THEORY

OF

DIFFERENTIAL EQUATIONS.

THEORY

OF

DIFFERENTIAL EQUATIONS.

PART II.

ORDINARY EQUATIONS, NOT LINEAR.

BY

ANDREW RUSSELL FORSYTH,

Sc.D. (Camb.), Hon. Sc.D. (Dubl.), F.R.S.,

SADLERIAN PROFESSOR OF PURE MATHEMATICS,
FELLOW OF TRINITY COLLEGE, CAMBRIDGE.

VOL. II.

CAMBRIDGE:

AT THE UNIVERSITY PRESS.

1900

CAMBRIDGE UNIVERSITY PRESS
Cambridge, New York, Melbourne, Madrid, Cape Town,
Singapore, São Paulo, Delhi, Tokyo, Mexico City

Cambridge University Press
The Edinburgh Building, Cambridge CB2 8RU, UK

Published in the United States of America by Cambridge University Press, New York

www.cambridge.org
Information on this title: www.cambridge.org/9781107640252

First published 1900
First paperback edition 2011

A catalogue record for this publication is available from the British Library

ISBN 978-1-107-64025-2 Paperback

PREFACE.

THE two volumes, now published as Part II, are my
second contribution towards the fulfilment of an old
promise. They deal almost entirely with the functional
character of the solutions of ordinary differential equations. At one time, I hoped to discuss the whole of this
theory in the present Part; the extent of the subject has
prevented me from realising this hope. Accordingly, I
have reserved the theory of linear differential equations
for another Part.

The revision of the proof-sheets has been made
lighter for me by the assistance of three friends.
Mr. E. T. Whittaker, M.A., Fellow of Trinity College,
Cambridge, has read both the volumes. Prof. W.
Burnside, M.A., F.R.S., Honorary Fellow of Pembroke
College, Cambridge, read a large part of the first volume.
Mr. R. W. H. T. Hudson, B.A., Scholar of St. John's
College, Cambridge, has read the whole of the second

volume. I wish to make a grateful acknowledgement of the help given me by these gentlemen.

I wish also to express my thanks to the Staff of the University Press, for the care and trouble they have taken, and the uniform consideration they have shewn me, during the progress of the printing.

A. R. FORSYTH.

Trinity College, Cambridge,
16 *December*, 1899.

CONTENTS.

CHAPTER I.

INTRODUCTORY.

CHAPTER II.

CAUCHY'S THEOREM ON THE EXISTENCE OF REGULAR INTEGRALS OF A SYSTEM OF EQUATIONS.

CHAPTER III.

CLASSES OF NON-ORDINARY POINTS CONNECTED WITH THE FORM OF THE EQUATION OF THE FIRST ORDER AND FIRST DEGREE IN THE DERIVATIVE.

CHAPTER IV.

INFLUENCE, UPON THE INTEGRAL, OF AN ACCIDENTAL SINGULARITY OF
THE FIRST KIND POSSESSED BY THE EQUATION.

CHAPTER V.

REDUCTION OF THE DIFFERENTIAL EQUATION TO FINAL TYPICAL FORMS,
VALID IN THE VICINITY OF AN ACCIDENTAL SINGULARITY OF THE
SECOND KIND.

CHAPTER VI.

THE CHARACTER OF THE INTEGRALS POSSESSED BY THE RESPECTIVE REDUCED FORMS OF THE ORIGINAL EQUATION IN THE VICINITY OF THE ACCIDENTAL SINGULARITY OF THE SECOND KIND.

CHAPTER VII.

CHAPTER VIII.

CHAPTER I.

1. THE theory of the solution of differential equations has been developed along several distinct lines of research. One of the many problems of the subject is the determination of those classes of differential equations which possess solutions expressible in terms of the functions already known in analysis. The most notable example of such a class is that of linear differential equations with constant coefficients: these can be solved by means of the exponential function. Another problem is the determination of those classes of differential equations which can be integrated by quadratures, that is, can be transformed so as to depend on the integration of equations of the form

$$\frac{dy}{dz} = Z,$$

where Z is a function of z. The solution of such equations involves only those transcendents which occur in the Integral Calculus. Examples of this kind are equations of the first order and the first degree in which the variables can be separated, and equations of the first order from which one of the variables is explicitly absent.

Equations of such classes were at one time the chief object of study in the theory of differential equations. They are somewhat limited in character and range. Many of the simpler results which

---

* In regard to the contents of this chapter, the following works may be consulted:—

Jordan, *Cours d'Analyse*, t. III, ch. I;

Königsberger, *Lehrbuch der Theorie der Differentialgleichungen*, Kap. I, Abschn. I, II.

have been obtained are given in introductory text-books on differential equations and therefore will not be developed here.

In the modern general theory, the problem of solution is considered from a different standpoint. It is proved that, within a suitably chosen region, a converging series of powers of the independent variable can be found which satisfies the differential equation. When, as is often the case, the function thus obtained is not included among the functions previously known in analysis, it is regarded as defined by the equation, and its properties are deduced as far as possible from the characteristic properties of the equation. Thus it may be possible to determine from the equation whether its integral is a uniform function or a multiform function; what are the places and the nature of its zeros, its singularities, its branch-points: and so on. In this way, new classes of functions are introduced to analysis, and the classes of differential equations, which can be solved by means of them, can be constructed.

It is to the consideration of this aspect of the theory of ordinary differential equations that nearly the whole of the present Part of this work is devoted. It will be seen that many of the investigations have regard to existence-theorems and are concerned with the character of the integral function in the vicinity of singularities. When all the singularities are known and the general character of the integral function in the immediate vicinity is determined, the further explicit determination of the integral is frequently taken for granted. The reason of this omission is that, as the respective investigations shew the kind of analytical expression which the integral acquires in the domain of any point, the actual substitution of an appropriately constructed expression and the determination of the coefficients, so as to make the equation identically satisfied, are matters of direct algebra. Such a process may be laborious but is not intrinsically difficult, and therefore only a few instances of it are carried through to their end; in several cases, it is omitted because its details are sufficiently obvious.

Moreover, the investigations will be restricted mainly to the analytical character of the solution of the equation. Some incidental illustrations may be given; but theories, that are concerned with descriptive and other properties of the equations considered, will be omitted.

2.   Simple considerations shew how little can be regarded at the outset as established knowledge and indicate that practically all the accepted propositions of a general character require to be reviewed so that their meaning and range may be clear and determinate. For instance, the complete solution of an ordinary equation of the first order is known to contain an arbitrary constant; and it is customary to declare that, in order to satisfy the conditions of a special problem associated with the equation, the value of the constant can be determined by any assigned relation.   On this basis, a solution of the equation

$$\frac{dy}{dx} = - \frac{y^2}{x}$$

might be required that would make y vanish when x vanishes. The complete solution is

$$\frac{1}{y} = A + \log x,$$

where A is left arbitrary by the equation; the datum is insufficient to determine A in the absence of information as to the mode in which x and y vanish. A precise solution cannot in this case be obtained; and a question is suggested as to possible limitations on data that serve to determine solutions. Further, the difficulty indicated has arisen after the general solution of the equation has been obtained; at least as grave a doubt must occur in the case of equations of which an explicit solution cannot be written down. In consequence, it is necessary to consider the fundamental question as to whether an integral exists; when the existence is established, some investigation must obtain the conditions by which it is limited and must deduce the characteristics of the function in the vicinity of ordinary and of critical values within and upon the boundary of its region of existence.

The existence-theorem for a system of equations

$$\frac{dw_i}{dz} = \phi_i(w_1, \ldots, w_n, z), \qquad (i = 1, 2, \ldots, n),$$

requires :

(i)   the establishment of integrals in the vicinity of values for which the functions ϕ_i are regular* : the range of existence of the integrals must also be considered :

* The term *regular* is applied, in accordance with Weierstrass's definition, *Ges. Werke*, t. II, p. 154, to a uniform function, (or to a uniform branch of a function),

1—2

(ii)   a proof of the uniqueness within the range of existence;
this gives rise to various questions connected with the
appropriate determining conditions, and also leads to a
discussion of some classes of singularities:

(iii)   (connected partially with (ii) in fact, though substantially
a quite independent investigation), a discussion of integrals
in the vicinity of values for which the functions ϕ_i cease to
be regular.

There is another class of investigations, distinct from those
just indicated: they are suggested by the corresponding class of
questions that arise in connection with ordinary linear equations.
When the path of the variable between z and some initial value z_0
is deformed in the plane, how is a particular set of simultaneous
solutions of the system affected? Or when the variable returns to
the initial point z_0 after describing any circuit in the plane, what
is the effect on the composite integral?

The various investigations thus suggested will, as far as
possible, be taken into successive account: the last class is,
however, discussed only briefly, for reasons adduced later.

3.   All the variables that occur are supposed to be complex
quantities with initially unlimited variation. As is usual, a
separate region is associated with each of them so that the
variation can be represented geometrically; the region of any
variable is generally a plane and, being so, is referred to as the
plane of the variable.

In most of the succeeding investigations, there is only a single
independent variable. The system of equations determining the
dependent variables is regarded as being constituted of simultaneous equations, which are independent of one another in the
sense that no one of them can be derived from the others by any
combination of algebraical and analytical processes. The number
of equations in such a system is the same as the number of
dependent variables. The dependent variables are generally
denoted by u, v, w, ..., the independent variable by z.

in a region of the variables at every point of which it can be represented in the form
of a converging power-series: it is finite and continuous for all values of the
variables included in such a region.

The equations may contain differential coefficients of any orders; but a transformation can be effected after which the only differential coefficients that occur are of the first order, it being sufficient for this purpose to associate appropriate equations of the type

$$\frac{dw}{dz} = w_1, \quad \frac{dw_1}{dz} = w_2, \ldots$$

with the system, which in its changed form will still be composed of as many equations as there are dependent variables. All the equations discussed will be algebraical in each of the derivatives of the dependent variables, and they will usually be algebraical also in these variables themselves, any deviation being indicated when it is of importance; but no such limitation as to functional occurrence is imposed as regards the occurrence of the independent variable.

It may happen that some equations of the system are free from derivatives: or it may be possible to construct such an equation from the system without integration or any equivalent process. Let such an equation be

$$g(u, v, w, \ldots, z) = 0,$$

so that

$$\frac{\partial g}{\partial u}\frac{du}{dz} + \frac{\partial g}{\partial v}\frac{dv}{dz} + \frac{\partial g}{\partial w}\frac{dw}{dz} + \ldots + \frac{\partial g}{\partial z} = 0.$$

By means of the latter, some one of the derivatives can be eliminated from all the equations of the system; by means of $g = 0$, the corresponding variable can be eliminated from each of the modified equations in turn. In this form, the number of equations is greater by unity than the number of dependent variables, so that one equation is satisfied in virtue of the remainder; this equation is therefore superfluous and should be removed. The original system is thus replaced by another, containing one dependent variable less and one equation less; the solution of the original system is compounded of the solution of the new system and of the equation $g = 0$. It would thus be sufficient to obtain the solution of the modified system; all the further processes necessary to solve the original system are algebraical in character.

For example, in the system

$$
\left.
\begin{aligned}
P_1 \frac{du}{dz} + P_2 \frac{dv}{dz} + P_3 \frac{dw}{dz} &= P_4 \\
Q_1 \frac{du}{dz} + Q_2 \frac{dv}{dz} + Q_3 \frac{dw}{dz} &= Q_4 \\
R_1 \frac{du}{dz} + R_2 \frac{dv}{dz} + R_3 \frac{dw}{dz} &= R_4
\end{aligned}
\right\},
$$

the coefficients P, Q, R, supposed to be functions of the variables, may be such that the determinant $(P_1 Q_2 R_3)$, say Δ, vanishes identically. In order that the derivatives of u, v, w, may not have infinite values only, it is necessary that the equations

$$(P_2 Q_3 R_4) = 0, \quad (P_1 Q_3 R_4) = 0, \quad (P_1 Q_2 R_4) = 0$$

be satisfied. They cannot all be identities, for the original system would then contain only two independent equations; properties of determinants shew that, as Δ vanishes identically, they are satisfied in virtue of a single new equation, say $S = 0$. This equation $S = 0$ would be used to transform the system into one involving only two dependent variables.

Systems of equations which can be transformed so as to yield, merely by processes of algebraical elimination, one or more equations free from derivatives, are called reducible; systems which do not admit of such a transformation are called irreducible. The process of modifying a reducible system, so that ultimately an irreducible system in a smaller number of variables shall remain, has been indicated; the properties of reducible systems and the tests of reducibility must be sought elsewhere. For the present purpose, it will be assumed that all the systems of equations under consideration are irreducible; and manifestly there is no loss of generality in assuming that each equation in a system is rationally irresoluble.

CONSTRUCTION AND PREPARATION OF NORMAL FORMS.

4. Before undertaking the discussion of the integral equivalent of a system of equations, it is desirable to select some typical form for the equations as one in which any given system can be expressed.

When a system is composed of only a single equation and when therefore only one dependent variable is involved, it is of the form

$$f \left(\frac{dw}{dz}, \, w, \, z \right) = 0,$$

where (after preceding explanations) f is rational and integral in dw/dz, is usually rational as regards w, and is unlimited as regards z.

When a system is composed of two equations and when therefore two dependent variables are involved, it will initially in the most general case have the form

$$f\left(\frac{du}{dz}, \frac{dv}{dz}, u, v, z\right) = 0, \qquad g\left(\frac{du}{dz}, \frac{dv}{dz}, u, v, z\right) = 0,$$

where both f and g are rational and integral in du/dz and in dv/dz; so that the two equations may be regarded temporarily as algebraical equations expressing du/dz and dv/dz in terms of u, v, z. To select a typical form of reference, the simultaneous roots of the two equations are to be found; and, for this purpose, Sylvester's dialytic process of elimination can be used. If f be of degree m_1 and g of degree m_2 in du/dz, then $m_1 + m_2$ equations are constructed, being in fact

$$\left(\frac{du}{dz}\right)^n f = 0, \qquad \left(\frac{du}{dz}\right)^{n'} g = 0,$$

for $n = 0, 1, \ldots, m_2 - 1$ and $n' = 0, 1, \ldots, m_1 - 1$. When all the $m_1 + m_2 - 1$ powers of du/dz are eliminated, the result is an equation

$$F\left(\frac{dv}{dz}, u, v, z\right) = 0$$

which is rational and integral in dv/dz; and any $m_1 + m_2 - 1$ of the equations *, solved linearly for du/dz, lead to an equation of the form

$$\frac{du}{dz} = G\left(\frac{dv}{dz}, u, v, z\right),$$

where G is algebraical and generally fractional† in $\dfrac{dv}{dz}$. Moreover,

---

* This is true in a case of complete generality; but nugatory forms may arise for particular cases. A full discussion of alternatives would require much of the algebra associated with the discussion of the intersections of algebraical plane curves.

† It can easily be made integral as follows. Let V denote dv/dz and suppose that the fractional form of G is

$$\frac{p\,(V, u, v, z)}{q\,(V, u, v, z)},$$

so that for any root, say V_1, of $F = 0$, we have

$$\frac{du}{dz} = \frac{p\,(V_1, u, v, z)}{q\,(V_1, u, v, z)}.$$

G is rational in $\dfrac{dv}{dz}$ when the first power of $\dfrac{du}{dz}$ can be deduced; it is a root of a rational function in $\dfrac{dv}{dz}$ when not the first power of $\dfrac{du}{dz}$ but some higher power is directly deduced from the $m_1 + m_2 - 1$ equations. The original system is thus equivalent to

$$F\left(\frac{dv}{dz},\, u,\, v,\, z\right) = 0, \qquad \frac{du}{dz} = G\left(\frac{dv}{dz},\, u,\, v,\, z\right),$$

which can be taken as a typical form for a system determining two dependent variables.

When a system is composed of three equations and when therefore three dependent variables are involved, it will initially in the most general case have the form

$$f\left(\frac{du}{dz},\, \frac{dv}{dz},\, \frac{dw}{dz},\, u,\, v,\, w,\, z\right) = 0,$$

$$g\left(\frac{du}{dz},\, \frac{dv}{dz},\, \frac{dw}{dz},\, u,\, v,\, w,\, z\right) = 0,$$

$$h\left(\frac{du}{dz},\, \frac{dv}{dz},\, \frac{dw}{dz},\, u,\, v,\, w,\, z\right) = 0,$$

where f, g, and h are rational and integral in each of the three derivatives. To obtain the modified form for this system, the process of dialytic elimination for several simultaneous equations is used *: it is similar in kind to the modification in the preceding case, but the details are more complicated. The result is that an equation of the form

$$F\left(\frac{dw}{dz},\, u,\, v,\, w,\, z\right) = 0$$

Let the other roots of $F = 0$ be V_2, V_3, ...; then

$$\frac{du}{dz} = \frac{p\,(V_1,\, u,\, v,\, z)\; q\,(V_2,\, u,\, v,\, z)\; q\,(V_3,\, u,\, v,\, z)\ldots}{q\,(V_1,\, u,\, v,\, z)\; q\,(V_2,\, u,\, v,\, z)\; q\,(V_3,\, u,\, v,\, z)\ldots}.$$

The denominator is a symmetric function of V_1, V_2, V_3, ... and therefore, by means of $F = 0$, can be made a function of u, v, z; the numerator is a symmetric function of V_2, V_3, ... which, by means of the same equation, can be made a function of V_1, u, v, z which is integral in V_1; hence the new form of $\dfrac{du}{dz}$ in terms of $\dfrac{dv}{dz}$ is integral and no longer fractional. Moreover, by means of $F = 0$, its degree in $\dfrac{dv}{dz}$ can be made lower than that of $F = 0$.

* Salmon's *Higher Algebra*, 3rd edn., §§ 91—94; Faà de Bruno, *Théorie générale de l'élimination*, 3me partie, chap. II, § iv; Cayley, *Coll. Math. Papers*, vol. I, pp. 370—374.

subsists, obtained as the eliminant of a number of equations linearly involving powers of du/dz, dv/dz and their combinations; and when all but one of these equations are treated simultaneously as giving the powers and products of du/dz and dv/dz, they generally lead to equations of the form

$$\frac{du}{dz} = G\left(\frac{dw}{dz}, u, v, w, z\right),$$

$$\frac{dv}{dz} = H\left(\frac{dw}{dz}, u, v, w, z\right),$$

where G and H are algebraical and fractional* in $\dfrac{dw}{dz}$. The two latter, with $F = 0$, can be taken as a typical form for a system determining three dependent variables.

When a system is composed of n equations and when therefore n dependent variables are involved, it will initially in the most general case have the form

$$f_s\left(\frac{du_1}{dz}, \frac{du_2}{dz}, \dots, \frac{du_n}{dz}, u_1, u_2, \dots, u_n, z\right) = 0$$

for $s = 1, 2, \dots, n$. The application of a similar dialytic process leads to the typical form for the system, as constituted by the equations

$$F_n\left(\frac{du_n}{dz}, u_1, u_2, \dots, u_n, z\right) = 0,$$

$$\frac{du_s}{dz} = G_s\left(\frac{du_n}{dz}, u_1, u_2, \dots, u_n, z\right),$$

for $s = 1, 2, \dots, n-1$; the equation $F_n = 0$ is rational and integral in du_n/dz, and all the expressions† G_s are algebraical and generally fractional (which can be made integral) in du_n/dz.

A very special example of such a system occurs in the case of

---

* Both G and H can be made integral in $\dfrac{dw}{dz}$ by the same process as in the preceding instance.

† It may be remarked that these expressions are usually rational; they cease to be rational only when the algebraical solution of the simultaneous equations used in the dialytic process leads to a nugatory result for du_s/dz, and then some finite integral power of the expression G_s is rational. The same remark applies to the expression G in the system with two variables, and to the expressions G, H in the system with three variables.

the ordinary linear differential equation of order n in a single variable. When it is

$$\frac{d^n y}{dx^n} + P_1 \frac{d^{n-1} y}{dx^{n-1}} + \ldots + P_{n-1} \frac{dy}{dx} + P_n y = 0,$$

the equivalent system of the preceding type is

$$\frac{dy}{dx} = y_1, \quad \frac{dy_1}{dx} = y_2, \ldots, \frac{dy_{n-2}}{dx} = y_{n-1},$$

$$\frac{dy_{n-1}}{dx} + P_1 y_{n-1} + \ldots + P_{n-1} y_1 + P_n y = 0.$$

In the transformation of the system involving n variables, there is no special reason for retaining du_n/dz as a derivative of reference in the typical form; any other of the derivatives might similarly have been retained. If each be retained in turn, there would be an equation

$$F_r \left(\frac{du_r}{dz}, u_1, \ldots, u_n, z \right) = 0,$$

$r = 1, \ldots, n$; these n equations would be distributed through the n reduced typical systems.

The aggregate of these n equations $F_r = 0$ is sometimes regarded as the normal form for the original system. It undoubtedly includes the original system; but it includes more. For a simultaneous solution of the aggregate would be given by combining any root of $F_1 = 0$, with any root of $F_2 = 0$, with any root of $F_3 = 0$, and so on; but only a limited number of such combinations would satisfy the original system*. The aggregate therefore cannot be regarded as a proper equivalent of an original system in which each equation involves all the derivatives.

It has been assumed that, in the original system of equations, all the derivatives occur in each equation or at least so many occur as to make the purely algebraical transformation possible. But sets of equations, that are less general, may be propounded. For example, in the system

$$f \left(z, u, v, w, \frac{du}{dz} \right) = 0,$$

$$g \left(z, u, v, w, \frac{dv}{dz}, \frac{dw}{dz} \right) = 0, \quad h \left(z, u, v, w, \frac{dv}{dz}, \frac{dw}{dz} \right) = 0,$$

---

* An analogous case would arise in finding the coordinates of the intersections of two curves, if they were determined from the x-eliminant and the y-eliminant alone.

the transformation, which can only partially be effected, leads to

$$\left.\begin{aligned}
f\left(z, u, v, w, \frac{du}{dz}\right) &= 0 \\[4pt]
G\left(z, u, v, w, \frac{dv}{dz}\right) &= 0 \\[4pt]
\frac{dw}{dz} &= H\left(z, u, v, w, \frac{dv}{dz}\right)
\end{aligned}\right\}.$$

In the system

$$f_r\left(z, u_1, \ldots, u_n, \frac{du_r}{dz}\right) = 0,$$

given as an original system, the transformation cannot be effected at all.

5. A somewhat different form of reference is selected as typical for the last class of equations; it can be constructed as follows.

Let $U, V, W, \ldots$ denote $du/dz, dv/dz, dw/dz, \ldots$; and in the first place let the system to be considered be

$$f(z, u, v, w, \ldots, U) = 0$$
$$g(z, u, v, w, \ldots, V) = 0$$
$$h(z, u, v, w, \ldots, W) = 0$$
$$\cdots\cdots\cdots\cdots\cdots$$

where f is of degree l in U and has $f_0 U^l$ for the term involving the highest power of U when the coefficients of all the powers of U are made integral functions; m and $g_0 V^m$ are corresponding quantities for the second equation, n and $h_0 W^n$ for the third, and so on. Let N denote the product $lmn\ldots.$

Denote the l roots of the first equation by $U_1, \ldots, U_l$; the m roots of the second equation by $V_1, \ldots, V_m$; and so on. If t denote

$$\lambda U + \mu V + \nu W + \ldots,$$

where $\lambda, \mu, \nu, \ldots$ are arbitrary constants, then, on substitution of the possible roots of the equations, the quantity t assumes values $t_{pqr\ldots}$ representing

$$\lambda U_p + \mu V_q + \nu W_r + \ldots,$$

which are N in number; and the equation which has these values $t_{pqr\ldots}$ for its roots is

$$F(t) = f_0^{\frac{N}{l}} g_0^{\frac{N}{m}} h_0^{\frac{N}{n}} \ldots \prod_{p,\,q,\,r,\,\ldots} (t - t_{pqr\ldots}) = 0.$$

The coefficients of the powers of t are symmetric functions of the various combinations of the roots of the equations $f = 0$, $g = 0$, $h = 0, \ldots$; when they are expressed in terms of the coefficients in these equations, all of them are rational integral functions of z, u, v, w, $\ldots$, and the coefficient of t^N is independent of λ, μ, ν, $\ldots$. Thus $F(t)$ has the form

$$F(t) = \theta_0 t^N + \theta_1 t^{N-1} + \ldots + \theta_N,$$

where all the coefficients are rational integral functions of z, u, v, w, $\ldots$; all of them except θ_0 involve the arbitrary constants λ, μ, ν, $\ldots$, being algebraical and integral in these constants; and the roots of $F(t)$ are the N values $t_{pqr\ldots}$. Now

$$\frac{\partial F(t)}{\partial t} = f_0^{\frac{N}{l}} g_0^{\frac{N}{m}} h_0^{\frac{N}{n}} \ldots \prod_{p,\,q,\,r,\,\ldots} (t - t_{pqr\ldots}) \sum_{p,\,q,\,r,\,\ldots} \frac{1}{t - t_{pqr\ldots}},$$

and therefore

$$\left[\frac{\partial F(t)}{\partial t} \right]_{t = t_{p'q'r'\ldots}} = f_0^{\frac{N}{l}} g_0^{\frac{N}{m}} h_0^{\frac{N}{n}} \ldots \prod_{p,\,q,\,r,\,\ldots} (t_{p'q'r'\ldots} - t_{pqr\ldots}).$$

Also

$$\frac{\partial F(t)}{\partial \lambda} = f_0^{\frac{N}{l}} g_0^{\frac{N}{m}} h_0^{\frac{N}{n}} \ldots \prod_{p,\,q,\,r,\,\ldots} (t - t_{pqr\ldots}) \sum_{p,\,q,\,r,\,\ldots} \frac{-U_p}{t - t_{pqr\ldots}},$$

and therefore

$$\left[\frac{\partial F(t)}{\partial \lambda} \right]_{t = t_{p'q'r'\ldots}} = - U_{p'} f_0^{\frac{N}{l}} g_0^{\frac{N}{m}} h_0^{\frac{N}{n}} \ldots \prod_{p,\,q,\,r,\,\ldots} (t_{p'q'r'\ldots} - t_{pqr\ldots})$$

$$= - U_{p'} \left[\frac{\partial F(t)}{\partial t} \right]_{t = t_{p'q'r'\ldots}},$$

provided the t-discriminant of $F(t)$ with regard to t does not vanish; and this condition is generally satisfied because the constants λ, μ, ν, $\ldots$ are arbitrary. In the same way, it can be deduced that

$$\left[\frac{\partial F(t)}{\partial \mu} \right]_{t = t_{p'q'r'\ldots}} = - V_{q'} \left[\frac{\partial F(t)}{\partial t} \right]_{t = t_{p'q'r'\ldots}},$$

$$\left[\frac{\partial F(t)}{\partial \nu} \right]_{t = t_{p'q'r'\ldots}} = - W_{r'} \left[\frac{\partial F(t)}{\partial t} \right]_{t = t_{p'q'r'\ldots}},$$

and so on; and it should be noticed that the quantities $U_{p'}$, $V_{q'}$, $W_{r'}$, ... are those which lead to the value $t_{p'q'r'...}$ of t. It therefore follows that the original system can be replaced by the system

$$\frac{du}{dz} = \frac{F_\lambda\,(t,\,z,\,u,\,v,\,w,\,...)}{\dfrac{\partial}{\partial t}F\,(t,\,z,\,u,\,v,\,w,\,...)},$$

$$\frac{dv}{dz} = \frac{F_\mu\,(t,\,z,\,u,\,v,\,w,\,...)}{\dfrac{\partial}{\partial t}F\,(t,\,z,\,u,\,v,\,w,\,...)},$$

$$\frac{dw}{dz} = \frac{F_\nu\,(t,\,z,\,u,\,v,\,w,\,...)}{\dfrac{\partial}{\partial t}F\,(t,\,z,\,u,\,v,\,w,\,...)},$$

$$.........................$$

where t is determined by the equation

$$F\,(t,\,z,\,u,\,v,\,w,\,...) = 0,$$

and the functions F_λ, F_μ, F_ν, ..., being respectively $-\partial F/\partial\lambda$, $-\partial F/\partial\mu$, $-\partial F/\partial\nu$, ..., are algebraical functions of t of degree one less than F. All the possible systems of values of du/dz, dv/dz, dw/dz, ... are obtained by taking all the values of t in turn as the roots of $F = 0$.

No substantial difference arises according as $F=0$ is a reducible equation or an irreducible equation; for, in the former case, it would reduce as in the form

$$F = G \cdot H,$$

where G and H are rational integral algebraical functions of t, and it is easy to see that for the roots of $G=0$ the system would be composed of equations such as

$$\frac{du}{dz} = \frac{-\,\partial G/\partial\lambda}{\dfrac{\partial}{\partial t}G}, \quad G=0\,;$$

while for the roots of $H=0$ the system would be composed of equations such as

$$\frac{du}{dz} = \frac{-\,\partial H/\partial\lambda}{\dfrac{\partial}{\partial t}H}, \quad H=0.$$

Both systems would be retained in order to secure the full equivalent of the original system. Thus the difference between the case when F is reducible and that when it is irreducible is a simplification in expression for the former; but the form is the same.

The form thus obtained* is Weierstrass's *normal* or *canonical* form.

In the second place, let the system to be considered be of the type

$$f_r\left(z, u_1, u_2, \ldots, u_n, u_{n+1}, \ldots, u_{n+m}, \frac{du_r}{dz}\right) = 0,$$

$$\frac{du_{n+s}}{dz} = f_{n+s}\left(z, u_1, u_2, \ldots, u_n, \ldots, u_{n+m}, \frac{du_1}{dz}, \ldots, \frac{du_n}{dz}\right),$$

for $r = 1, \ldots, n$ and $s = 1, \ldots, m$; the functions f_r are rational integral functions of the derivatives that occur, and the functions f_{n+s} generally are, or can be made, also rational integral functions of the derivatives that occur. Introducing a variable t defined by the relation

$$t = \mu_1 \frac{du_1}{dz} + \mu_2 \frac{du_2}{dz} + \ldots + \mu_n \frac{du_n}{dz},$$

the set of n equations $f_r = 0$ can, by the preceding investigation, be replaced by the equivalent system

$$G(t, z, u_1, \ldots, u_{n+m}) = 0,$$

$$\frac{du_r}{dz} = \frac{G_r(t, z, u_1, \ldots, u_{n+m})}{\frac{\partial}{\partial t} G(t, z, u_1, \ldots, u_{n+m})},$$

where G is practically the eliminant of the equation defining t and the n equations $f_r = 0$ for $r = 1, \ldots, n$; and the remaining m equations, after substitution is made of the various expressions just obtained for du_1/dz, du_2/dz, $\ldots$, du_n/dz, become

$$\frac{du_{n+s}}{dz} = H_{n+s}(t, z, u_1, u_2, \ldots, u_{n+m}).$$

The form thus obtained will be regarded as the normal form.

It thus appears that any system of equations can be so transformed so as to be equivalent either to a system of the type

$$\left.\begin{array}{l} F\left(\dfrac{du_n}{dz}, u_1, \ldots, u_n, z\right) = 0 \\[2ex] \dfrac{du_r}{dz} = F_r\left(\dfrac{du_n}{dz}, u_1, \ldots, u_n, z\right) \end{array}\right\},$$

---

* Königsberger, *Theorie der Differentialgleichungen*, p. 11; Biermann, *Theorie der analytischen Functionen*, p. 248.

for $r = 1, \ldots, n-1$; or to a system of the type

$$F(t, u_1, \ldots, u_n, z) = 0 \left.\vphantom{\frac{du_r}{dz}}\right\}$$
$$\frac{du_r}{dz} = H_r(t, u_1, \ldots, u_n, z) \left.\vphantom{\frac{du_r}{dz}}\right\},$$

for $r = 1, \ldots, n$. The former is less symmetrical than the latter, but it dispenses with the introduction of the other variable t. A comparison of the investigations shews that the former type could be changed into the latter, but that the latter cannot be changed into the former.

6.   In all the forms to which the system of equations has been reduced, the completion of the process of expressing each of the derivatives in terms only of the variables depends upon the solution of an algebraical equation defining one of the derivatives as an implicit function. That the implicit function does exist when so defined is an inference from Weierstrass's theorem* on the resolution of a function of several variables: a theorem which, for the present purpose, may be stated as follows:

Let $z = c$, $w_1 = a_1, \ldots, w_n = a_n$ be simultaneous values of variables giving a zero value to a uniform analytical function $\phi(z, w_1, \ldots, w_n)$; and let the changes $z = c + x$, $w_r = a_r + x_r$ for $r = 1, \ldots, n$, transform $\phi(z, w_1, \ldots, w_n)$ into $F(x, x_1, \ldots, x_n)$ so that F vanishes when each of its variables acquires a zero value. Then it is always possible to choose non-vanishing quantities $\rho, r_1, \ldots, r_n$ such that, for values of the variables within the regions

$$|x| \leqslant \rho, \;\; |x_1| \leqslant r_1, \;\; \ldots\ldots, \;\; |x_n| \leqslant r_n,$$

$F(x, x_1, \ldots, x_n)$ *can be expressed in the form*

$$f(x, x_1, \ldots, x_n)\, e^{g(x, x_1, \ldots, x_n)},$$

where $g(x, x_1, \ldots, x_n)$ is finite for the range of variables indicated and $f(x, x_1, \ldots, x_n)$ is an algebraical polynomial in x having, for the coefficients of powers of x, analytical functions of $x_1, \ldots, x_n$ which are regular within the range. Moreover, when the expansion of $F(x, 0, \ldots, 0)$, supposed not to vanish identically, begins with Cx^m, the polynomial is of degree m; so that if, in

---

* *Abhandlungen aus der Functionenlehre*, pp. 107—114; *Ges. Werke*, t. II, pp. 135—142: a proof is given at the end of the present chapter. It need hardly be pointed out that, as the theorem holds for uniform analytical functions in general, it holds for those that are merely algebraical in some of their variables.

particular, $\partial F(x, x_1, \ldots, x_n)/\partial x$ does not vanish for zero values of $x, x_1, \ldots, x_n$, the value of m is unity and then f takes the form

$$x - \text{analytical function of } x_1, \ldots, x_n.$$

It thus appears that if $\partial F/\partial x$ is not zero for $x = 0$, $x_1 = 0$, $\ldots$, $x_n = 0$, the equation $F = 0$ is satisfied, for values of the variable in the vicinity, solely in virtue of the equation

$$z - c = x = h(x_1, \ldots, x_n) = h(w_1 - a_1, \ldots, w_n - a_n),$$

where h is an analytical function of $x_1, \ldots, x_n$ that vanishes when all the variables are zero and is regular for values of the variables within the assigned regions.

7. In the first of the typical forms to which a system of equations can be reduced, the central equation

$$F\left(\frac{du_n}{dz}, u_1, \ldots, u_n, z\right) = 0$$

is algebraical in du_n/dz. If du_n/dz have a value b for $z = c$, $u_1 = a_1, \ldots, u_n = a_n$, and if

$$F(b + U, a_1, \ldots, a_n, c)$$

when expanded in powers of U begin with the first power of U or—what is the same thing—if $\partial F/\partial U$ be not zero for the values specified, then there are finite regions of the variables defined by

$$|z - c| \leqslant \rho, \; |u_1 - a_1| \leqslant r_1, \ldots, |u_n - a_n| \leqslant r_n,$$

for which the above equation is satisfied by

$$\frac{du_n}{dz} - b = h(z - c, u_1 - a_1, \ldots, u_n - a_n)$$

where h is a function regular for all values of the variables within those regions. When this value is substituted in

$$\frac{du_r}{dz} = F_r\left(\frac{du_n}{dz}, u_1, \ldots, u_n, z\right),$$

the new expression for the derivative is a function of $z, u_1, \ldots, u_n$ which is an analytical function of $z - c, u_1 - a_1, \ldots, u_n - a_n$ unless the set of values $c, a_1, \ldots, a_n$ constitute a singularity of one or more of the coefficients of powers of $\dfrac{du_n}{dz}$ in F_r. This form holds for each of the derivatives; and thus *the final reduced form of the system is*

$$\frac{du_s}{dz} = f_s(u_1, \ldots, u_n, z), \qquad (s = 1, \ldots, n),$$

where the functions f_s are analytical functions of the variables $u_1, \ldots, u_n, z$. In discussing the integral equivalent of these equations, when there is an integral equivalent, it will be necessary to take account of the singularities of the analytical functions f_s.

8. In the second of the typical forms to which a given system can be reduced, the central equation is

$$F(t, u_1, \ldots, u_n, z) = 0,$$

algebraical in the subsidiary variable t. If τ be a value of t when $u_1, \ldots, u_n, z$ acquire the values $a_1, \ldots, a_n, c$ respectively, an argument similar to that which applies to the first case leads to the inference that the equation $F = 0$ is satisfied by

$$t - \tau = h(z - c, u_1 - a_1, \ldots, u_n - a_n),$$

where h is a function regular for all values of the variables within certain finite regions. The substitution of this value in the remaining equations of the system shews, as before, that the final reduced form of the system is of the same character as in the preceding case. This final reduced form also may be called a *normal form*.

It would have been possible also to infer the final reduced form of a system of equations, without the construction of the intermediate forms, by using a theorem on the existence of a number of implicit functions given by the same number of algebraical equations: the theorem is the generalisation of Weierstrass's theorem quoted in the process which has been adopted.

It was assumed that $\partial F(b + U, a_1, \ldots, a_n, c)/\partial U$ does not vanish with U; and there is a corresponding assumption in effecting the modification of the alternative typical form. The consideration of the equations when the assumptions are not justified is reserved until a later stage, when branch-points connected with equations will be discussed.

Further, cases occur in which the value of U_n, as determined by

$$F(U_n, u_1, \ldots, u_n, z) = 0,$$

satisfies

$$\frac{\partial}{\partial U_n} F(U_n, u_1, \ldots, u_n, z) = 0$$

not solely for systems of particular values of $u_1, \ldots, u_n, z$ but in general. This possibility will be discussed when the singular integrals of systems of equations are considered.

9. The purpose of the preceding investigation has been the derivation of a normal form equivalent to the original system ; the form has been chosen as one that will be found convenient in discussing the existence of the integral. It may however be pointed out that, in practice, the solution of a system given in a normal form can be made to depend upon the solution of a single equation of order equal to the number of dependent variables, when the analytical functions are algebraical in the dependent variables. Thus let the system be

$$\frac{dw}{dz} = W(w, u_1, \ldots, u_{n-1}, z) = W,$$

$$\frac{du_r}{dz} = U_r(w, u_1, \ldots, u_{n-1}, z) = U_r, \qquad (r = 1, 2, \ldots, n-1);$$

and denote the operation

$$\frac{\partial}{\partial z} + W\frac{\partial}{\partial w} + U_1\frac{\partial}{\partial u_1} + \ldots + U_{n-1}\frac{\partial}{\partial u_{n-1}}$$

by Θ. Then

$$\frac{dw}{dz} = W,$$

$$\frac{d^2w}{dz^2} = \Theta W,$$

$$\frac{d^3w}{dz^3} = \Theta^2 W,$$

$$\vdots$$

$$\frac{d^nw}{dz^n} = \Theta^{n-1} W.$$

In these n equations the variables $u_1, \ldots, u_{n-1}$ occur algebraically: consequently they can be eliminated by ordinary processes, and the resulting equation is of the form

$$\phi\left(z, w, \frac{dw}{dz}, \frac{d^2w}{dz^2}, \ldots, \frac{d^nw}{dz^n}\right) = 0,$$

the order n being the number of dependent variables in the former system. If the complete integral of this equation be known in the form

$$\psi(z, w, a_1, \ldots, a_n) = 0,$$

involving n arbitrary constants, the solution of the original system can be deduced. For from $\psi = 0$, the values of $\frac{dw}{dz}, \frac{d^2w}{dz^2}, \ldots, \frac{d^{n-1}w}{dz^{n-1}}$ can be derived ; and the n equations

$$\psi = 0, \quad \frac{dw}{dz} = W, \ldots, \frac{d^{n-1}w}{dz^{n-1}} = \Theta^{n-2} W,$$

will then constitute the practical solution of the original system.

The similar inference, that the solution of any given system, which involves only one independent variable with any number of dependent variables and their derivatives in any order, can be made to depend upon the solution of a single equation of higher order, is true when the system is not given in a normal form. The derivation of that single equation and even the investigation of the order of the highest derivative that occurs are somewhat complicated questions, which will be omitted as not contributing towards the development of the theories to be considered; they have been discussed by Jacobi *.

NOTE.

Weierstrass's theorem on the form of a regular function of several variables in the vicinity of a zero value.

This theorem † has been used in §§ 6—8 when the number of variables is general. It will be used several times in succeeding chapters, particularly in the case when the number of variables is two.

It is proved as follows by Weierstrass (*l. c.*). Two cases arise for consideration according as $F(x, 0, ..., 0)$ does not, or does, vanish for all values of x.

In the former case, $F(x, 0, ..., 0)$, when it does not vanish for all values of x, is a uniform analytical function of x, vanishing when $x = 0$; let the lowest exponent of x which it contains be m. Denote it by $F_0(x)$; and introduce a function $F_1(x, x_1, ..., x_n)$ defined by the equation

$$F(x, x_1, ..., x_n) = F_0(x) - F_1(x, x_1, ..., x_n),$$

so that F_1 vanishes for all values of x when $x_1, ..., x_n$ vanish, and is a uniform analytical function within the region of convergence of the power-series for F. Because F_0 is independent of $x_1, ..., x_n$ and does not vanish for all values of x, we can choose points, in the vicinity of $0, 0, ..., 0$ and lying within the region of convergence of F_1, such that

$$|F_0| > |F_1|.$$

But F_0 vanishes when $x = 0$, so that there may be some limit of $|x|$ other than zero below which the inequality does not hold: suppose that the range for which the inequality does hold is given by

$$\rho_0 < |x| < \rho, \quad |x_s| < |r_s|, \qquad (s = 1, ..., n).$$

---

* *Ges. Werke*, t. v, pp. 191—216; *ib.* pp. 483—513. See also Jordan, *Cours d'Analyse*, t. III, pp. 5—7.

† See *Ges. Werke*, t. II, pp. 135 sqq. The theorem is of great importance in the theory of functions of several variables; some of the applications are given by Weierstrass.

Another proof, when the number of variables is two,-has been obtained by Simart; it is reproduced by Picard, *Traité d'Analyse*, t. II, pp. 243—245.

For all such values we obtain, on forming logarithmic derivatives of the equation

$$F = F_0 \left(1 - \frac{F_1}{F_0} \right),$$

the relation

$$\frac{1}{F} \frac{\partial F}{\partial x} = \frac{1}{F_0} \frac{dF_0}{dx} - \frac{\partial}{\partial x} \sum_{\lambda=1}^{\infty} \frac{1}{\lambda} \frac{F_1^\lambda}{F_0^\lambda}.$$

Since F_0 is a uniform analytical function of x, the lowest exponent of which is m, we have

$$\frac{1}{F_0} \frac{dF_0}{dx} = \frac{m}{x} + G(x),$$

where $G(x)$ is a converging series of integral powers. Similarly

$$\frac{F_1^\lambda}{F_0^\lambda} = \sum_{\mu=0}^{\infty} G_{\lambda, \mu} \, x^{-m\lambda + \mu},$$

where the coefficients $G_{\lambda, \mu}$ are converging series of integral powers of $x_1, \ldots, x_n$, all of them vanishing with these variables; and accordingly, collecting the terms that involve the same power of x, we have

$$\sum_{\lambda=1}^{\infty} \frac{1}{\lambda} \frac{F_1^\lambda}{F_0^\lambda} = \sum_{p=-\infty}^{p=\infty} G_p x^p,$$

where the coefficients G_p are converging series of integral powers of $x_1, \ldots, x_n$, all of them vanishing with these variables. Hence

$$\frac{1}{F} \frac{\partial F}{\partial x} = \frac{m}{x} + G(x) - \frac{\partial}{\partial x} \sum_{p=-\infty}^{\infty} G_p x^p.$$

This result shews that, for values of $x_1, \ldots, x_n$ given by

$$|x_s| < r_s, \qquad\qquad (s = 1, \ldots, n),$$

there are m values of x such that F vanishes, each zero being counted in its proper multiplicity. For if within the ranges indicated there be values $a_1, \ldots, a_n$ such that no root of $F(x, a_1, \ldots, a_n)$ occurs within the included range of x, then

$$\frac{1}{F(x, a_1, \ldots, a_n)} \frac{\partial F(x, a_1, \ldots, a_n)}{\partial x},$$

when expanded as a converging series of powers of x, would contain only positive integral powers. This expansion, however, ought to be the same as

$$\frac{m}{x} + G(x) - \frac{\partial}{\partial x} \sum_{p=-\infty}^{\infty} G_p x^p;$$

an identity in form that does not exist owing to the presence of $\frac{m}{x}$ in the latter, the only term with an exponent -1. Accordingly, the hypothesis that $F(x, a_1, \ldots, a_n)$ is rootless within the range is untenable.

Suppose therefore that $\xi_1, \ldots, \xi_r$ are all the zeros of $F(x, a_1, \ldots, a_n)$ for values of x such that $|x| < \rho$, repetition of a value ξ accounting for possible multiplicity; then

$$\frac{1}{F}\frac{\partial F}{\partial x} - \sum_{s=1}^{r}\frac{1}{x-\xi_s}$$

is finite for all values of x within the range and it can therefore be expanded in a converging series of positive integral powers, say $P(x)$, so that

$$\frac{1}{F}\frac{\partial F}{\partial x} = P(x) + \sum_{s=1}^{r}\frac{1}{x-\xi_s}.$$

Now choose values of x such that $|x|$, while still less than ρ, is greater than the greatest of the quantities $|\xi_s|$; for all such values, the fractions on the right-hand side can be expanded in descending powers of x, and we have

$$\frac{1}{F}\frac{\partial F}{\partial x} = P(x) + \frac{r}{x} + \sum_{\kappa=1}^{\infty} S_\kappa x^{-\kappa-1},$$

where

$$S_\kappa = \xi_1^\kappa + \xi_2^\kappa + \ldots\ldots + \xi_r^\kappa.$$

Consequently, comparing the expansions

$$\frac{m}{x} + G(x) - \frac{\partial}{\partial x}\sum_{p=-\infty}^{\infty} G_p x^p$$

and

$$\frac{r}{x} + P(x) + \sum_{\kappa=1}^{\infty} S_\kappa x^{-\kappa-1},$$

and writing $x_1, \ldots, x_n$ for $a_1, \ldots, a_n$ in the latter, so that the two must now be identical, we have

$$r = m,$$
$$S_\kappa = \kappa G_{-\kappa}.$$

The first result shews that there are m roots of F within the range. The other expresses the sums of the powers of those roots as converging series of positive powers of $x_1, \ldots, x_n$; and therefore, when we write

$$f(x, x_1, \ldots, x_n) = (x - \xi_1)(x - \xi_2)\ldots(x - \xi_m)$$
$$= x^m + f_1 x^{m-1} + f_2 x^{m-2} + \ldots + f_m,$$

the coefficients $f_1, f_2, \ldots, f_m$ are power-series in $x_1, \ldots, x_n$, converging within the specified ranges*, and vanishing at $0, \ldots, 0$.

Further, the identity of the two expansions leads to the equation

$$P(x) = G(x) - \sum_{p=0}^{\infty} (p+1) G_{p+1} x^p,$$

so that, if

$$\Gamma(x, x_1, \ldots, x_n) = \int_0^x G(x)\, dx - \sum_{p=0}^{\infty} G_{p+1} x^{p+1},$$

where Γ is obviously a regular function of $x, x_1, \ldots, x_n$, we have

$$P(x) = \frac{\partial}{\partial x} \Gamma(x, x_1, \ldots, x_n).$$

Thus
$$\frac{1}{F} \frac{\partial F}{\partial x} = P(x) + \sum_{s=1}^{m} \frac{1}{x - \xi_s}$$
$$= \frac{\partial}{\partial x} \Gamma(x, x_1, \ldots, x_n) + \frac{\partial}{\partial x} \{\log f(x, x_1, \ldots, x_n)\},$$

and therefore

$$F = Uf(x, x_1, \ldots, x_n)\, e^{\Gamma(x, x_1, \ldots, x_n)},$$

where U is independent of x.

In order to determine U, it may be noticed that, when $x_1, \ldots, x_n$ all vanish, the value of U must be C: but it does not follow that U is equal to C for non-zero values of $x_1, \ldots, x_n$. The function F is a uniform analytical function of $x, x_1, \ldots, x_n$; the function f is a polynomial in x and is regular in $x_1, \ldots, x_n$; also Γ is regular; consequently U, if it be variable, is a regular function of $x_1, \ldots, x_n$, and therefore

$$U = C(1 + \text{positive powers of } x_1, \ldots, x_n)$$
$$= Ce^u,$$

where u is a regular function of the n variables. This function u may be absorbed into the function $\Gamma(x, x_1, \ldots, x_n)$ which still will be regular after the change; denoting the new function by $G(x, x_1, \ldots, x_n)$, we have

$$F = Cf(x, x_1, \ldots, x_n)\, e^{G(x, x_1, \ldots, x_n)}.$$

* If there be only one variable x_1, then $f_1, f_2, \ldots, f_m$ are regular functions of x_1 which vanish when $x_1 = 0$.

The coefficients in the regular functions f and G are independent of ρ, r_1, ..., r_n; the application of the principle of continuation shews that the equality holds throughout the domain of convergence round $0, 0, ..., 0$ that is common to F, f, G.

The alternative case is that in which the function F vanishes with x_1, ..., x_n for all values of x, so that no function $F_0(x)$ exists. We then transform the variables by a substitution

$$x, x_1, ..., x_n = \left(\begin{array}{cccc} c_{00}, & c_{01}, & ..., & c_{0n} \\ c_{10}, & c_{11}, & ..., & c_{1n} \\ .... \\ c_{n0}, & c_{n1}, & ..., & c_{nn} \end{array}\right) \!\!\left(y, y_1, ..., y_n\right);$$

the constant coefficients c of the matrix of substitution are arbitrary, subject solely to two conditions of inequality (i) that the determinant of the matrix does not vanish, (ii) that the quantity $g(c_{00}, c_{10}, ..., c_{n0})$ does not vanish, where $g(x, x_1, ..., x_n)$ denotes the aggregate of the terms of lowest dimensions, say l, in F. Manifestly with $(n+1)^2$ arbitrary constants at our disposal, the two conditions can be satisfied in an infinitude of ways. When the substitution is effected upon $F(x, x_1, ..., x_n)$, the new function, say $\bar{F}(y, y_1, ..., y_n)$, is an analytical function of y, y_1, ..., y_n; and clearly

$$\bar{F}(y, 0, ..., 0) = g(c_{00}, c_{10}, ..., c_{n0})\, y^l + \text{higher powers of } y,$$

so that $\bar{F}(y, 0, ..., 0)$ does not vanish for all values of y.

The preceding analysis may now be applied; and we infer that

$$\bar{F}(y, y_1, ..., y_n) = g(c_{00}, c_{10}, ..., c_{n0})\, \bar{f}(y, y_1, ..., y_n)\, e^{\bar{G}(y, y_1, ..., y_n)}.$$

The regular function $\bar{G}(y, y_1, ..., y_n)$ can, by the inverse substitution, be changed into a regular function

$$G(x, x_1, ..., x_n).$$

The function $\bar{f}(y, y_1, ..., y_n)$ is an algebraical polynomial in y of degree l of the form

$$y^l + \bar{f}_1 y^{l-1} + \bar{f}_2 y^{l-2} + ... + \bar{f}_l,$$

the coefficients $\bar{f}_1$, $\bar{f}_2$, ..., $\bar{f}_l$ being regular functions of y_1, ..., y_n, vanishing with y_1, ..., y_n.

Manifestly the zeros of the uniform analytical function $F(x, x_1, ..., x_n)$ in the domain of $0, 0, ..., 0$ are given by

$$f(x, x_1, ..., x_n) = 0, \quad \bar{f}(y, y_1, ..., y_n) = 0,$$

in the respective cases.

In the case when the number of variables is only two, we denote them by x, y. The function $F(x, y)$ then vanishes at 0, 0 and is regular in the immediate vicinity of those values; consequently it can be expanded in a converging power-series.

If $F(x, 0)$ does not vanish for all values of x, let Cx^m be the lowest power of x which it contains; then a function $f(x, y)$ of the form

$$f(x, y) = x^m + f_1 x^{m-1} + f_2 x^{m-2} + \ldots + f_m,$$

where $f_1, \ldots, f_m$ are regular functions of y vanishing when $y = 0$ exists such that

$$F(x, y) = Cf(x, y) e^{G(x, y)},$$

where G is a regular function of x and y, vanishing at 0, 0.

If $F(x, 0)$ does vanish for all values of x, then a transformation of variables

$$x = \lambda u + \mu v, \quad y = \lambda' u + \mu' v$$

is effected, subject to the limitations

$$\lambda \mu' - \lambda' \mu \neq 0, \quad F_m(\lambda, \lambda') \neq 0,$$

where $F_m(x, y)$ is the aggregate of terms of lowest dimensions in $F(x, y)$. Then $F(x, y)$ becomes a regular function of u and v, say $\bar{F}(u, v)$, such that $\bar{F}(u, 0)$ does not vanish for all values of u. If Au^m be the lowest power of u which $\bar{F}(u, 0)$ contains, then a function $\bar{f}(u, v)$ of the form

$$\bar{f}(u, v) = u^m + \bar{f}_1 u^{m-1} + \bar{f}_2 u^{m-2} + \ldots + \bar{f}_m,$$

where $\bar{f}_1, \bar{f}_2, \ldots, \bar{f}_m$ are regular functions of v vanishing when $v = 0$, exists such that

$$F(x, y) = A\bar{f}(u, v) e^{G(x, y)},$$

where G is a regular function of x and y, vanishing at 0, 0.

In this particular case when the number of variables is two, there is an alternative expression for F which is equally effective. The form that has been obtained has been made special in the variable x so that if, for instance, the zeros of F in the vicinity of 0, 0 are wanted, they are given by m values of x as functions of y from the equation

$$f(x, y) = 0.$$

It may, however, be desirable to have these zeros given by values of y as functions of x; they would ·be obtained most simply as follows.

If $F(0, y)$ does not vanish for all values of y, let By^p be the lowest power of y which it contains; then a function $g(x, y)$ of the form

$$g(x, y) = y^p + g_1 y^{p-1} + g_2 y^{p-2} + \ldots + g_p,$$

where $g_1, \ldots, g_p$ are regular functions of x vanishing when $x = 0$, exists such that

$$F(x, y) = Bg(x, y) e^{G(x, y)},$$

where G is a regular function of x and y, vanishing at $0, 0$. The zeros of F in the required form would be given by $g(x, y) = 0$.

Similarly, if $F(0, y)$ does vanish for all values of y, we transform the variables to

$$x = \lambda u + \mu v, \quad y = \lambda' u + \mu' v,$$

where $\lambda \mu' - \lambda' \mu \neq 0$, $F_p(\mu, \mu') \neq 0$; and we obtain a corresponding expression

$$F(x, y) = D\bar{g}(u, v) e^{G(x, y)};$$

here $\bar{g}(u, v)$ is algebraical in v of degree equal to the lowest power of v in $\bar{F}(0, v)$, where $F(x, y)$ becomes $\bar{F}(u, v)$, and the coefficients of the powers of v are regular functions of u which vanish when $u = 0$.

CHAPTER II.

10. It has been shewn that the normal form of a system of equations involving one independent variable is

$$\frac{du_r}{dz} = f_r(u_1, \ldots, u_n, z), \qquad (r = 1, \ldots, n),$$

where the functions f_r are analytical functions. To secure the existence of integrals of this system, an obvious preliminary condition is that the regions of existence of the functions $f_1, f_2, \ldots, f_n$, regarded as analytical functions of $u_1, \ldots, u_n, z$, must have some common range that is not infinitesimal. In this common range let $z = c$, $u_1 = a_1, \ldots, u_n = a_n$ denote a place at which all the functions are regular; each of them can therefore be expressed in the form of a power-series converging absolutely for all values of the variables defined by

$$|z - c| \leqslant \rho, \quad |u_1 - a_1| \leqslant r_1, \ldots, |u_n - a_n| \leqslant r_n.$$

Then, if r be the smallest of the magnitudes $r_1, \ldots, r_n$, the series certainly converge absolutely for values of the variables defined by

$$|z - c| \leqslant \rho, \quad |u_m - a_m| \leqslant r, \qquad (m = 1, \ldots, n).$$

The fundamental theorem relating to the integrals of the system of equations, when these various conditions are satisfied, is due to Cauchy; it is as follows:—

In a region of the z-variable that is not infinitesimal in extent, functions $u_1, u_2, \ldots, u_n$ of z exist satisfying the system of differential equations and assuming values $a_1, \ldots, a_n$ respectively when $z = c$.

---

* References to a number of authorities are given in a bibliographical note at the end of this chapter.

Let M denote the maximum among the values of the moduli of all the functions $f_1, \ldots, f_n$ for all possible places within the common range of existence; and let a new function, denoted by $F(v_1, \ldots, v_n, z)$ and defined as

$$\frac{M}{\left(1 - \dfrac{z - c}{\rho}\right)\left(1 - \dfrac{v_1 - a_1}{r}\right)\left(1 - \dfrac{v_2 - a_2}{r}\right) \ldots \left(1 - \dfrac{v_n - a_n}{r}\right)},$$

be constructed. It is a known proposition* that

$$\left| \frac{\partial^{m_1 + m_2 + \ldots + m_n + p} f_r}{\partial u_1^{m_1} \ldots \ldots \partial u_n^{m_n} \partial z^p} \right| < \left| \frac{\partial^{m_1 + \ldots + m_n + p} F}{\partial v_1^{m_1} \ldots \ldots \partial v_n^{m_n} \partial z^p} \right|$$

$$< m_1! \, m_2! \ldots m_n! \, p! \, \frac{M}{r^{m_1 + \ldots + m_n} \rho^p},$$

for all positive or zero values of the integers $m_1, \ldots, m_n, p$, when in the former the values $u_1 = a_1, \ldots, u_n = a_n$, $z = c$ are substituted and in the latter the values $v_1 = a_1, \ldots, v_n = a_n$, $z = c$. The function F is called a *dominant* function†.

First, consider the system of equations

$$\frac{dv_1}{dz} = \frac{dv_2}{dz} = \ldots = \frac{dv_n}{dz} = F(v_1, \ldots, v_n, z).$$

From

$$\frac{dv_1}{dz} = \frac{dv_m}{dz}$$

it follows that

$$v_1 - a_1 = v_m - a_m,$$

the constant of integration being determined by the assignment of $a_1, \ldots, a_n$ as simultaneous values of $v_1, \ldots, v_n$ respectively; and therefore the quantities $v_m - a_m$, for $m = 1, \ldots, n$, are equal to one another. Thus any one of them, say $v_1 - a_1$, is determined by the equation

$$\frac{dv_1}{dz} = \frac{M}{\left(1 - \dfrac{z - c}{\rho}\right)\left(1 - \dfrac{v_1 - a_1}{r}\right)^n},$$

whence

$$\left(1 - \frac{v_1 - a_1}{r}\right)^{n+1} = A + (n+1)\frac{\rho}{r} M \log\left(1 - \frac{z - c}{\rho}\right),$$

---

* See the author's *Theory of Functions* (which, in subsequent references, will be denoted by *Th. Fns.*), § 22.

† Poincaré calls such a function *majorante*.

A being a constant of integration to be determined. Let that branch of the logarithmic function be chosen which is zero for $z = c$; the branch is regular for all values of z such that $|z - c| < \rho$, that is, for all values of z at present under consideration. Now let c be the value of z when a_1 is the value assigned to v_1; it appears that $A = 1$, and therefore

$$\frac{v_1 - a_1}{r} = 1 - \left\{ 1 + (n + 1)\frac{\rho}{r} M \log \left(1 - \frac{z - c}{\rho} \right) \right\}^{\frac{1}{n+1}},$$

on choosing that branch of the radical which is unity when $z = c$.

The value of v_1 thus obtained is regular for values of $|z - c|$ which, being less than ρ, exclude the branch-points of the radical from a simply-connected field of variation in the z-plane. The radius σ of the circle within which v_1 is uniform is therefore given by

$$1 + (n + 1)\frac{\rho}{r} M \log \left(1 - \frac{\sigma}{\rho} \right) = 0,$$

so that
$$\sigma = \rho \left\{ 1 - e^{-\frac{r}{(n+1)M\rho}} \right\};$$

evidently σ is a finite quantity less than ρ, so that v_1 is regular within a finite circle. Consequently, functions $v_1, \ldots, v_n$ exist satisfying the system of equations

$$\frac{dv_1}{dz} = \frac{dv_2}{dz} = \ldots = \frac{dv_n}{dz} = F(v_1, \ldots, v_n, z),$$

and acquiring the values $a_1, \ldots, a_n$ when $z = c$; and each of them is a function of z, regular for all values of z such that

$$|z - c| < \rho \left\{ 1 - e^{-\frac{r}{(n+1)M\rho}} \right\}.$$

For the purpose of establishing the existence of integrals of the equations

$$\frac{du_s}{dz} = f_s(u_1, \ldots, u_n, z), \qquad (s = 1, \ldots, n),$$

(which will be called the original system), it is convenient to compare them with the integrals of the system just considered

$$\frac{dv_s}{dz} = F(v_1, \ldots, v_n, z), \qquad (s = 1, \ldots, n),$$

(which will be called the dominant system). The integrals of the latter have been proved to be functions of z that are regular for all points z lying within the circle $|z - c| = \sigma$; and therefore for such values of z they can be expressed in the form

$$v_s - a_s = (z - c)\frac{dv_s}{dz_c} + \frac{(z - c)^2}{2!}\frac{d^2v_s}{dz_c^2} + \dots,$$

where the coefficients of the powers of $z - c$ are the values of the derivatives of v_s when $z = c$. The values of the successive derivatives of all the functions v_s when $z = c$ can be obtained by differentiating the equations of the dominant system any number of times, substituting after each differentiation the general values of the first derivatives, and then inserting the values $z = c$, $v_1 = a_1, \dots, v_n = a_n$. They all clearly are real positive quantities.

Now if integrals of the original system exist which are regular and acquire the values $a_1, \dots, a_n$ for the value c of z, then power-series of the form

$$u_s - a_s = (z - c)\frac{du_s}{dz_c} + \frac{(z - c)^2}{2!}\frac{d^2u_s}{dz_c^2} + \dots$$

define the integrals within some region of existence in the vicinity of the point c, the coefficients of the powers of $z - c$ being the values of the derivatives of the functions $u_1, \dots, u_n$ when $z = c$. And the values of these derivatives can be formally deduced from the equations of the original system by precisely the same process as the one by which, as explained above, it is possible to deduce the values of the derivatives of v from the equations of the dominant system.

From the definition of the function $F(v_1, \dots, v_n, z)$, it at once follows that, when $z = c$,

$$\left|\frac{du_s}{dz}\right| < \frac{dv_s}{dz}, \qquad\qquad (s = 1, \dots, n).$$

Again,

$$\frac{d^2u_s}{dz^2} = \frac{\partial f_s}{\partial z} + \sum_{p=1}^{n} f_p \frac{\partial f_s}{\partial u_p},$$

so that, when $z = c$,

$$\left|\frac{d^2u_s}{dz^2}\right| < \left|\frac{\partial f_s}{\partial z}\right| + \sum_{p=1}^{n} |f_p| \left|\frac{\partial f_s}{\partial u_p}\right|,$$

the values $z = c$, $u_1 = a_1, \dots, u_n = a_n$ being substituted in the right-hand side. But for these values

$$\left|\frac{\partial f_s}{\partial z}\right| < \frac{\partial F}{\partial z}, \qquad \left|\frac{\partial f_s}{\partial u_p}\right| < \frac{\partial F}{\partial v_p}, \qquad |f_p| < F,$$

when the values $z = c$, $v_1 = a_1$, ..., $v_n = a_n$ are substituted in F and its derivatives; and therefore

$$\left| \frac{d^2 u_s}{dz^2} \right| < \frac{\partial F}{\partial z} + \sum_{p=1}^{n} F \frac{\partial F}{\partial v_p}$$

for such values. By proceeding from the dominant system in the same manner, it is evident that

$$\frac{d^2 v_s}{dz^2} = \frac{\partial F}{\partial z} + \sum_{p=1}^{n} F \frac{\partial F}{\partial v_p}$$

for the values considered; and therefore, when $z = c$,

$$\left| \frac{d^2 u_s}{dz^2} \right| < \frac{d^2 v_s}{dz^2}.$$

For the values of the third derivatives at $z = c$, the equations

$$\frac{d^2 u_s}{dz^2} = \frac{\partial f_s}{\partial z} + \sum_{p=1}^{n} f_p \frac{\partial f_s}{\partial u_p}, \qquad \frac{d^2 v_s}{dz^2} = \frac{\partial F}{\partial z} + \sum_{p=1}^{n} F \frac{\partial F}{\partial v_p}$$

would be treated in the same manner as above: and a similar argument leads to the conclusion that

$$\left| \frac{d^3 u_s}{dz^3} \right| < \frac{d^3 v_s}{dz^3}$$

at $z = c$. And it can now be seen that, for all values of m, the inequality

$$\left| \frac{d^m u_s}{dz^m} \right| < \frac{d^m v_s}{dz^m}$$

can be established; for the processes that occur in using the equations of the original system are differentiation, multiplication, addition and the replacement of the modulus of any term by the greatest possible value it can possess: and the completion of these processes leads to the same result as is obtained in using the equations of the dominant system.

The series $\quad (z - c) \dfrac{dv_s}{dz_c} + \dfrac{(z - c)^2}{2!} \dfrac{d^2 v_s}{dz_c^2} + \cdots,$

which defines the function $v_s - a_s$, converges absolutely for points within the circle $|z - c| = \sigma$. The moduli of its terms are greater than the moduli of the terms of the series

$$(z - c) \frac{du_s}{dz_c} + \frac{(z - c)^2}{2!} \frac{d^2 u_s}{dz_c^2} + \cdots;$$

and therefore the latter series also converges absolutely for points within the circle. We accordingly construct functions u_s', given by

$$u_s' - a_s = (z - c)\frac{du_s}{dz_c} + \frac{(z - c)^2}{2!}\frac{d^2 u_s}{dz_c^2} + \dots, \qquad (s = 1, \dots, n),$$

using the values of $\dfrac{du_s}{dz_c}, \dfrac{d^2 u_s}{dz_c^2}, \dots$ at $z = c$ as deduced in the preceding formal calculations. These functions u_s' are regular within the circle $|z - c| = \sigma$; consequently*, all their derivatives also exist with the same character within that circle. The function $|u_s' - a_s|$ is continuous, and its value is zero when $z = c$; being always positive, it begins to increase (save in one case to be noticed hereafter) with increasing values of $|z - c|$, though the increase may not be persistent. It is finite for all values of z such that $|z - c| < \sigma$, and it cannot exceed the value r, as was proved; and therefore $|u_s' - a_s|$ either will never attain a value r for values of z within the circle $|z - c| = \sigma$ or, if it attains the value r, it will first attain that value when $|z - c| = \alpha_s \sigma$, where α_s is a proper fraction that does not vanish. Of all the fractions α_s, if any arise, let α be the smallest; then either

$$|u_s' - a_s| < r \text{ for } |z - c| < \sigma, \qquad (s = 1, \dots, n),$$

or

$$|u_s' - a_s| \leqslant r \text{ for } |z - c| \leqslant \alpha\sigma, \qquad (s = 1, \dots, n).$$

Also $\sigma < \rho$, and therefore also $\alpha\sigma < \rho$; hence there is a region within which du_s'/dz has been proved to exist, this region is finite, and it is certainly included for all the variables in the region in which the functions $f_1, \dots, f_n$ are supposed regular.

Lastly, from the manner in which the coefficients in the power-series giving the values of $u_s' - a_s$ have been obtained, it is evident that du_s'/dz and $f_s(u_1', \dots, u_n', z)$ have the same value at $z = c$ and that any derivative of either of them has the same value at $z = c$ as the derivative of the other of the same order. They have been shewn to possess a common region of existence, within which they can be expressed as power-series; and their derivatives are the same at $z = c$. Hence

$$\frac{du_s'}{dz} = f_s(u_1', \dots, u_n', z), \qquad (s = 1, \dots, n),$$

at all points within this common region of existence: that is to

* *Th. Fns.*, § 21.

say, functions $u_1{}'$, ..., $u_n{}'$ exist, satisfying the differential equations and acquiring values a_1, ..., a_n for $z = c$.

Cauchy's theorem is thus established under the assigned conditions.

The case of exception referred to, in which the value of $|u_s - a_s|$ does not begin to increase from zero with values of $|z - c|$ increasing from zero, is that in which the power-series for $u_s - a_s$ is evanescent by reason of zero values of all of the derivatives of u_s when $z = c$; and this may occur for more than one of the quantities u. The subsequent argument is not valid in this case: and the inference is that the corresponding functions u_s, defined by the equations and the assigned conditions jointly, are merely constants for values of z such that $|z - c| < \rho$. On the other hand, the inference can only be made for values of the variables occurring in the regions where the functions f_1, ..., f_n are regular; other investigations will be required when values of the variables are considered that give rise to branchings, or infinities, or discontinuities, or other irregularities, of the functions f_1, ..., f_n.

A simple example will illustrate the last remark. Let the equation

$$\frac{du}{dz} = \frac{u^2}{1-z}$$

be considered; a general solution is

$$\frac{1}{u} = \log(1-z) + A.$$

If u is to have the value 0 when $z = 0$, and if z is restricted so that $|z| < 1$, then A is infinite; and for all such values of z, the value of u is zero. If other values of z be included and it be possible to have $z = 1$, the value of u cannot be asserted to be zero; but $z = 1$ is an infinity of $u^2/(1-z)$ for non-zero values of u, and $u = 0$, $z = 1$ make $u^2/(1-z)$ indeterminate, and thus the initial conditions of the theorem cease to be satisfied.

And more generally, it is not difficult to see that the case of exception can occur only when at least one equation in the system has the form

$$\frac{du_s}{dz} = (u_1 - a_1)g_{1s} + (u_2 - a_2)g_{2s} + \ldots + (u_n - a_n)g_{ns}$$

for $s = 1$, ..., n, the various coefficients g_{11}, ..., g_{nn} being analytical functions of u_1, ..., u_n, z that are regular for the regions of variation considered; in such regions, the equations determine the corresponding functions u_s as constants.

11. In the same way* as in the case of functions defined by power-series, it may be proved that, if two paths ACB and ADB in the z-plane enclose a space for every point of which all the conditions necessary for the establishment of the existence-theorem are satisfied, then the set of regular integrals at B are the same whether the passage of z from A to B be made by the path ACB or by the path ADB. Hence also the values of the set of integrals at any point, as depending upon the values assigned at any other point, are unaffected when the path of variation of the independent variable is deformed continuously within the region of existence of the integrals. Further, it at once follows that a path $ACBDA$, returning to the initial point and enclosing a simply-connected space every point of which is ordinary for all the functions concerned, restores at A the values of the set of integrals as initially assigned.

Unique Determination of the Regular System.

12. The preceding investigation establishes the existence of integrals of the equations

$$\frac{du_s}{dz} = f_s(u_1, \dots, u_n, z), \qquad (s = 1, \dots, n),$$

the integrals being regular functions of z which assume values $a_1, \dots, a_n$ when $z = c$; and, save in the case of exception indicated, the range of variation of the variables is given by equations of the form

$$|z - c| \leqslant \rho', \quad |u_s - a_s| \leqslant \rho'',$$

where ρ' and ρ'' are finite quantities.

Moreover, a review of the investigation shews that at no stage has there occurred any possibility of deviation from uniqueness of value; and it might therefore be concluded that the system of regular integrals is unique, a conclusion which can be established by the following considerations.

The theorem of Cauchy proves the existence of a set of quantities $w_1, w_2, \dots, w_n$, which are regular functions of z near the point $z = c$, take the values $\alpha_1, \alpha_2, \dots, \alpha_n$, respectively at that point, and satisfy the differential equations.

* *Th. Fns.*, § 90.

It remains to be seen whether more than one set of quantities $w_1, w_2, \ldots, w_n$, exists, satisfying these conditions.

If possible, let $w_1', w_2', \ldots, w_n'$ and $w_1, w_2, \ldots, w_n$, be two different sets of quantities satisfying the conditions.

Since $w_1', w_2', \ldots, w_n'$; $w_1, w_2, \ldots, w_n$, are functions of z, regular near $z = c$, they can in the vicinity of this point be expanded in series of ascending powers of $z - c$. Let the series be represented by

$$w_r = \alpha_r + b_r (z - c) + c_r (z - c)^2 + \ldots, \qquad (r = 1, 2, \ldots, n),$$
$$w_r' = \alpha_r + b_r' (z - c) + c_r' (z - c)^2 + \ldots, \qquad (r = 1, 2, \ldots, n).$$

Then since the differential equations are satisfied, we have

$$\frac{dw_r}{dz} = f_r (w_1, w_2, \ldots, w_n, z).$$

Differentiating this,

$$\frac{d^2 w_r}{dz^2} = \frac{\partial f_r}{\partial w_1} f_1 + \frac{\partial f_r}{\partial w_2} f_2 + \ldots + \frac{\partial f_r}{\partial w_n} f_n + \frac{\partial f_r}{\partial z} = f_r' (w_1, w_2, \ldots, w_n, z)$$

say. Therefore

$$\frac{d^3 w_r}{dz^3} = \frac{\partial f_r'}{\partial w_1} f_1 + \frac{\partial f_r'}{\partial w_2} f_2 + \ldots + \frac{\partial f_r'}{\partial w_n} f_n + \frac{\partial f_r'}{\partial z} = f_r'' (w_1, \ldots, w_n, z)$$

say. Similarly the higher derivatives can be found in terms of

$$w_1, w_2, \ldots, w_n, z.$$

Hence

$$\left(\frac{dw_r}{dz}\right)_{z=c} = f_r (\alpha_1, \alpha_2, \ldots, \alpha_n, c); \qquad \left(\frac{d^2 w_r}{dz^2}\right)_{z=c} = f_r' (\alpha_1, \alpha_2, \ldots, \alpha_n, c);$$
$$\left(\frac{d^3 w_r}{dz^3}\right)_{z=c} = f_r'' (\alpha_1, \alpha_2, \ldots, \alpha_n, c); \ldots.$$

But in the same way it can be shewn that

$$\left(\frac{dw_r'}{dz}\right)_{z=c} = f_r (\alpha_1, \alpha_2, \ldots, \alpha_n, c); \qquad \left(\frac{d^2 w_r'}{dz^2}\right)_{z=c} = f_r' (\alpha_1, \alpha_2, \ldots, \alpha_n, c);$$
$$\left(\frac{d^3 w_r'}{dz^3}\right)_{z=c} = f_r'' (\alpha_1, \alpha_2, \ldots, \alpha_n, c); \ldots.$$

Hence

$$\left(\frac{dw_r}{dz}\right)_{z=c} = \left(\frac{dw_r'}{dz}\right)_{z=c}; \quad \left(\frac{d^2 w_r}{dz^2}\right)_{z=c} = \left(\frac{d^2 w_r'}{dz^2}\right)_{z=c}; \ldots;$$

that is, $\qquad\qquad b_r = b_r', \quad c_r = c_r', \ldots.$

Therefore the series expressing w_r and w_r' are identical; that is, w_r' is not distinct from w_r.

So there cannot be two distinct sets of quantities $w_1, w_2, \ldots, w_n$ satisfying the given conditions; the set given by Cauchy's existence-theorem is the only possible set of regular integrals.

13. Another proof, establishing this important result, is as follows*.

Suppose that it is possible that two different sets of regular solutions of the equations

$$\frac{du_r}{dz} = f_r(u_1, \ldots, u_n, z), \qquad (r = 1, \ldots, n),$$

can exist having the same values $a_1, \ldots, a_n$ for $z = c$; and let them be denoted by $u_1, \ldots, u_n$; $u_1 + v_1, \ldots, u_n + v_n$ respectively. All the quantities u, $u + v$, f are regular functions of z for the respective regions in which they exist, all these regions being included within the region of existence of the n functions f. It therefore follows that the quantities v_s, being $(u_s + v_s) - u_s$, also are regular functions of z for the region of existence common to the quantities u and $u + v$; and they all have a zero value for $z = c$.

The region of simultaneous existence of the two sets of integrals may be more restricted than the region of existence of the n functions f; but it is not infinitesimal. Within this region let a portion be defined by $|z - c| \leqslant b$; and suppose that, for $|z - c| = b$, the value of $|u_s - a_s|$ is equal to or less than μ_s, with the further assumption that, as the value of $|z - c|$ increases from 0, this value b is the first for which $|u_s - a_s|$ acquires the value μ_s: an assumption justified for values μ_s, b that are not infinitesimal, because the continuous quantities $|u_s - a_s|$ begin to increase from zero with values of $|z - c|$ increasing from zero.

Similarly, let $|u_s + v_s - a_s| \leqslant \lambda_s$ for $|z - c| = b$, and suppose that, as the value of $|z - c|$ increases from zero, this value b is the first for which $|u_s + v_s - a_s|$ acquires the value λ_s: this assumption is justified as before, by noting that the quantities $|u_s + v_s - a_s|$ begin to increase from zero as $|z - c|$ increases from zero, and by tracing simultaneously the values of $|u_s - a_s|$ and $|u_s + v_s - a_s|$ with increasing values of $|z - c|$. Then as the region of existence considered lies within the region of existence of the functions $f_1, \ldots, f_n$, which was defined by

$$|z - c| \leqslant \rho, \qquad |u_m - a_m| \leqslant r,$$

* It is based upon Jordan's method, *Cours d'Analyse*, t. III, pp. 94—101.

it follows that $b < \rho$, $\lambda_s < r$, $\mu_s < r$. Moreover as all the quantities considered are regular functions, so that (§ 11) they acquire at z the same value whatever be the path by which the variable passes from c to z, it will be assumed that the z-path is a straight line so that, if at any point $|z - c| = s$, the value of s varies from 0 to b and $|dz| = ds$.

Let t denote a real variable lying between 0 and 1, and let

$$w_s = u_s + tv_s,$$

so that, for all values of t, w_s lies within the region of existence common to u_s and $u_s + v_s$, is regular in that region, and acquires the value a_s for $z = c$; it is easy to see that the greatest value of $|w_s - a_s|$ lies between λ_s and μ_s; so that, denoting it by $\theta_s|$, this value $\theta_s < r$. It follows that all possible values of w_s for $|z - c| \leqslant b$ are included within the region of existence of the functions $f_1, \ldots, f_n$, and that consequently $f_r(w_1, \ldots, w_n, z)$, as well as all its partial derivatives with regard to $w_1, \ldots, w_n$, for $r = 1, \ldots, n$, can be expanded in absolutely converging power-series of $w_1 - a_1$, $w_2 - a_2$, $\ldots$, $w_n - a_n$, $z - c$.

Hence

$$\frac{\partial f_s}{\partial w_1} = \frac{\partial f_s}{\partial a_1} + \frac{\partial^2 f_s}{\partial a_1 \partial c}(z - c) + \frac{\partial^2 f_s}{\partial a_1{}^2}(w_1 - a_1) + \frac{\partial^2 f_s}{\partial a_1 \partial a_2}(w_2 - a_2) + \ldots,$$

where

$$\frac{\partial^{r + r_1 + r_2 + \ldots} f_s}{\partial c^r \partial a_1{}^{r_1} \partial a_2{}^{r_2} \ldots}$$

denotes the result of substituting the values c, a_1, a_2, $\ldots$ for z, w_1, w_2, $\ldots$ respectively, in

$$\frac{\partial^{r + r_1 + r_2 + \ldots} f_s}{\partial z^r \partial w_1{}^{r_1} \partial w_2{}^{r_2} \ldots};$$

and therefore

$$\left| \frac{\partial f_s}{\partial w_1} \right| \leqslant \left| \frac{\partial f_s}{\partial a_1} \right| + |z - c| \left| \frac{\partial^2 f_s}{\partial a_1 \partial c} \right| + |w_1 - a_1| \left| \frac{\partial^2 f_s}{\partial a_1{}^2} \right| + \ldots$$

$$\leqslant \frac{M}{r} + |z - c| \frac{M}{r\rho} + |w_1 - a_1| \, 2! \, \frac{M}{r^2} + \ldots$$

(as in § 10), so that

$$\left| \frac{\partial f_s}{\partial w_1} \right| \leqslant \frac{M}{r} \frac{1}{1 - \dfrac{|z - c|}{\rho}} \frac{1}{\left\{ 1 - \dfrac{|w_1 - a_1|}{r} \right\}^2} \frac{1}{1 - \dfrac{|w_2 - a_2|}{r}} \cdots \frac{1}{1 - \dfrac{|w_n - a_n|}{r}}$$

$$\leqslant \frac{M}{r} \frac{1}{1 - \dfrac{b}{\rho}} \frac{1}{\left(1 - \dfrac{\theta_1}{r} \right)^2} \frac{1}{1 - \dfrac{\theta_2}{r}} \cdots \frac{1}{1 - \dfrac{\theta_n}{r}}.$$

Hence if θ be the greatest of the quantities $\theta_1, \ldots, \theta_n$ so that $\theta < r$, it follows that

$$\left|\frac{\partial f_s}{\partial w_1}\right| \leqslant \frac{M}{r} \frac{1}{1 - \dfrac{b}{\rho}} \frac{1}{\left(1 - \dfrac{\theta}{r}\right)^{n+1}}$$

for all values of the variables within the portion of the region of existence under consideration.

In the same way it may be established that

$$\left|\frac{\partial f_s}{\partial w_m}\right| \leqslant N, \qquad (m = 1, \ldots, n;\ s = 1, \ldots, n),$$

for all values of the variables within the portion of the region of existence under consideration, N denoting the quantity

$$\frac{1}{r} M \left(1 - \frac{b}{\rho}\right)^{-1} \left(1 - \frac{\theta}{r}\right)^{-n-1},$$

which is finite.

Now

$$f_s(u_1 + v_1,\ u_2 + v_2,\ \ldots,\ u_n + v_n,\ z) - f_s(u_1,\ u_2,\ \ldots,\ u_n,\ z)$$

$$= \int_0^1 \frac{d}{dt} f_s(u_1 + v_1 t,\ u_2 + v_2 t,\ \ldots,\ u_n + v_n t,\ z)\, dt$$

$$= \int_0^1 \left(v_1 \frac{\partial}{\partial w_1} + v_2 \frac{\partial}{\partial w_2} + \ldots + v_n \frac{\partial}{\partial w_n}\right) f_s(w_1,\ \ldots,\ w_n,\ z)\, dt$$

$$= v_1 \int_0^1 \frac{\partial f_s}{\partial w_1}\, dt + v_2 \int_0^1 \frac{\partial f_s}{\partial w_2}\, dt + \ldots + v_n \int_0^1 \frac{\partial f_s}{\partial w_n}\, dt.$$

Also

$$\left|\frac{\partial f_s}{\partial w_m}\right| \leqslant N$$

for all possible values of the variables that occur in the integral, and therefore

$$\left|\int_0^1 \frac{\partial f_s}{\partial w_m}\, dt\right| \leqslant N \int_0^1 dt \leqslant N.$$

Consequently

$$|f_s(u_1 + v_1,\ u_2 + v_2,\ \ldots,\ u_n + v_n,\ z) - f_s(u_1,\ \ldots,\ u_n,\ z)|$$
$$\leqslant N\{|v_1| + |v_2| + \ldots + |v_n|\}.$$

Because $u_1, \ldots, u_n$ and $u_1 + v_1, \ldots, u_n + v_n$ are two sets of solutions of the given system of differential equations, we have

$$\frac{dv_m}{dz} = \frac{d(u_m + v_m)}{dz} - \frac{du_m}{dz}$$

$$= f_m(u_1 + v_1,\ u_2 + v_2,\ \ldots,\ u_n + v_n,\ z) - f_m(u_1,\ \ldots,\ u_n,\ z)$$

for $m = 1, \ldots, n$; and therefore

$$\left| \frac{dv_m}{dz} \right| \leqslant N \{ |v_1| + |v_2| + \ldots + |v_n| \}$$

for the n values of m, so that

$$|dv_m| \leqslant N \{ |v_1| + |v_2| + \ldots + |v_n| \} \, ds.$$

Now for any complex variable w

$$d |w| \leqslant |dw|.$$

Hence if $V = |v_1| + |v_2| + \ldots + |v_n|,$

it follows that $dV = d|v_1| + d|v_2| + \ldots + d|v_n|$

$$\leqslant |dv_1| + |dv_2| + \ldots + |dv_n|$$

$$\leqslant nNV \, ds,$$

or $$\frac{dV}{ds} \leqslant nNV.$$

Therefore $$\frac{dV}{ds} = nNV - p,$$

where p is a real quantity that may be zero but cannot be negative.

The variables $v_1, \ldots, v_n$ all have a zero value for $z = c$; consequently V has a zero value when $|z - c| = 0$, that is, when $s = 0$. As n and N are independent of s, the above equation gives

$$Ve^{-nNs} = A - \int pe^{-nNs} \, ds,$$

where A is an arbitrary constant, which can be determined by the condition that $V = 0$ when $s = 0$; so that

$$Ve^{-nNs} = - \int_0^s pe^{-nNs} \, ds.$$

The real quantity p may be zero but cannot be negative; hence, as s is real and not infinitesimal, the integral is positive except only in the case when p is zero for all values of s. But V is necessarily not a negative quantity, so that p must be zero; and therefore

$$\frac{dV}{ds} = nNV,$$

with the condition that $V = 0$ when $s = 0$. This equation gives

$$Ve^{-nNs} = B,$$

where B is an arbitrary constant. It is determined by the condition that $V = 0$ when $s = 0$, and by the fact that only a finite region is under consideration, so that s has only finite values; and the equation becomes

$$V e^{-nNs} = 0,$$

so that V must vanish for all the values of s considered. Hence each of the quantities $|v_1|, |v_2|, ..., |v_n|$ must vanish; and therefore also the quantities $v_1, ..., v_n$ vanish for the range of variables considered.

Each of these quantities v is a regular function; each of them has been proved to be zero along a finite continuous line in the z-plane; consequently*, each of them is zero everywhere in the part of the region of its existence that lies in the domain of $z = c$. It therefore follows that the two sets of regular solutions, having the same values $a_1, ..., a_n$ for $z = c$, are the same; consequently, *the regular integrals of the differential equations having the values* $a_1, ..., a_n$ *for $z = c$ are uniquely determined by these conditions.*

It remains to take account of the possibility that the set of integrals $u_1, ..., u_n$, determined as having the values $a_1, ..., a_n$ for $z = c$, have those constant values for a finite range of variation of z, say, for $|z - c| \leqslant \sigma$.

If another set $u_1 + v_1, ..., u_n + v_n$ exist within that range determined by the same initial values for $z = c$, then $v_1, ..., v_n$ have zero values for $z = c$; and they are regular within a finite range of variation. The sole difference between the more general case, when the quantities u are variable, and the present, when they are constant, is that all the quantities μ_s in the preceding discussion now become zero. This change is easily seen to leave both the course of the argument and the inferences unaffected; and therefore it is concluded, in the same way as before, that all the quantities v are zero for the range of variation of z; that is, that the set of solutions determined by their values for $z = c$ are a unique set for a range of variation of z such as to make the functions $f_1, ..., f_n$ regular functions.

It has been proved that, if the region of existence of the analytical functions $f_1, ..., f_n$ within which they are regular, be defined by the conditions

$$|z - c| \leqslant \rho, \quad |u_m - a_m| \leqslant r,$$

* *Th. Fns.*, § 37.

then solutions of the equations

$$\frac{du_s}{dz} = f_s(u_1, \ldots, u_n, z), \qquad\qquad (s=1, \ldots, n),$$

are obtainable as power-series in $z - c$, which certainly converge within the range $|z - c| \leqslant \rho_1$, where

$$\rho_1 < \rho \left\{ 1 - e^{-\frac{r}{(n+1)M\rho}} \right\},$$

a range included within the range of the functions f. But though these series for $u_s - a_s$ converge, no indication is given as to the greatest value of $|u_s - a_s|$ within the range of z, except that it is less than a quantity which itself is less than r; as has been seen, it may even be a constant zero. Manifestly, the greatest value of $|u_s - a_s|$, within the region of existence of the functions f occurring in the differential equations that define the quantities u in a general case, must depend upon the quantities $a_1, \ldots, a_n, c, \rho$; the more precise determination of this dependence requires to be effected.

Note.

Several methods have been devised for the establishment of the existence of integrals of differential equations.

The earliest of them appears to be due to Cauchy* who in the first instance applied the calculus of limits, as it is called, to the discussion of a system of differential equations in which all the variables are real; an exposition of the method is given by Lipschitz†, by Picard‡, and by others, in a form that is simpler and is more precise in definition than Cauchy's. Cauchy afterwards extended the application of the calculus of limits to systems of ordinary equations in which the variables are complex and to systems of partial differential equations§; and his methods have been improved and amplified by Méray‖, and by Riquier¶, who has applied Méray's methods to partial differential equations.

The most conspicuous additions after Cauchy's memoirs are those contained in the memoir of Briot and Bouquet**, who discussed in much detail the properties of the integral of a single equation; some limitations on their proof are pointed out by Picard in his exposition†† of the method, and Fuchs has devoted some memoirs to the discussion of difficulties and omitted cases‡‡.

* It is reproduced by Moigno, *Calcul Différentiel et Intégral*, (1844).

† *Lehrbuch der Analysis.*

‡ *Traité d'Analyse*, t. II, ch. XI, §§ 1–4.

§ *Œuvres complètes de Cauchy*, 1re sér., t. VII.

‖ A full statement of M. Méray's position as regards the development of the theory of the existence of integrals is given in his *Leçons nouvelles sur l'analyse infinitésimale*, t. I.

¶ *Annales de l'École Norm. Supér.*, 3me Sér., t. X, (1893), pp. 65–86, 123–150, 167–181.

** *Journal de l'École Polytechnique*, t. XXI, (1856), pp. 133–198.

†† *l. c. supra*, § 10.

‡‡ References will be found later, §§ 28–34, where a discussion of these matters will be found.

Briot and Bouquet's proof of the uniqueness of a system of solutions defined by suitably chosen initial conditions has been improved by Jordan*; and the whole method is applied by him to the establishment of the existence of integrals of a system of equations. And an extension has been made by Poincaré† to the case when the equations involve an arbitrary parameter, in powers of which it may be desired to expand the functions that occur.

An independent method, having some analogy with Cauchy's, has been constructed by Weierstrass; it is slightly sketched by Mme v. Kowalevsky‡ who extended it to partial differential equations: and it is expounded in fuller detail by Königsberger§, who proceeds to the establishment of the integrals directly from Weierstrass's canonical form.

There is also another process due to Weierstrass‖ constructed at a time when he was unacquainted with Cauchy's investigation (*l. c.*, p. 85).

Picard uses a method of successive approximations to the integrals, applying it in the first instance to the case when the variables are real; by generalising this method so that it may be applied to the case when the variables are complex, he is able to shew that the series obtained as integrals of the equations have a circle of convergence that is larger than the circle obtained above in § 10. He also uses the existence-theorem for partial differential equations to deduce the existence of integrals of ordinary differential equations¶.

---

* *Cours d'Analyse*, t. III, §§ 77–81.

† *Les méthodes nouvelles de la mécanique céleste*, t. I, §§ 23–27; it had been given previously in the *Acta Mathematica*, vol. XIII, (1890), pp. 15 sqq.

‡ *Crelle's Journal*, vol. LXXX, (1875), pp. 1–5.

§ *Lehrbuch der Theorie der Differentialgleichungen*, Kap. I, Sect. III.

‖ *Ges. Werke*, t. I, pp. 75–84.

¶ *Ib.* t. II, chap. XI; the whole of the chapter will well repay perusal.

CHAPTER III.

Continuations of a Regular Integral.

14. The equations under discussion, viz.,

$$\frac{du_s}{dz} = f_s (u_1, \dots, u_n, z), \qquad (s = 1, \dots, n),$$

were obtained as a normal form of a general system of differential equations; the functions f_s being regular in the vicinity of assigned values of the variables, when certain conditions are satisfied. The equations may, however, be propounded independently of any such origin of existence; in that case, there need not be a limitation to the particular region initially considered and the functions f may, for values of the variables that arise, cease to possess some at least of the properties of regular functions. We therefore shall take account of this wider possibility.

The existence of integrals of the differential equations has been established for finite domains of a set of values of the variables within which the functions f occurring in the equations are regular; and these integrals, obtained as power-series, are regular for some domain of the variable z. Within the whole of

---

* Reference can be made to Briot and Bouquet's memoir quoted in the bibliographical note at the end of chapter II. A different method of reduction of the differential equation to typical forms, valid in the vicinity of the respective exceptional points, is adopted in this chapter; it is based upon Weierstrass's theorem (Note to chap. I) upon functions of more than a single variable.

In regard to non-regular integrals, only a few preliminary remarks are made in this chapter; they are discussed more fully in later chapters.

this domain, the derivatives are also regular and the differential equations are satisfied.

Let c' be a point in the domain and let $a_1', \ldots, a_n'$ be the values of the integrals at that point; so that $c', a_1', \ldots, a_n'$ constitute an ordinary combination of variables for the functions f. When the existence-theorem is applied with this set of values as initial values, there is a domain of c' at all points of which there exist integrals satisfying the differential equations, these integrals being regular in the domain. By a repeated application of the process, the region of existence of the integrals can be constructed gradually.

It is evident that this process is equivalent to the process of continuation* applied to power-series; and that, for each one of the integrals, the series obtained for the first domain is the initial element. It may happen that parts of the z-plane cannot thus be included, as not belonging to the region of existence; regular integrals exist determined by initial values, but they have been established only so long as the variable moves in a simply-connected part of the plane leading to values of the variables which, with the variable z, leave the functions f regular.

It has been seen (§ 11) that, within the domain of a point $z = c$, the path of variation of the independent variable from c to any other point c' can be deformed without affecting the values of the integrals at c' as determined by the assigned initial values at c. But the result may not necessarily hold when, by the process of continuation, domains of successive points and corresponding successive elements of the integrals are obtained. For when this process of continuation is completely effected, it may happen that a closed path of the variable z, returning to the initial point c, does not lead to initial values of the integrals. From what has already been established, it must then be inferred that the path of the variable encloses points in the z-plane corresponding to values of the variables for which some of the functions occurring in the differential equations cease to be regular.

Points of this character, when taken in the aggregate, will, for the sake of brevity and to distinguish them from ordinary points, be called *exceptional* (or *critical*†) points of the differential

---

* *Th. Fns.*, §§ 84, 90.

† Critical points are, by some writers, restricted to the class of branch-points round which a finite number of branches circulate. When f is everywhere a uniform function, the exceptional points are frequently called *singularities*.

equations. It is therefore necessary to consider the system of differential equations for values of the variables in the immediate vicinity of the exceptional points: further, it will be convenient to obtain the various classes of points included in this general aggregate.

On the Possibility of Non-regular Integrals.

15. Thus far, only those solutions which are characterised by the possession of the properties of regular functions have been taken into account; they are the only set established by Cauchy's theorem. But Cauchy's theorem does not exclude the possibility of other non-regular functions which may be determined by the same conditions; its sequel declares that, if the solution is regular, it is unique. Instances of the actual occurrence of non-regular solutions satisfying the same conditions as regular solutions can be constructed, and one such will be given: the general investigation is difficult.

The investigations will, in the first instance, be undertaken in connection with only a single equation, so that there is only one dependent variable. And further, for the sake of simplifying the argument, the equation will in the initial stage be taken of the first degree in the derivative, so as to avoid complications caused by concurrence of different classes of exceptional points; the substantially important critical points of a single equation, that initially is not of the first degree in the derivative, will be considered subsequently.

Accordingly for the immediate purpose, the equation to be considered is

$$\frac{dw}{dz} = f(w, z).$$

When Cauchy's theorem is applied to this equation, it establishes the existence of a regular solution acquiring the value α when $z = c$, provided that $f(w, z)$ is regular within a finite domain in the vicinities of α and c; the solution is obtained in the form of a converging series of powers of $z - c$ and, being regular, it is known to be unique.

There is one case of exception, so far at least as concerns the variability of the regular solution, though $f(w, z)$ is still a regular function. It arises when the value assigned to w makes f vanish without regard to the range of variation of z; the equation then has the form

$$\frac{dw}{dz} = (w - \alpha)^m g(w, z),$$

where m is a positive integer greater than zero and $g(w, z)$ is regular in the domain considered. The equation is undoubtedly satisfied by $w = \alpha$: the possibility of variation in the solution is precluded by the conditions imposed which are manifestly quite special in connection with the form of the equation.

The proof as to the uniqueness of the regular solution obtained depends upon the fundamental assumption, tacitly made and actually used in the analysis, that the solution is uniform. When once the possibility of multiform integrals is admitted, an attempt to apply the proof to them, or to each branch of them, must fail; the proposition, that the path of the independent variable could be deformed at pleasure and could therefore be taken a straight line, can no longer be applied. Fuchs* and Picard† object to the validity of the argument for the case of integrals that are not regular, on the ground that the finiteness of the path from the initial point c to a point z is of the essence of the argument. It is true that the full variation of the variable is excluded, if a path infinite in length is not permissible, though the proposition is not therefore wrong; but so long as the possible multiformity of a solution can be entertained, the full variation of the variable is not admissible, in this sequence of ideas.

Consider the example

$$\frac{dw}{dz} = \frac{(w - \alpha)^2}{z - \zeta},$$

a special form of an equation instanced by Fuchs in another connection; the complete solution of the equation is

$$w - \alpha = \frac{1}{A - \log(z - \zeta)}.$$

This solution is one possessing an unlimited number of branches: they can be deduced from one another by adding to any one branch of the logarithmic function (positive or negative) integer

---

* *Berliner Sitzungsber.*, 1886 (I), p. 283..

† *Cours d'Analyse*, t. II, (1893), p. 314.

multiples of $2\pi i$; and those branches for which the integer multiples are infinite reduce the solution to the form $w = \alpha$, though they reduce it to this form whether z be equal to c or not.

In this mode of stating the result, there is no question of the variation of the path of z. Each branch of the solution is uniform in a simply-connected region in the finite part of the plane that does not enclose the point ζ; there is an unlimited number of branches of the function each of which has reduced the solution to the assigned form. The point $z = \zeta$ is a singularity of the function $\dfrac{dw}{dz}$, and therefore it is excluded from the range of variation of z contemplated in Cauchy's theorem; if therefore all the conditions in Cauchy's theorem are to be definitely maintained, it is not permitted to make the variable z describe a path round the point ζ. Were it otherwise, it would be possible to make the variable pass from c, move round ζ an unlimited number of times * and then move to the final value z: the final value of w would be α,—in effect, the same as that obtained from the branch of the logarithmic function chosen above.

On the other hand, it is to be noticed that each of the branches of the multiform function under consideration coincides, within the region of variation retained, with the uniform solution known to exist. So far as concerns the particular equation under discussion, the difference between the two ways of regarding the solution, which is equal to α when z is c, may be resolved into a difference of mere statement. Yet when we regard a multiform function as composed of all the branches that do not isolate themselves into uniform functions over the whole range of their existence, the more exact way of stating the result would appear to be, in the present case, that some branches of the multiform integral satisfy the conditions that determine the unique uniform integral. Of course, the exceptional points of the functions in the differential equations are excluded from the range of variation; but critical points of the integral may be introduced by a process of integration and some of them might be included within the specified range.

The conclusion therefore to which the argument and the

---

* This could be secured by a finite path, for instance, along an equiangular spiral having its pole at ζ and passing through c.

example, limited as the latter is, lead, is that further investigation is necessary before the proposition, that the unique regular integral of the equation is the only integral of any kind satisfying the conditions within the range of variation assigned by Cauchy's theorem, can be regarded as definitely established *. The actual resolution of the question will be possible after the singularities of the equation have been considered in their effect upon the integral in their immediate vicinity (§ 34).

EXCEPTIONAL POINTS OF THE EQUATION.

16. Having adverted to the possibility of integrals of an equation that are multiform in character and determinable in part by initial conditions, we now proceed to the consideration of the exceptional points of the equation of the first degree in the derivative. So far as is suggested by the preceding investigation in the case of a single equation

$$\frac{dw}{dz} = f(w, z),$$

all these exceptional points are included among those near or at which the function $f(w, z)$ ceases to be regular and therefore can be

      (i)   not finite, or

      (ii)   not determinate, or

      (iii)   not uniform.

Further a function $f(w, z)$ can cease to be continuous from various causes. One such cause may be the occurrence of an infinite value; it will be deemed to be included in the first of the above

---

* The proof given by Picard, *Cours d'Analyse*, t. II, pp. 314—318, based upon the corresponding existence-theorem for partial differential equations, seems to me to be incomplete. The solutions indicated by the existence-theorem are regular and it can be proved, as in the earlier part of this chapter, that the regular solutions are the only regular solutions determined by the assigned conditions; but it is not proved that they are the only solutions so determined and, indeed, there is the same kind of difficulty in establishing their uniqueness as arises in the case of ordinary equations.

A proof, which is outlined by Painlevé in his *Stockholm Lectures*, pp. 19—20, virtually assumes that any branch of a possible non-regular solution is regular within a circle of half the radius of that over which the regular solution is known to exist: the assumption requires to be justified before the proposition can be regarded as established.

classes. Another may be the occurrence of one or more points of essential singularity; it will be deemed to be included in the second of the above classes. Further, it will be assumed that the function $f(w, z)$ has no lines or spaces of essential singularity for variations of z in its plane. Hence so far as regards the cessations of continuity of $f(w, z)$ that are admitted as possible, they give rise to no classes of exceptional points other than those already retained.

It does not follow that all such points, which lead to deviations from regularity in the function $f(w, z)$, are exceptional points of the integral; all that can be stated before investigation is that, taking account of Cauchy's theorem, all possible exceptional points of the integral that arise through the form of the differential equation have been retained.

On the other hand, it equally does not follow that all the exceptional points of the integral, regarded as a function of the independent variable, are thus secured. For example, the integral of the equation

$$nw^{n-1}\frac{dw}{dz} = mz^{m-1}$$

is
$$w = (z^m - a^m)^{\frac{1}{n}},$$

where a is arbitrary: manifestly the m points given by $z^m = a^m$ are parametric branch-points of the function w. The integral of the equations

$$\frac{dw}{dz} = w_1,$$

$$\frac{dw_1}{dz} = \frac{w_1^2 + 2iw^{\frac{1}{2}}w_1^{\frac{3}{2}}}{w},$$

is
$$w = ae^{\frac{1}{z-c}},$$

where a and c are arbitrary; the parametric point c is an essential singularity of the integral. In the former instance, the parametric branch-points would be found to occur naturally in the discussion of the infinities and the indeterminate values of the expression for $\frac{dw}{dz}$, provision for which has already been made. In the latter instance, the parametric singularity does not obviously arise in connection with the exceptional points of the two simultaneous equations. In fact, the obvious singularities to take account of are a zero value of w and an infinite value of w_1; but the form of the solution shews that, for the parametric singularity c, the value of w is not definite nor is its variation in the immediate vicinity definite: and indefiniteness in value of one of the variables gives rise to exceptional points distinct from specific singularities of functions that occur. Though it is to be noted in passing that the latter instance is not that of a single equation, still it manifestly is desirable that account should be taken of the possible occurrence of parametric singularities of the integral which do not arise through the exceptional points of the equation or equations under consideration.

SINGULARITIES OF THE FUNCTION $f(w, z)$, WHEN UNIFORM.

17. It thus is necessary to consider the various classes of values for which $f(w, z)$ is not regular. Let $w = \alpha$, $z = c$ be such a combination of values; then there may be some power-series $P_0(w - \alpha, z - c)$, vanishing for the combination $w = \alpha$, $z = c$, and such that the product

$$P_0(w - \alpha, z - c) f(w, z)$$

is a regular function in the vicinity of α and c; the combination is called an *accidental singularity* of $f(w, z)$.

Of accidental singularities, there are two kinds. If the above product $P_0 f$ can, by appropriate choice of P_0, be made to have at α, c a value different from zero, then the singularity is said to be of the first kind. If however all the power-series, which render the above product regular, are such as to make the product vanish at α, c, the singularity is said to be of the second kind.

If there be no power-series vanishing at the values α, c such that the product $P_0 f$ is regular at and in the immediate vicinity of α, c, then the combination α, c is said to be an *essential singularity* of the function[*]. This class of values manifestly includes those for which $w = \alpha$ is an essential singularity of f without regard to the value of z, and those for which $z = c$ is an essential singularity of f without regard to the value of w.

---

[*] Weierstrass, *Crelle*, LXXXIX (1880), p. 3; *Ges. Werke*, t.-II, p. 128. See also the memoir in the latter volume, pp. 135 et seq.

If a function $f(w, z)$ be such that $w = \alpha$ is an essential singularity of f without regard to the values of z, and if all the essential singularities of f also arise for values of w without regard to the values of z, then, as they are also essential singularities of $\frac{1}{f}$, by taking the equation in the form

$$\frac{dz}{dw} = \frac{1}{f(w, z)},$$

we have an equation in which all the essential singularities for the value of the derivative arise for values of the independent variable.

If a function $f(w, z)$ be such that all the essential singularities arise for values of z without regard to the value of w, then the equation

$$\frac{dw}{dz} = f(w, z)$$

is of the same character as in the preceding case.

If a function be such that some essential singularities of f arise for values of w without regard to the values of z and some arise for values of z without regard to the values of w, the equation cannot be transformed, as in the first case, so that the essential singularities of the derivative arise for values of the independent variable.

Manifestly neither the first case nor the third case can arise if $f(w, z)$ is a rational function of w.

18.   In order to discriminate exactly between the two classes of accidental singularity, we use a special form of Weierstrass's theorem already quoted*, taking the case in which the number of variables is two.

As regards the power-series $P_0(w - \alpha, z - c)$, if $P_0(w - \alpha, 0)$ do not vanish for all values of w, and if the lowest power of $w - \alpha$ which it contains have an exponent m, it is possible to determine a function $p(w, z)$, where

$$p(w, z) = (w - \alpha)^m + p_1 (w - \alpha)^{m-1} + p_2 (w - \alpha)^{m-2} + \ldots + p_m,$$

and $p_1, \ldots, p_m$ denote functions of $z - c$ regular in the vicinity of $z = c$, such that

$$P_0(w - \alpha, z - c) = p(w, z) e^{G(w, z)},$$

where G is a regular function of w and z in the immediate vicinity

---

* See Note, chap. i, p. 19; in particular, pp. 24, 25.

of α and c. But if $P_0(w-\alpha, 0)$ vanishes for all values of w, then, effecting a transformation

$$w - \alpha = \lambda u + \mu v, \quad z - c = \lambda' u + \mu' v,$$

where λ, μ, λ', μ' are arbitrary constants subject to the conditions

$$\lambda \mu' - \lambda' \mu \neq 0, \quad p_0(\lambda, \lambda') \neq 0,$$

where $p_0(x, y)$ is the aggregate of terms of the lowest dimensions, say l, in $P_0(x, y)$, it is possible to determine a function $p(u, v)$, where

$$p(u, v) = u^l + p_1 u^{l-1} + p_2 u^{l-2} + \dots + p_l,$$

and $p_1, \dots, p_l$ are analytical functions of v regular in the vicinity of $v = 0$, such that

$$P_0(w - \alpha, z - c) = p(u, v) e^{H(w, z)},$$

where H is a regular function of w and z in the immediate vicinity of α and c. In the former of these resolutions, the functions p_r (for $r = 1, \dots, m$) are regular in the vicinity $z = c$ and vanish there; in the latter, the functions p_s (for $s = 1, \dots, l$) are regular in the vicinity $v = 0$ and vanish there.

When α, c is an accidental singularity of $f(w, z)$, then the product

$$P_0(w - \alpha, z - c) f(w, z)$$

is a regular function, say $Q_0(w - \alpha, z - c)$, in the vicinity of α, c.

If the accidental singularity is of the first kind, Q_0 does not vanish there; and therefore the term of lowest dimension in $Q_0(w - \alpha, 0)$ is of order zero. Accordingly, applying Weierstrass's theorem, we have

$$Q_0(w - \alpha, z - c) = e^{J(w, z)},$$

where J is a regular function in the vicinity of α, c.

If the accidental singularity is of the second kind, Q_0 does vanish at it: the corresponding alternatives for $Q_0(w - \alpha, z - c)$ must be taken as they were taken for P_0. If $Q_0(w - \alpha, 0)$ does not vanish for all values of w and if the lowest power of $w - \alpha$ which it contains has an exponent n, it is possible to determine a function $q(w, z)$, where

$$q(w, z) = (w - \alpha)^n + q_1(w - \alpha)^{n-1} + \dots + q_n,$$

and $q_1, \dots, q_n$ denote functions of $z - c$ regular in the vicinity of $z = c$, such that

$$Q_0(w - \alpha, z - c) = q(w, z) e^{\Gamma(w, z)},$$

where Γ is a regular function of w and z in the immediate vicinity of α and c. But if $Q_0 (w - \alpha, 0)$ vanishes for all values of w, then effecting the same transformation as before, viz.

$$w - \alpha = \lambda u + \mu v, \quad z - c = \lambda' u + \mu' v,$$

and imposing a further exclusive limitation by the condition

$$q_0 (\lambda, \lambda') \neq 0,$$

where $q_0 (x, y)$ is the aggregate of terms of lowest dimensions, say k, in $Q_0 (x, y)$, it is possible to determine a function $q (u, v)$, where

$$q (u, v) = u^k + q_1 u^{k-1} + q_2 u^{k-2} + \ldots + q_k,$$

and $q_1, \ldots, q_k$ are analytical functions of v regular in the vicinity of $v = 0$, such that

$$Q_0 (w - \alpha, z - c) = q (u, v) e^{\Upsilon (w, z)},$$

where Υ is a regular function of w and z in the immediate vicinity of α, c. In the former of these resolutions, the functions q_s (for $s = 1, \ldots, n$), regular in the vicinity of $z = c$, vanish there ; in the latter, the functions q_s (for $s = 1, \ldots, k$), regular in the vicinity of $v = 0$, vanish there.

In each of these cases we have

$$f (w, z) = \frac{Q_0 (w - \alpha, z - c)}{P_0 (w - \alpha, z - c)}.$$

19.   When the accidental singularity is of the first kind, then $f (w, z)$ has either the form

$$\frac{1}{p (w, z)} e^{J (w, z) - G (w, z)},$$

or the form

$$\frac{1}{p (u, v)} e^{J (w, z) - H (w, z)}$$

For the first of these, $J (w, z) - G (w, z)$ is a regular function in the region considered : for the second of them, $J (w, z) - H (w, z)$ is likewise a regular function.   It is manifest that $f (w, z)$ is infinite at α, c ; and the infinity is a determinate infinity.   We have

$$\frac{dw}{dz} = \frac{1}{p (w, z)} e^{-G_1 (w, z)},$$

or

$$\frac{dw}{dz} = \frac{1}{p (u, v)} e^{-G_2 (w, z)} ;$$

and therefore

$$\frac{dz}{dw} = p (w, z) e^{G_1 (w, z)},$$

or

$$\frac{dz}{dw} = p (u, v) e^{G_2 (w, z)} ;$$

where the functions on the right-hand side are regular in the immediate vicinity of $w = \alpha$, $z = c$.

20. When the accidental singularity is of the second kind, then $f(w, z)$ has one of four forms: viz.

$$\frac{q(w, z)}{p(w, z)} e^{\Gamma(w,z) - G(w,z)},$$

$$\frac{q(u, v)}{p(w, z)} e^{\Upsilon(w,z) - G(w,z)},$$

$$\frac{q(w, z)}{p(u, v)} e^{\Gamma(w,z) - H(w,z)},$$

$$\frac{q(u, v)}{p(u, v)} e^{\Upsilon(w,z) - H(w,z)}.$$

Each of the functions in the index of the exponential is regular in the region considered; and each of the functions $q(w, z)$, $p(w, z)$, $q(u, v)$, $p(u, v)$ vanishes at α, c. Moreover, owing to the source of the functions in connection with $f(w, z)$, it may be assumed that the numerator function q and the denominator function p have no common factor*.

Consider the first of these forms. At α, c, both q and p vanish. For other values of w and z, both q and p can vanish only if the resultant of the two functions, obtained either as an eliminant in z or as an eliminant in w, should vanish, for some value or values of z or for some value or values of w; and these values have a finite modulus and are isolated values, because q and p have no common factor. Accordingly if we take a small, not infinitesimal, region round $w = \alpha$, $z = c$, then for no values in that region other than the respective centres can q and p vanish together.

Now q and p are algebraical in w. Hence within this region there will be an infinitude of simultaneous values of w and z for which q can vanish alone without p vanishing for any of the combinations: for each of them $\dfrac{dw}{dz}$ is zero. There will be another infinitude of simultaneous values for which p can vanish alone without q vanishing for any of the combinations: for each of them

---

* There is one exception, of slight functional importance, to this statement in regard to the last of the four forms: it will be noticed in its proper connection (§ 35, p. 86).

dw/dz is infinite. There will be another infinitude of simultaneous values for which

$$p - Aq = 0,$$

where A is any constant, without either p or q vanishing for any of the combinations: for each of them dw/dz has the value A. Hence dw/dz can have any value in the immediate vicinity of α, c: and at α, c its value is not determinate.

A similar argument applies in connection with each of the other three forms, when they are transformed as in § 35; the indeterminateness of the value of the derivative at and near the point is the general characteristic of the accidental singularity of the second kind.

Hence in the immediate vicinity of such a singularity of $f(w, z)$, the differential equation has one of the forms

$$\frac{dw}{dz} = \frac{q(w, z)}{p(w, z)} e^{G_2(w, z)},$$

$$\frac{dw}{dz} = \frac{q(u, v)}{p(w, z)} e^{G_2(w, z)},$$

$$\frac{dw}{dz} = \frac{q(w, z)}{p(u, v)} e^{G_3(w, z)},$$

$$\frac{dw}{dz} = \frac{q(u, v)}{p(u, v)} e^{G_2(w, z)},$$

where $w - \alpha = \lambda u + \mu v$, $z - c = \lambda' u + \mu' v$, with merely exclusive limitations on the constants $\lambda, \mu, \lambda', \mu'$; in each of the forms, q and p are algebraical functions of w and u respectively: and both q and p vanish at α, c.

It is clear that accidental singularities of $f(w, z)$, depending upon only one particular value of one of the variables and without regard to the value of the other, are included in these two classes. Thus let $z = c$ be such a point; then there is some positive integer κ such that

$$(z - c)^\kappa f(w, z)$$

is regular in the vicinity of c; if its value at c is not zero for a value of w, that accidental singularity belongs to the first kind; if its value at c is zero for a value of w, that accidental singularity belongs to the second kind.

21. No limitation, external to the equation, has been imposed

upon the arbitrary values assigned to the variable w and the variable z as initial values ; but the forms considered depend upon a tacit assumption that both variables initially are finite. It is necessary to take account of infinite values, which need not be singularities of the functions in the equations ; it is convenient to take account of them separately from the finite values. For this purpose, let $ww_1 = 1$, so that

$$\frac{dw_1}{dz} = - w_1{}^2 f\left(\frac{1}{w_1}, z\right)$$

$$= f_1(w_1, z),$$

say. The initial value of w_1 to be taken is zero; for an assigned value of z, the combination may be an ordinary combination for f_1, or it may be an exceptional combination. In the former case, the existence-theorem applies; in the latter case, the combination belongs to one of the classes considered for the earlier form and the corresponding reduction to a typical form should be effected.

Similarly, for infinite values of z, we change the variable to ζ by assuming $z\zeta = 1$.

22. As regards the various classes of singularities, one property should be noticed.

In the case of essential singularities, the combination α, c represents a combination of isolated points in the respective planes. If c alone, without respect to the value of w, be an essential singularity of the function $f(w, z)$, then the function is determinate and finite for points in the z-plane in the immediate vicinity of c. If f be transcendental in w, it is possible that isolated values of w alone could be essential singularities of f: as pointed out before, a transformation making z the dependent variable and w the independent variable brings the equation under the case last considered, in which the essential singularity is a value of the independent variable without regard to the dependent variable. If f be rational in w, then isolated values of w alone cannot be essential singularities of f; points in the z-plane, being isolated points, can provide this kind of singularity alike for special values of w and without regard to the value of w. Essential singularities of the function $f(w, z)$ are isolated points in the z-plane, when f is a rational function of w; preceding assumptions require f to be merely a uniform function, quà function of z.

In the case of accidental singularities of the second kind, say of the combination α, c, the function behaves like

$$\frac{q\,(w-\alpha,\ z-c)}{p\,(w-\alpha,\ z-c)}$$

in the immediate vicinity; for the actual values α, c, the function is indeterminate in value: and for any region round α, c however small, the function can acquire any value. But in the infinitesimal region indicated, there are no other simultaneous values of w and z where the function can be represented in a corresponding analytical form; so that the combination α, c represents a combination of isolated points in the respective planes. Accordingly, accidental singularities of the second kind possessed by the function $f(w, z)$ are isolated points in the z-plane.

In the case of accidental singularities of the first kind, say of the combination α, c, the function behaves like

$$\frac{1}{p\,(w-\alpha,\ z-c)}$$

in the immediate vicinity. For the actual values α, c, the function is infinite in value; for any small region round α, c, the value of the function is everywhere determinate and, by making the region sufficiently small, the modulus of the function can be made as large as we please. Not merely so, but for every point z in such a region there are values of w which make the modulus of the function actually infinite; for writing $z-c=\zeta$, where $|\zeta|$ is to be taken small, and $w-\alpha=W$, the equation

$$p\,(W,\ \zeta)=0$$

is an algebraical equation in W which has zero roots when $\zeta=0$ and, by the known theory as to continuity of the roots of an algebraical equation, has one or more roots of small modulus when $|\zeta|$ is small. It therefore appears that, in the case of a combination α, c proving to be an accidental singularity of the first kind, the point c is not an isolated point in the z-plane; at all points in the immediate vicinity of c, there are some associable values of w in the immediate vicinity of α which give actually infinite values for f, and all values of w in the immediate vicinity of α leave $|f|$ larger than any finite quantity by taking the vicinity sufficiently small. We may proceed from a point on the boundary of a small vicinity of c in a direction outside that vicinity and obtain another small vicinity for every point of which there can arise one or more combinations of w and z which give determinately infinite values to f. Accordingly, accidental singularities of the first kind possessed by the function f form a continuous region in the z-plane. In fact, for any value of Z, we have a definite infinity given by

$$\frac{1}{f(w,\, Z)} = 0\,;$$

because f is rational in w, this is generally a definite algebraical equation in w, known therefore to possess roots each of which, associated with Z, is an accidental singularity of the first kind.

It may happen, however, that an isolated value of z is a pole of some coefficient in $f(w, z)$ without regard to the value of w; there then is no algebraical equation for values of w. Similarly, it may happen that a value of w makes f infinite without regard to the value of z. These are merely very special cases, arising as limiting forms of the general case.

Corresponding considerations arise and should be applied when, for the singularities in question, either w is infinite, or z is infinite, or both are infinite.

It thus is clear that, for a function $f(w, z)$ which is rational in w and uniform in z, the mode of occurrence of points in the z-plane for essential singularities and for accidental singularities of the second kind is markedly different from that for accidental singularities of the first kind; in the two former classes, the points are isolated points in general*: in the latter class, the points belong to a continuous region or continuous regions. When therefore a curve is drawn in the z-plane, it will always be possible to draw the curve continuously so that no point of it is at only an infinitesimal distance from a point of singularity, either essential or accidental of the second kind; the explanations adduced shew that it is not possible to exclude accidental singularities of the first kind from its course.

---

* This would not apply to families of functions which possess lines of essential singularity or lacunary spaces in the z-plane; at the present stage, such families of functions are tacitly excluded from discussion.

BRANCHES OF $f(w, z)$.

23. It still remains to take account of points, or combinations
of values, where the function $f(w, z)$ does not possess only a single
value, the points themselves not belonging to any of the preceding classes. In the vicinity of such points the function is not
uniform ; the number of values for any point in that vicinity may
be limited or unlimited. By taking continuations of the function
$f(w, z)$ so as to make closed paths round all such points in the
respective planes of w and z, (or, if the branching is due to the
mode of occurrence of only one variable, in the plane of that
variable alone), and by noting the various branches of the function
that thus arise solely from continuation of $f(w, z)$ and without
regard to the fact that f is to be equal to the derivative of w, we
shall find that the total number of branches is either limited or
unlimited.

If the number is unlimited, then either f is not determined by
one equation or, if it is so determined, the equation is transcendental in f. When f is replaced by $\dfrac{dw}{dz}$, the former case implies
that no differential equation of the first order is satisfied by w;
the latter case implies that a differential equation of the first
order is satisfied by w, the equation being transcendental in the
derivative, the coefficients being uniform functions of w and z.
As our attention will be limited to equations that are algebraical
in the derivative, both the possibilities that thus have arisen will
be ignored.

When the number of branches of the function f is limited and
equal to n, say, then f is the root of an algebraical equation

$$F(f, w, z) = 0,$$

which, when made integral in f, is of degree n in f; the coefficients
are uniform functions of w and z. Accordingly, we shall have a
differential equation

$$F\left(\frac{dw}{dz}, w, z\right) = 0$$

of the first order and of degree n when made integral in powers of
the derivative ; the coefficients are uniform functions of w and z.

Such combinations of values of w and z, as make the branches of f circulate for contours round them in their respective planes, are undoubtedly exceptional points for the equation

$$\frac{dw}{dz} = f(w, z).$$

This however is only one of n equations to which F is equivalent, and the exceptional combinations of points of f are exceptional combinations for $F = 0$, being the branch-points of f considered as a function of w and z. They accordingly will be discussed later in connection with equations of the first order when the degree of the equation is higher than the first: there is moreover an added advantage in this reservation of branch-points for the present, because some of the branch-points of the equation

$$F\left(\frac{dw}{dz}, w, z\right) = 0$$

may be also singularities of the derivative.

CHAPTER IV.

WE now proceed to the consideration of the integral of the equation

$$\frac{dw}{dz} = f(w, z)$$

in the vicinity of the various singularities which it possesses; and the various forms will be considered in turn. These forms have arisen in connection with special combinations of values; so that, for the immediate purpose, what is desired is a knowledge of the behaviour of an integral or integrals (if any) in the vicinity of α, c, such that $w = \alpha$ when $z = c$.

But in an earlier instance, when certain forms were obtained as reduced canonical forms of a system of differential equations, these forms having certain restrictions upon their properties to serve that purpose, it was pointed out (p. 42) that such forms might be propounded initially without such restrictions: so here also, it may be that equations are propounded with the same analytical expressions as the canonical forms, though not restricted by the same initial conditions and the same limiting range of functional existence.

---

* In connection with the subject of this chapter, reference may be made to the memoir by Briot and Bouquet (cited on p. 40) and particularly to the examples which they discuss. Their method is followed by Picard, *Cours d'Analyse*, t. II, ch. XII. Also the memoir by Fuchs, "Ueber die Werte welche die Integrale einer Differentialgleichung erster Ordnung in singulären Punkten annehmen können," *Berl. Sitzungsber.* (1886, 1), pp. 279—300, should be consulted; it is in this memoir that the notion of points of indeterminateness of a function is first introduced into the discussion of the integrals of differential equations.

Accidental Singularities of the First Kind.

24. We shall first deal with the class of accidental singularities of $f(w, z)$ which are the simplest in regard to the analytical expression of f in their vicinity—viz. those of the first kind. Taking $w = \alpha$, $z = c$ to denote such a singularity, we know that the function

$$\frac{1}{f(w, z)}$$

is regular in the vicinity of the combination of values.

In the first place, if $\dfrac{1}{f(y + \alpha, c)}$ does not vanish for all values of y, then (Note, ch. I) there is a function $p(w, z)$, where

$$p(w, z) = (w - \alpha)^m + p_1 (w - \alpha)^{m-1} + \ldots + p_m,$$

and the functions $p_1, \ldots, p_m$ are regular functions of $z - c$ vanishing at c, such that

$$\frac{1}{f(w, z)} = p(w, z) e^{G_1(w, z)}.$$

To consider the differential equation, write

$$w = \alpha + y, \quad z = c + x.$$

Also, since p_i is regular in the vicinity of the x-origin and vanishes there, write

$$p_i = x^{n_i} P_i,$$

where n_i is an integer $\geqslant 1$ and P_i is a regular function of x which does not vanish with x; and let

$$e^{G_1(a, c)} = A,$$

so that $e^{G(w, z)}$, being a regular function of w and z, is a regular function of x and y, equal to A when $y = 0$, $x = 0$. Thus the differential equation becomes

$$\frac{dx}{dy} = A \left\{ y^m + x^{n_1} P_1 y^{m-1} + x^{n_2} P_2 y^{m-2} + \ldots + x^{n_m} P_m \right\} + \ldots,$$

the unexpressed terms being of moduli smaller than the expressed terms in the immediate vicinity of the origin; or it may be written

$$\frac{dx}{dy} = A y^m \Omega + x \Upsilon,$$

where Ω may be unity but more generally is a regular function of x and y, which is equal to unity when $x = 0$, $y = 0$; and Υ is a similar function, without the restriction of being unity when $x = 0$, $y = 0$. As the right-hand side of the equation is a regular function of x and y, the existence-theorem leads to the conclusion that there is an integral of the equation, uniquely determined by the condition of acquiring a zero value when $y = 0$; and it is known that this integral can be expanded in a converging series in powers of y. The process adopted in the proof of Cauchy's theorem leads to the expansion. Evidently $d^s x/dy^s$ vanishes for $x = 0$, $y = 0$ when $s = 1, 2, \ldots, m$, but its value for $x = 0$, $y = 0$ is $m! A$ when $s = m + 1$; and the derivatives of higher order can be deduced by making the same use of the equation as in § 10. The expansion takes the form

$$x = \frac{y^{m+1}}{(m+1)!} \left(\frac{d^{m+1} x}{dy^{m+1}} \right)_0 + \frac{y^{m+2}}{(m+2)!} \left(\frac{d^{m+2} x}{dy^{m+2}} \right)_0 + \cdots$$

$$= \frac{A}{m+1} y^{m+1} + b y^{m+2} + c y^{m+3} + \cdots,$$

where A, b, c, ... are finite constants: and this value of x is the desired solution. To obtain w as a function of z, this series must be reversed; the result of the reversion gives

$$w - \alpha = y$$

$$= \alpha' \xi + \beta \xi^2 + \gamma \xi^3 + \cdots,$$

where $\alpha' = \{(m+1)/A\}^{\frac{1}{m+1}}$ and $\xi = (z-c)^{\frac{1}{m+1}}$, and the series converges[*] for sufficiently small values of ξ. It thus follows[†] that $z = c$ is an algebraical branch-point of order m, that is, a branch-point round which $m + 1$ branches of the function w interchange in a cycle and at which they have a common value α. The results may be summarized as follows :—

If the function $f(w, z)$ become infinite for $w = \alpha$, $z = c$ in such a way that its reciprocal is regular in the immediate vicinity of those values, and if m be the order of the first of the partial derivatives of $1/f$ with regard to w that does not vanish for $w = \alpha$, $z = c$, then there exists an integral of the equation

$$\frac{dw}{dz} = f(w, z)$$

---

[*] Chrystal's *Algebra*, vol. ii, p. 354.
[†] Briot et Bouquet, (l. c., p. 40), p. 148.

acquiring the value α *when* $z = c$; *this integral has* $m + 1$ *branches in the immediate vicinity of* $z = c$, *forming a single cycle for interchange round the branch-point* c.

The point is sometimes called an *algebraical critical point* for the function w.

25.   In the second place, suppose that

$$\frac{1}{f(w - \alpha, \, 0)}$$

does vanish for all values of w; then there is a function $p(u, v)$, given by

$$p(u, v) = u^m + p_1 u^{m-1} + \ldots + p_m,$$

where $p_1, \ldots, p_m$ are regular functions of v vanishing with v, such that

$$\frac{1}{f(w - \alpha, \, z - c)} = p(u, v) \, e^{G_1(w, \, z)},$$

where G_1 is a regular function of $w - \alpha$, $z - c$, and where

$$w - \alpha = \lambda u + \mu v, \quad z - c = \lambda' u + \mu' v.$$

Since $\dfrac{1}{f(w - \alpha, \, 0)}$ vanishes for all values of w, it follows that $p(u, v)$ vanishes when $\lambda' u + \mu' v = 0$.   Now $p(u, v)$ is an integral algebraical function of u, the coefficients of which are regular functions of v vanishing with v; hence $\lambda' u + \mu' v$ is a factor of p. Let the sth power of this linear factor be the highest contained in p, and let

$$p(u, v) = (\lambda' u + \mu' v)^s \, p_1(u, v),$$

where p_1 now does not vanish with $\lambda' u + \mu' v$.   Now write

$$w - \alpha = y, \quad z - c = x;$$

then $$p(u, v) \, e^{G_1(w, \, z)} = x^s g(y, x),$$

where $g(y, x)$ is a regular function of x, y in the immediate vicinity of 0, 0, and $g(y, 0)$ does not vanish for all values of x.

The differential equation then becomes

$$\frac{dx}{dy} = x^s g(y, x),$$

where $g(y, x)$ is a regular function of y and x in the vicinity of $0, 0$, such that $g(y, 0)$ is not zero, and s is a positive integer.

So long as the variables remain in the domain of the new origin, that is, so long as w and z are confined to a region round α and c that includes no other exceptional combination, there is only a single regular integral of this equation which vanishes when $y = 0$; and it is given by $x = 0$. Hence for this relation between w and z, it appears that z remains steadily equal to c while w varies from α within a domain of α and c that contains no other exceptional combination of the original equation; and therefore, with this limitation on the range of the variables, w cannot be regarded as a function of z. In other words, no regular integral of the equation can be determined by the assigned conditions of values α and c, if the range of the variables is restricted to a small domain round those points.

This kind of relation between w and z is sufficiently illustrated by the equation

$$\frac{dw}{dz} = a\,(z - c)^{-n}\,(w - \alpha)^{-m},$$

where m and n are integers and a is a constant; the general integral can be obtained explicitly and its evanescence be observed when $w = \alpha$, $z = c$ are assigned as conditions for the determination of the constant of integration.

But if this restriction as regards the range of variation of the variables is not imposed, the result is no longer valid. As an example, consider

$$\frac{dx}{dy} = \frac{x^2}{a + y},$$

which satisfies all the assigned conditions; the complete integral is

$$x = \frac{1}{A - \log (a + y)},$$

that is,

$$z - c = \frac{1}{A - \log (w - \alpha + a)}.$$

If then w, starting from the value α, passes an unlimited number of times round the point $\alpha - a$, then undoubtedly any number of branches of the function on the right-hand side give $z = c$ on the return of w to α. When w is expressed as a function of z, we find

$$w - \alpha + a = \epsilon^A e^{\frac{1}{z - c}};$$

the point $z = c$ is an essential singularity of w. Now this function is uniform; it is known that, in the immediate vicinity of an essential singularity, a function can assume any value*, and therefore w can assume the value α. The difference from the preceding case arises from the fact that, at the essential singularity $z = c$, the value of the function is not determinate: but the solution may not be excluded.

The difference between the two cases, according as the range of variation of the variables is restricted or is not restricted, recalls the difference between the two views regarding the uniqueness of the regular function determined by initial conditions, as stated in Cauchy's existence-theorem and discussed in §§ 14, 15. In connection with a given equation

$$\frac{dw}{dz} = f(w, z),$$

for which α, c is an accidental singularity of the first kind, the modified expression for f is only known to be valid within the immediate vicinity of α, c (though it may be valid in a less limited vicinity), and there could then be a justification for restricting the variation of the variable. In connection with a given equation

$$\frac{dz}{dw} = p(w, z) e^{G(w, z)},$$

where the region of validity of the expression on the right-hand side is not thus limited, there then is not the corresponding justification for restricting the variation of the variable.

ASSIGNMENT OF INFINITE VALUES.

26. These inferences have been deduced on the implied supposition that α is not infinite; the possibility of infinite values must be considered.

The equation is

$$\frac{dw}{dz} = f(w, z)$$
$$= \frac{p_n + p_{n-1}w + \ldots + p_1 w^{n-1} + p_0 w^n}{q_m + q_{m-1}w + \ldots + q_1 w^{m-1} + q_0 w^m},$$

* *Th. Fns.*, §§ 33, 62.

where $f(w, z)$ is expressed as a rational function of w. Then $n > m$, because, by hypothesis, $w = \infty$ must make $f(w, z)$ infinite for some assigned value or values of z; and both p_0, q_0 may be assumed different from zero for a value of z, say $z = c$, this value being an ordinary point for all the other functions p and q.

Let $n = m + 1$; then if $w_1 w = 1$, the equation for w_1 is

$$\frac{dw_1}{dz} = - w_1 \frac{p_0 + p_1 w_1 + \dots}{q_0 + q_1 w_1 + \dots}$$

$$= w_1 h(w_1, z - c),$$

where h is regular for sufficiently small values of $|w_1|$ and $|z - c|$. From § 10, p. 32, it follows that, as an initial value zero is assigned to w_1, the only regular integral of the equation is that which arises by giving a persistent zero value to w_1; and therefore the integral of the equation

$$\frac{dw}{dz} = \frac{p_{m+1} + p_m w + \dots + p_1 w^m + p_0 w^{m+1}}{q_m + q_{m-1} w + \dots + q_1 w^{m-1} + q_0 w^m},$$

to which is assigned an initial infinite value for a value of z that is an ordinary point for all the coefficients p and q and is not a zero of either p_0 or q_0, is infinite for all values of z in the vicinity of c; the integral is not properly a function of z.

Next, let $n = m + 2$. On making the same substitution for the dependent variable, the equation for w_1 is

$$\frac{dw_1}{dz} = - \frac{p_0 + p_1 w_1 + \dots}{q_0 + q_1 w_1 + \dots}$$

$$= k(w_1, z - c),$$

where k is regular for sufficiently small values of $|w_1|$ and $|z - c|$, and does not vanish for $w_1 = 0$, $z = c$. The values $w_1 = 0$, $z = c$, constitute an ordinary set of values for the new equation; and therefore there exists a unique integral w_1 which is a regular function of $z - c$ and acquires the value 0 for $z = c$. Hence also there exists a unique integral of the equation

$$\frac{dw}{dz} = \frac{p_{m+2} + p_{m+1} w + \dots + p_1 w^{m+1} + p_0 w^{m+2}}{q_m + q_{m-1} w + \dots + q_1 w^{m-1} + q_0 w^m},$$

determined by the condition that it has an infinite value for any given value of z, which is an ordinary point of each of the co-efficients p and q and is not a zero of p_0 or q_0. The unique integral w of course changes as the selected point z is changed subject to the governing conditions. It is of the form

$$\frac{1}{w} = \text{uniform analytical function of } z - c,$$

the term independent of $z - c$ being absent.

Lastly, let $n > m + 2$. As before, the equation in w_1 is constructed; it can be taken in the form

$$\frac{dz}{dw_1} = - w_1{}^{-n+m+2} \frac{q_0 + q_1 w_1 + \dots + q_m w_1{}^m}{p_0 + p_1 w_1 + \dots + p_n w_1{}^n}$$

$$= w_1{}^{-n+m+2} \, \Omega,$$

where Ω is a uniform analytical function of w_1 and $z - c$ which does not vanish when $w_1 = 0$, $z = c$. By § 24, an integral of this equation exists, determined by the condition that $w_1 = 0$ when $z = c$; the integral has $n - m - 1$ branches in the immediate vicinity of $z = c$, these branches forming a single cycle for interchange round the branch-point c. Consequently there exists a corresponding integral of the original equation determined by the condition that it acquires an infinite value when $z = c$: and this integral has $n - m - 1$ branches which interchange in a single cycle round $z = c$.

Hence it appears that, *with three general exceptions, the accidental singularities of the first kind are algebraical branch-points of the integrals which certainly exist in the vicinity of these points.* These three exceptions are:

(i) when $\dfrac{1}{f(\alpha + y, \, c + x)}$ takes the form $x^s g(y, x)$, where g is regular: there is no regular function of z, satisfying the equation and acquiring the value α when $z = c$, when the range of variation is limited to a small region;

(ii) when $f\left(\dfrac{1}{w_1}, \, z - c\right)$ takes the form $\dfrac{1}{w_1} h(w_1, z - c)$, where h is regular and does not vanish for $w_1 = 0$, $z = c$: there is no regular function of z satisfying the equation and acquiring the value ∞ when $z = c$;

(iii)   when $f\left(\dfrac{1}{w_1}, z - c\right)$ takes the form $\dfrac{1}{w_1{}^2} k\,(w_1,\ z - c)$, where k is regular and does not vanish for $w_1 = 0$, $z = c$; there is a unique integral satisfying the equation and acquiring the value ∞ when $z = c$: its reciprocal is a regular function of $z - c$ for sufficiently small values of $|\,z - c\,|$.

The first two of these cases will be considered immediately in regard to the possibility of non-regular integrals.

27.   *Ex.* 1.   In illustration of these results, consider the integral relation between w and z which is equivalent to

$$\frac{dw}{dz} = \frac{w\,(w+1)}{w-1}\,.$$

There exists an integral w, which is a regular function of z and is uniquely determined by acquiring a value a for some value of z, say c; but a may not be 0 or -1 or ∞.   In case an initial value 0, or an initial value -1, or an initial value ∞, be assigned to w, then w cannot vary when determined by the equation although z is capable of variation.   The function

$$\frac{w-1}{w\,(w+1)}$$

is regular in the vicinity of $w = 1$, and the integer s is unity; hence if ζ be the value of z when $w = 1$,

$$z - \zeta = \tfrac{1}{4}\,(w-1)^2 + \dots,$$

and ζ is a simple branch-point of the function w.   These results are established by the preceding theorems: they can be verified as follows.

The equation can be integrated; for

$$dz = \frac{w-1}{w\,(w+1)}\,dw = d \log \frac{(w+1)^2}{w},$$

so that

$$\frac{(w+1)^2}{w} = \frac{(a+1)^2}{a}\,e^{z-c},$$

assuming that $w = a$ when $z = c$.   The value of a cannot be -1, for w would then be -1 for all values of z: and it cannot be 0, for w would then be 0 for all values of z: and it cannot be infinite, for w would then be infinite for all values of z: in none of these cases is w a function of z.   For all values of a, other than 0, -1, and ∞, the function $w - a$ can be expressed as a regular function of $z - c$ vanishing when $z = c$.   A branch-point for the functional relation between w and z is given by $w = 1$: if ζ be the corresponding value of z, then

$$4 = \frac{(a+1)^2}{a}\,e^{\zeta - c},$$

so that

$$\frac{(w+1)^2}{w} = 4e^{z - \zeta};$$

and therefore

$$z - \zeta = \tfrac{1}{4}\,(w-1)^2 + \dots.$$

in the immediate vicinity of $z=\zeta$, $w=1$, shewing that ζ is a simple branch-point.

Ex. 2. Discuss in a similar manner the properties of the integral equivalents of

$$\text{(i)} \quad \frac{dw}{dz} = \frac{w-a}{w-b} \,;$$

$$\text{(ii)} \quad \frac{dw}{dz} = \frac{a}{(w+z)^n}, \text{ for } n=1, 2 \,;$$

$$\text{(iii)} \quad \frac{dw}{dz} = \frac{a+\beta w+\gamma z w^2}{\delta+\epsilon\,(z-1)\,w},$$

in particular, considering the integrals in the vicinity of $z=1$, $z=0$.

Ex. 3. Discuss the equations considered in § 26, when $z=c$ is a zero of p_0 or of q_0 or of both.

Ex. 4. Discuss the function satisfying the equation

$$\frac{dw}{dz} = aw^2 + bz^m,$$

determined by an assigned value of w (finite or infinite) when $z=0$; the integer m being positive.

NON-REGULAR INTEGRALS.

28. It now remains to consider the possibility of non-regular integrals of the equation in the general case, when the combination of values initially assigned to the original equation is an accidental singularity of the first kind; and therefore we proceed to consider the possibility of non-regular integrals of

$$\frac{dx}{dy} = x^s g\,(y, x),$$

where g is a regular function of y and x, such that $g\,(y, 0)$ is not zero for all values of y, and s is a positive integer.

Following Fuchs[*], we introduce the notion of a point of *indeterminateness* (Unbestimmtheit), as associated with functions; it is a point (it may be an essential singularity, but is not necessarily so) at which a function can assume one of a series of values, the value assumed depending upon the path by which the independent variable approaches the point. Moreover, the point may be a branch-point; if it is a branch-point, the branching can be one of two distinct kinds.

[*] *Berl. Sitzungsber.*, (1886, i), pp. 279—300.

In the case of the first kind, it may be possible to surround such a point c with a circle, not infinitesimal in size yet small enough to exclude every other branch-point; the branching is then called *definite*. Thus if ϕ be a function which is uniform in the immediate vicinity of c but actually indeterminate at c, then for

$$(z-c)^\lambda \phi,$$

where λ is a real positive quantity other than an integer, c is a point of indeterminateness with definite branching.

In the case of the second kind, it is not possible to surround the point c with a circle of such character, because there is an infinitude of points in the immediate vicinity of c at which branching takes place; the branching is then called *indefinite*. For example, let ζ be any point other than c, and let

$$e^{\frac{1}{\zeta-c}} = C,$$

where manifestly C is determinate and not zero; then all the points defined by

$$\frac{1}{z'-c} = \frac{1}{\zeta-c} + 2n\pi i,$$

for integer values of n from $-\infty$ to ∞, give

$$e^{\frac{1}{z'-c}} = C.$$

Of this series of points there is an unlimited number within any circle, however small, having c as its centre; each of them is a branch-point of the function

$$(e^{\frac{1}{z-c}} - C)^{\frac{1}{m}},$$

where m is an integer. For this function, the point c is a point of indeterminateness with indefinite branching.

It is necessary to consider the effect of such points upon the possibility of the expansion of a function in their vicinity.

When the point of indeterminateness is one without any branching whatever, the function can be expanded in a series of positive and negative powers of $z - c$, according to Laurent's theorem, the series converging in a region of the plane between two circles. Thus the function

$$\frac{1}{e^{\frac{1}{z-c}} - C}$$

can be expanded in such a series valid over a ring between two circles, having c for a common centre and such that the ring-space contains no zero of the denominator; but the circles may not be infinitesimal in radius, in order that this condition may be satisfied.

When the point of indeterminateness is one with definite branching, in many cases the expansion is possible by means of a subsidiary variable; sometimes the region of expansion is a strip, sometimes a Laurent ring, the actual region depending on the subsidiary variable adopted.

But when the point of indeterminateness is one with indefinite branching, the expansion is not, in general, possible; thus the function instanced above, for which the point c is of the class specified, cannot be expanded in positive and negative powers of $z - c$.

29. Turning to the equation

$$\frac{dx}{dy} = x^s g (y, x),$$

or rather to

$$\frac{dx}{dw} = x^s g (w - \alpha, x),$$

from which it is derived, the function g is a regular function in the immediate vicinity of $w = \alpha$, $x = 0$, on the assumption that the variables themselves are definite and specific at each point in their respective planes; the expansion of g then has the form

$$A_0 + A_1 x + A_2 (w - \alpha) + \dots,$$

and it is definite. Now w, if it exist at all as a solution, is a function of x. If $x = 0$ is not a point of indeterminateness for w, then $w = \alpha$ is a definite and specific value of w; the foregoing expansion is definite and possible, and the existence-theorem then shews that $w = \alpha$ is the solution under the assigned conditions. But if $x = 0$ is a point of indeterminateness (it might even be with indefinite branching) for w, then w cannot be regarded as determinately equal to α at $x = 0$: in the vicinity of $x = 0$, the variable $w - \alpha$ is not definite and specific, so that $g (w - \alpha, x)$ cannot thus uniquely be expanded as for the other case: and the inference in the existence-theorem cannot then be drawn.

It therefore is necessary to settle this particular relation of $x = 0$ to the functionality of w.

For this purpose, we introduce a new variable t having $x = 0$ for a point of indeterminateness and then, if possible, discuss the dependent variable w as a function of the new variable.

When $s > 1$, we define the new variable t by the relation

$$t = e^{\frac{1}{1-s}\frac{1}{x^{s-1}}},$$

so that

$$\frac{1}{t}\frac{dt}{dx} = \frac{1}{x^s}.$$

The value $x = 0$ is a point of indeterminateness for t: the value of t in the immediate vicinity of $x = 0$ depends upon the path of x in its approach to zero and it can have any magnitude. Taking this relation together with the differential equation, we have

$$\frac{dt}{t} = \frac{dx}{x^s} = \frac{dy}{\dfrac{1}{g(y,\,x)}};$$

with these relations in differential elements, we associate the partial differential equation

$$t\frac{\partial U}{\partial t} + x^s\frac{\partial U}{\partial x} + \frac{1}{g(y,\,x)}\frac{\partial U}{\partial y} = 0.$$

If $U(t,\,x,\,y)$ is an integral of this equation, then we know (from the elementary properties of partial differential equations of this type) that

$$U(t,\,x,\,y) = \text{constant}$$

is an integral of the equations in the differential elements. Accordingly when, with this integral, we associate the integral

$$t = e^{\frac{1}{1-s}\frac{1}{x^{s-1}}}$$

which is independent of it, we have the integral of the original differential equation. Now the equation

$$U(t,\,x,\,y) = \text{constant},$$

if it involves y at all, may be regarded as determining y as a function of x and t. As x approaches the value zero, the value of t depends upon the mode of approach; consequently, in general, y (which is a function of x and t) will also have a value depending upon the method of approach. In that case, $x = 0$ is a point of indeterminateness.

It may happen that, for special values of the constant to which U is equal, y ceases to depend upon t and is a function of x alone; $x = 0$ then ceases to be a point of indeterminateness for the integral.

It should be noted that, when $s = 1$, the variable t, as defined above in the form

$$\frac{1}{t}\frac{dt}{dx} = \frac{1}{x},$$

ceases to be effective for the purpose of discriminating the character of $x = 0$ as a point of the integral.

30. As a simple example of a point of indeterminateness, take the equation

$$\frac{dx}{dy} = \frac{x^s}{a_0 + a_1 y + a_2 y^2},$$

where a_0, a_1, a_2 are non-zero constants; we are to consider the relation of $x = 0$ to values of y. We have

$$-\frac{1}{s-1}\frac{1}{x^{s-1}} = \frac{1}{a_2}\frac{1}{\lambda - \mu}\log\frac{y - \lambda}{y - \mu} + C,$$

where λ and μ are the roots of

$$a_0 + a_1 \xi + a_2 \xi^2 = 0,$$

and C is arbitrary. It is evident that unless $\lambda = \mu$, that is, unless $a_1^2 = 4a_0 a_2$, $x = 0$ is a point of indeterminateness for y.

As a more general example, consider the equation

$$\frac{dx}{dy} = \frac{x^s}{p_0 + p_1 y + p_2 y^2},$$

where p_0, p_1, p_2 are regular functions of x in the vicinity of $x = 0$ and no one of them is zero there. Let

$$y = -\frac{x^s}{p_2}\frac{1}{\eta}\frac{d\eta}{dx}:$$

then on substituting, we find

$$\frac{d^2\eta}{dx^2} - \left(\frac{p_1}{x^s} - \frac{s}{x} + \frac{1}{p_2}\frac{dp_2}{dx}\right)\frac{d\eta}{dx} + \frac{p_0 p_2}{x^{2s}}\eta = 0,$$

a linear equation of the second order. When $s > 1$, this equation has no integrals that are regular* in the vicinity of $x = 0$; it can be made to have one sub-regular* integral; and the point $x = 0$ is a point of indeterminateness for the variable y of the original equation.

This result depends upon the theory of ordinary linear differential equations and is established by the following analysis, which is suggested by that theory. Anticipating so far the discussion (in a future volume) of the

---

* In the sense in which these terms are used in theory of linear differential equations by Fuchs, Thomé, Cayley, and others.

process of dealing with the sub-regular integrals (if any) of a linear differential equation, take

$$\eta = e^{\Omega} Y;$$

then the equation for Y is

$$\frac{d^2 Y}{dx^2} + Q_1 \frac{dY}{dx} + Q_2 Y = 0,$$

on the removal of the factor e^{Ω}. The coefficients Q_1 and Q_2 are given by

$$Q_1 = 2\Omega' + \frac{s}{x} - \frac{p_1}{x^s} - \frac{1}{p_2} \frac{dp_2}{dx},$$

$$Q_2 = \Omega'' + \Omega'^2 + \Omega' \left(\frac{s}{x} - \frac{p_1}{x^s} - \frac{1}{p_2} \frac{dp_2}{dx} \right) + \frac{p_0 p_2}{x^{2s}} :$$

the quantity Ω being at our disposal. Now if the two independent integrals of the equation in Y are to be regular (in the sense adopted for linear differential equations), the lowest power in the expansion of Q_1 in ascending powers of x cannot have an index lower than -1, and the lowest power in the similar expansion of Q_2 cannot have an index lower than -2. And if some one integral of the equation in Y is to be regular, the preceding conditions are to be replaced by a condition that the lowest index in the expansion of Q_2 is less by unity than the lowest index in the expansion of Q_1—a result that may be achievable in two ways, each leading to a regular integral; if they give distinct regular integrals, they can be linearly combined for the equation in η, according to the general theory. It may be added that the conditions stated are necessary: but they do not form the aggregate of sufficient conditions to secure the result.

With our present aim in view, which is the derivation of at least one critical relation among the constants, it will suffice to take the second alternative, and apply it to a special instance, say $s = 2$. Moreover, let

$$p_i = a_i + b_i x + c_i x^2 + \ldots, \qquad\qquad (i = 0, 1, 2).$$

If we assume

$$\Omega' = \frac{\lambda}{x^2},$$

where λ is a constant, the assigned conditions of relation between the lowest exponents in the expansion of Q_1 and Q_2 are satisfied. We then have

$$Q_1 = \frac{2\lambda - a_1}{x^2} - \frac{b_1 - 2}{x} + P_1(x),$$

$$Q_2 = \frac{\lambda^2 - \lambda a_1 + a_0 a_2}{x^4} + \frac{a_2 b_0 + a_0 b_2 - \lambda b_1}{x^3} + \frac{1}{x^2} P_2(x),$$

where P_1 and P_2 are regular functions of x in the vicinity of $x = 0$ and do not necessarily vanish there.

Let the roots of the equation

$$\lambda^2 - \lambda a_1 + a_0 a_2 = 0$$

be different from one another, and denote them by λ_1, λ_2: and let

$$a_1^2 - 4a_0 a_2 = \epsilon^2,$$

so that

$$2\lambda - a_1 = \pm \epsilon.$$

The determining equation of the equation for Y is

$$(2\lambda - a_1) n = a_2 b_0 + a_0 b_2 - \lambda b_1$$

so that, if ϵ be different from zero, there are two different values for n according to the two values of λ. When $\lambda = \lambda_1$, let $n = n_1$; when $\lambda = \lambda_2$, let $n = n_2$.

Then n_1 is a characteristic index for a regular integral Y_1, say

$$Y_1 = x^{n_1} \psi_1,$$

where ψ_1 is a regular function of x in the vicinity of $x = 0$ and does not vanish there; the corresponding value of Ω is

$$\Omega = -\frac{\lambda_1}{x}.$$

Similarly, n_2 determines an integral Y_2, such that

$$Y_2 = x^{n_2} \psi_2,$$

where ψ_2 is a regular function of x in the vicinity of $x = 0$ and does not vanish there: the corresponding value of Ω is

$$\Omega = -\frac{\lambda_2}{x}.$$

Hence the general value of η is

$$\eta = A e^{-\frac{\lambda_1}{x}} x^{n_1} \psi_1 + B e^{-\frac{\lambda_2}{x}} x^{n_2} \psi_2,$$

where A and B are arbitrary constants; the corresponding value of y, which is a solution of the equation

$$\frac{dx}{dy} = \frac{x^s}{p_0 + p_1 y + p_2 y^2},$$

is

$$y = -\frac{x^2}{p_2} \frac{1}{\eta} \frac{d\eta}{dx}.$$

When the value of η is substituted, then (as λ_1 and λ_2 are unequal) the expression for y is of the form

$$y = \frac{e^{\frac{\lambda_2 - \lambda_1}{x}} U + V}{e^{\frac{\lambda_2 - \lambda_1}{x}} U_1 + V_1},$$

where U, V, U_1, V_1 are either uniform in the vicinity of $x = 0$, or merely multiform with specific branches. It is evident that the point $x = 0$ is a point of indeterminateness (it is an essential singularity) of y; and therefore, unless the roots of the equation

$$\lambda^2 - \lambda a_1 + a_0 a_2 = 0$$

are equal to one another, the value $x = 0$ is certainly a point of indeterminateness fo the equation

$$\frac{dx}{dy} = \frac{x^s}{p_0 + p_1 y + p_2 y^2},$$

where

$$p_i = a_i + b_i x + c_i x^2 + \dots,$$

for $i = 0$, 1, 2.

It must not however be assumed that, if this condition is satisfied, the integrals of the equation in Y are regular and consequently that $x=0$ is not a point of indeterminateness of the integral of the equation in y so determined. Thus let the roots of

$$\lambda^2 - \lambda a_1 + a_0 a_2 = 0$$

be equal, so that $a_1{}^2 = 4 a_0 a_2$: then $2\lambda = a_1$, and we have

$$Q_1 = \frac{2 - b_1}{x} + P_1(x).$$

Then $$Q_2 = \frac{a_2 b_0 + a_0 b_2 - \frac{1}{2} a_1 b_1}{x^3} + \frac{1}{x^2} P_2(x).$$

If $a_2 b_0 + a_0 b_2 - \frac{1}{2} a_1 b_1$ is different from zero, then no integral of the equation in Y is regular; but if

$$a_2 b_0 + a_0 b_2 - \frac{1}{2} a_1 b_1 = 0,$$

then Q_2 begins with a term in $\frac{1}{x^2}$. In that case, one independent integral of the equation in Y is certainly regular. A second integral will also be regular if the roots of the determining (quadratic) equation differ by a quantity which is not an integer: and then we have

$$Y_1 = x^{n_1} \psi_1, \quad Y_2 = x^{n_2} \psi_2,$$

where ψ_1 and ψ_2 are regular functions of x that do not vanish with x; and the value of η is

$$\eta = e^{-\frac{1}{2} \frac{a_1}{x}} (A Y_1 + B Y_2),$$

where A and B are arbitrary constants. Since

$$y = -\frac{x^2}{p_2} \frac{1}{\eta} \frac{d\eta}{dx},$$

it is manifest that $x=0$ is no longer a point of indeterminateness for the integral y thus determined.

If however the roots of the determining quadratic equation differ by an integer, then the two independent integrals of the equation in Y are

$$Y_1 = x^n \psi_1,$$
$$Y_2 = x^n (\psi_2 + \psi_1 \log x),$$

where ψ_1 and ψ_2 are regular functions of x, not vanishing with x. Then

and we have $$\eta = e^{-\frac{1}{2} \frac{a_1}{x}} (A Y_1 + B Y_2),$$

$$y = -\frac{x^2}{p_2} \frac{1}{\eta} \frac{d\eta}{dx}.$$

Manifestly the value of y depends (when B is different from zero) upon the way in which x approaches its origin: the point consequently is a point of indeterminateness for the integral so obtained.

The two conditions, viz.

$$a_1{}^2 = 4 a_0 a_2, \quad a_2 b_0 + a_0 b_2 = \frac{1}{2} a_1 b_1,$$

shew that $p_1{}^2 - 4 p_0 p_2$ is divisible by x^2.

It therefore appears that the point $x = 0$ is certainly a point of indeterminateness for the integral of the equation

$$\frac{dx}{dy} = \frac{x^2}{p_0 + p_1 y + p_2 y^2},$$

unless $p_1^2 - 4 p_0 p_2$ is divisible by x^2, a condition involving two relations among the coefficients ; but the point ceases to be a point of indeterminateness if these two relations between the coefficients are satisfied and if, in addition, the difference between the two roots of the (quadratic) determining equation for the linear equation in Y, obtained after the transformation

$$y = -\frac{a_1}{2p_2} - \frac{x^2}{p_2}\frac{1}{Y}\frac{dY}{dx},$$

is a real quantity distinct from an integer.

31.   In all the cases considered, it has been assumed that the value of w and the value of z for the singularity are finite. Corresponding investigations for the cases when either is infinite, and for the case when both are infinite, would be necessary to complete the discussion : they would follow the lines already laid down, and so they will not be detailed here. But in each individual instance, the infinite values must be considered so as to make the discussion complete.

32.   Another method suggested by Fuchs, as introducing a new independent variable of intermediary type for the discussion of the equation

$$\frac{dx}{dy} = x^s g\,(y, x), \qquad\qquad (s \geqslant 1),$$

depends upon the equation

$$\frac{dx}{d\eta} = x^s g\,(\eta, 0),$$

so that, if $s > 1$,

$$-\frac{1}{s-1}\frac{1}{x^{s-1}} = \int g\,(\eta, 0)\, d\eta,$$

and, if $s = 1$,

$$\log x = \int g\,(\eta, 0)\, d\eta.$$

The variable x is thus determined in terms of η. The original equation becomes

$$\frac{d\eta}{dy} = \frac{g\,(y, x)}{g\,(\eta, 0)},$$

in the right-hand side of which the value of x is to be substituted in terms of η: this form, however, is not necessarily simpler than the initial form.

Suppose that logarithmic terms occur in $\int g(\eta, 0)\, d\eta$, say

$$\int g(\eta, 0)\, d\eta = \Sigma A \log(\eta - a) + g_1(\eta) + \text{constant}.$$

First, suppose that $s > 1$. Then when the variable η describes a contour round a an unlimited number of times and completes its path by ending at any point in its plane, the modulus of the right-hand side becomes infinitely great; the corresponding value of x is zero and this holds for a region of continuous variation of η.

If $s = 1$, then we have a relation of the form

$$x = c_1 \Pi (\eta - a)^A e^{g_1(\eta)}.$$

If in any one of the indices A the imaginary part should be different from zero, say equal to $i\alpha$, (where α is positive), then when η describes a contour round a and returns to its initial value, this contour not including any other point such as a associated with a factor in the product Π, the new value of x is

$$x_0 e^{2\pi i A},$$

x_0 being some one value; and therefore the new value of $|x|$ is

$$|x_0|\, e^{-2\pi\alpha}.$$

Consequently after an unlimited number of contours round a, the value of $|x|$ tends to the limit zero: and therefore x tends to the limit zero, in consequence of such a path, whatever the final value of η may be.

As in the preceding method, we consider the system of equations

$$\frac{dx}{x^s} = \frac{dy}{\dfrac{1}{g(y, x)}} = \frac{d\eta}{\dfrac{1}{g(\eta, 0)}}$$

in differential elements, and with them we associate the partial differential equation

$$x^s \frac{\partial U}{\partial x} + \frac{1}{g(y, x)} \frac{\partial U}{\partial y} + \frac{1}{g(\eta, 0)} \frac{\partial U}{\partial \eta} = 0.$$

Any integral of this equation that involves y is an integral of the system of two equations and therefore, when taken in conjunction with the integral of

$$\frac{dx}{x^s} = \frac{d\eta}{\dfrac{1}{g(\eta, 0)}},$$

furnishes an integral of the original differential equation. Now the existence-theorem of partial differential equations shews that integrals exist; let one such be

$$U(y, x, \eta) = \text{constant}.$$

As x tends to the value zero, then in the preceding cases it has been seen that there is an unlimited number of values of η: and conversely, the values of η depend upon the path by which x approaches zero. When an integral y is derivable from $U = $ constant and when it depends upon η, there will generally (though not universally) in such cases be a value or a number of values of y depending upon η as well as upon x; that is, owing to the relation between x and η, an unlimited number of different values of y depending upon the path by which x approaches zero. When all these conditions are satisfied, whether for $s > 1$ or for $s = 1$, the point $x = 0$ is a point of indeterminateness for the integral.

Examples for the case $s > 1$ have been given in connection with the other method. A simple example, (for which the intermediary variable need not be used), is given by

$$\frac{dx}{dy} = ax\frac{y+1}{y^2+1},$$

where a is a real positive constant. The integral of this equation is

$$x = c\,(y - i)^{\frac{1}{2}a\,(1-i)}\,(y + i)^{\frac{1}{2}a\,(1+i)}.$$

As x approaches the value 0, there are an unlimited number of values of y satisfying the equation; each can be obtained by making that variable describe negatively an unlimited number of times a contour enclosing i and not $-i$, or by making it describe positively an unlimited number of times a contour enclosing $-i$ but not i; subject to this condition, the value of y can remain arbitrary. Thus the point $x = 0$ is manifestly a point of indeterminateness for the integral.

A corresponding investigation would shew that $x = 0$ is not a similar point of indeterminateness for the integral of the equation

$$\frac{dx}{dy} = \frac{axy}{y^2+1},$$

where a is a real commensurable constant.

33. We may summarise the results for the integral in the immediate vicinity of an accidental singularity of the first kind as follows.

Denoting the singularity by α, c, the reciprocal of the value of $\frac{dw}{dz}$ is regular in the vicinity of α, c, and vanishes there. If

W, $= \{f(w - \alpha, 0)\}^{-1}$, does not vanish for all values of w, the singularity is an algebraic branch-point for the integral, the branches circulating round the point form one system, and their number is greater by unity than the index of the lowest power of $w - \alpha$ in the expansion of W. If W does vanish for all values of w—so that $\{f(w - \alpha, z - c)\}^{-1}$, expanded in the vicinity of the singularity, is of the form $(z - c)^s g(w - \alpha, z - c)$, where s is a positive integer, and g is a regular function such that $g(w - \alpha, 0)$ does not vanish for all values of w—then there is no regular integral acquiring the value α at $z = c$, but there can be non-regular integrals; when s is greater than unity, the singularity is generally (but not universally) a point of indeterminateness for the non-regular integral; and when $s = 1$, the singularity may or may not be a point of indeterminateness for the non-regular integral. The discrimination between the cases depends upon the form of the function g.

With the appropriate changes, corresponding results hold if α, or if c, or if both α and c, be infinite.

On the Uniqueness of Cauchy's Integral.

34. The analytical results just obtained, in connection with the points of indeterminateness of the integral that can arise from an accidental singularity of the first kind belonging to the original differential equation, can be applied to settle the question, which was left unsettled (§ 15), as to whether the unique regular integral in Cauchy's existence-theorem is the only integral of the equation satisfying the conditions in that theorem.

Let u denote the unique regular integral of the equation

$$\frac{dw}{dz} = f(w, z),$$

determined by the condition that it assumes the value α when z is c, the function f being regular in the vicinity of α, c. Let $u + v$ denote any other integral (if any such exist) satisfying the same initial conditions; it will not be assumed that the integral is regular, and therefore it may not be assumed that v is regular, so that the argument of §§ 12, 13 is no longer valid: it is not possible by that means to prove that v is zero everywhere in the vicinity of $z = c$. Because u and $u + v$ are integrals of the equation, we have

$$\frac{dv}{dz} = f(u + v, z) - f(u, z).$$

When v is zero, the function on the right-hand side is zero; so that we may write

$$f(u + v, z) - f(u, z) = v^s g(u, z, v),$$

where the new function g is a regular function of u, z, v; or since u is a regular function of z, the function g is a regular function of z and v, say $G(z, v)$. Moreover, if the variable v be definite (§ 29), G can be expanded in power-series. The equation for v now is

$$\frac{dv}{dz} = v^s G(z, v).$$

But from what we have seen, the point $z = c$ is a point of indeterminateness in general for v when $s > 1$; and, when $s = 1$, it may be a point of indeterminateness according to the character of the function G. If then $z = c$ is a point of indeterminateness, v cannot be regarded as definite in its immediate vicinity; the expansion of the function G is therefore also not valid; and neither the argument of § 12 nor that of § 13 can be applied to prove that v is steadily zero, as indeed it is not when $z = c$ is a point of indeterminateness. We therefore infer the following result, due to Fuchs (*l. c.*):

The regular integral u of the differential equation

$$\frac{dw}{dz} = f(w, z),$$

determined by the initial conditions in the existence-theorem, is the sole integral of the equation determined by those conditions only if the initial value of z is not a point of indeterminateness for the equation

$$\frac{dv}{dz} = f(u + v, z) - f(u, z).$$

As a special example constructed by Fuchs, consider the equation

$$\frac{dw}{dz} = \frac{a(az + \beta w)^2 - a(az + bw)}{\beta(az + bw) - b(az + \beta w)^2},$$

where a, b, a, β are constants such that $a\beta - ba$ is not zero. Assign c as an initial value of z, $-\frac{a}{\beta} c$ as the initial value of w; then $\frac{dw}{dz}$ is determinate and finite for these initial values. It is easy to see that the corresponding integral is

$$w = -\frac{a}{\beta} z.$$

We thus take

$$w = -\frac{a}{\beta} z + y,$$

so that

$$\frac{dy}{dz} = \frac{\beta (a\beta - ab) y^2}{(a\beta - ab) z + b\beta y - b\beta^2 y^2};$$

and therefore*

$$\beta (a\beta - ab) \frac{dz}{dy} = \frac{1}{y^2} \{(a\beta - ab) z + b\beta y - b\beta^2 y^2\}.$$

The integral of this is

$$z (ab - a\beta) = b\beta y + A e^{-\frac{1}{\beta y}},$$

where A is an arbitrary constant. Manifestly y can be made to tend to the value zero (and can have an unlimited number of values in the vicinity of zero) so that z shall tend to the value c: all that is necessary for this purpose is to take

$$-\frac{1}{\beta y} + 2k\pi i = \log \{c (ab - a\beta)\},$$

and to give the integer k a succession of infinitely large values.

In fact, $w = -\frac{a}{\beta} c$ is an essential singularity of z regarded as a function of w; w is a non-regular function of z, having $z = c$ as a singularity; and some, among its infinitude of branches, tend to the value $-\frac{a}{\beta} c$ as z tends to c.

It is manifest that a whole class of cases, with similar results, is given by an equation

$$\frac{dw}{dz} = \gamma + (w - \gamma z)^s F(w, z),$$

where F is a regular function and the assigned condition is that $w = \gamma c$ when $z = c$.

Ex. Another example, being a case in which for $s = 1$ a point of indeterminateness arises, is given by Fuchs as follows.

The integral of the equation

$$\frac{dw}{dz} = -\frac{1}{4} \frac{1}{z(z-1)} - \frac{2z-1}{z(z-1)} w - w^2$$

is given by

$$w = \frac{A \dfrac{d\eta_1}{dz} + \dfrac{d\eta_2}{dz}}{A\eta_1 + \eta_2},$$

where A is an arbitrary constant, $\eta_1 = z^{-\frac{1}{2}} K$, $\eta_2 = z^{-\frac{1}{2}} K'$, and K, K' are the elliptic quarter-periods of modulus $z^{-\frac{1}{2}}$. It is proved that the number of integrals of the equation, converging to one value as z by selected variation is made to converge to any arbitrary value, is unlimited.

---

* The equation in question is evidently an example of the value $s = 2$ indicated on the preceding page.

CHAPTER V.

TYPICAL REDUCED FORMS OF THE DIFFERENTIAL EQUATION IN THE VICINITY OF AN ACCIDENTAL SINGULARITY OF THE SECOND KIND*.

ACCIDENTAL SINGULARITIES OF THE SECOND KIND: FORM OF THE EQUATION.

35. WHEN the combination α, c is an accidental singularity of the second kind for the function $f(w, z)$ in the differential equation

$$\frac{dw}{dz} = f(w, z),$$

it has been proved (§ 20) that the differential equation can be changed so as to have one or other of four definite forms.

Let

$$w - \alpha = y, \quad z - c = x;$$

then the first form is

$$\frac{dy}{dx} = \frac{q(y, x)}{p(y, x)} e^{G(y, x)},$$

---

* The more important cases of reduction were discussed by Briot and Bouquet in their memoir (cited on p. 40); in the present chapter, the discussion is rendered complete because the general functional expression due to Weierstrass (Note to ch. I) of a function of two variables is used as the basis of departure. The various results attained by Briot and Bouquet have been adopted or reproduced by many other writers but without any amplification; the only exception to this statement, so far as concerns the form of the equation (as distinct, that is to say, from the character of the integral), is to be found in a memoir by Horn, *Crelle*, t. CXIII (1894), pp. 50—57.

where G is a regular function of y, x in the immediate vicinity of 0, 0; also

$$q(y, x) = y^n + q_1 y^{n-1} + q_2 y^{n-2} + \ldots + q_n,$$

$$p(y, x) = y^m + p_1 y^{m-1} + p_2 y^{m-2} + \ldots + p_m,$$

m and n being positive integers, and all the coefficients q_r and p_r being functions of x which are regular in the vicinity of its origin and vanish there.

The second form is

$$\frac{dy}{dx} = \frac{q'(u, v)}{p(y, x)} e^{G_1(y, x)},$$

where G_1 is a regular function of y and x in the immediate vicinity of the origin, p has the same significance as before, and q' is given by

$$q'(u, v) = u^{n_1} + q_1' u^{n_1-1} + q_2' u^{n_1-2} + \ldots + q'_{n_1},$$

in which the coefficients q_s' are functions of v, regular in the vicinity of $v = 0$ and vanishing at $v = 0$. The variables u and v are connected with x and y by the relations

$$y = \lambda u + \mu v, \quad x = \lambda' u + \mu' v,$$

containing arbitrary constants λ, μ, λ', μ', subjected only to the two inequalities specified at the time of the reduction to the form. Moreover, the transformation of the variables was rendered desirable by the fact that, if $q'(u, v)$ is the transformed expression for $q(w - \alpha, z - c)$, the quantity $q(w - \alpha, 0)$ vanishes for all values of w.

Consequently, $q'(u, v)$ vanishes when $\lambda' u + \mu' v = 0$. Now $q'(u, v)$ is an algebraical function of u, the coefficients being regular functions of v which vanish with v; hence $\lambda' u + \mu' v$ is a factor of $q'(u, v)$. Let the sth power of this linear factor be the highest contained in $q'(u, v)$, and let

$$q'(u, v) = (\lambda' u + \mu' v)^s q_1(u, v), \qquad (s \geqslant 1),$$

where q_1 now does not vanish with $\lambda' u + \mu' v$. Effecting the reverse transformation so as to express $q_1(u, v)$ as a function of y and x, we obtain a function $Q_1(y, x)$ of y and x, which is regular in the vicinity of 0, 0 and vanishes there but does not vanish for all values of y when $x = 0$. To $Q_1(y, x)$ we again apply the theorem of Weierstrass proved in the Note to Chapter I; since $Q_1(y, 0)$

does not vanish for all values of y, let y^n be the lowest power of y which it contains; then

$$q_1(u, v) = Q_1(y, x)$$
$$= (y^n + q_1 y^{n-1} + q_2 y^{n-2} + \ldots + q_n) e^{G_2(y, x)}$$
$$= q(y, x) e^{G_2(y, x)}$$

say, where G_2 is regular and the coefficients $q_1, \ldots, q_n$ are functions of x which vanish at $x = 0$ and are regular in its vicinity. Accordingly

$$q'(u, v) = x^s q(y, x) e^{G_2(y, x)} \; ;$$

and therefore, writing

$$G(y, x) = G_1(y, x) + G_2(y, x),$$

so that G is regular in the vicinity of $0, 0$, the differential equation becomes

$$\frac{dy}{dx} = x^s \frac{q(y, x)}{p(y, x)} e^{G(y, x)},$$

where s is a positive integer.

In particular cases, it might happen that $(\lambda' u + \mu' v)^s$ is the whole of $q'(u, v)$; for each of them, the function $q(y, x)$ would be replaced by unity.

The third form is

$$\frac{dy}{dx} = \frac{q(y, x)}{p'(u, v)} e^{G_1(y, x)},$$

where the function p' has properties and character similar to those of q' in the last case. Similar modification of p' and transformation by Weierstrass's theorem lead to an expression

$$p'(u, v) = x^s p(y, x) e^{G_2(y, x)},$$

where

$$p(y, x) = y^m + p_1 y^{m-1} + p_2 y^{m-2} + \ldots + p_m;$$

the coefficients p_r are functions of x, which vanish when $x = 0$ and are regular in its vicinity. Writing

$$G(y, x) = G_1(y, x) - G_2(y, x),$$

so that G is a regular function of y, x in the vicinity of $0, 0$, the form of the differential equation is

$$\frac{dy}{dx} = x^{-s} \frac{q(y, x)}{p(y, x)} e^{G(y, x)},$$

where s is a positive integer.

In particular cases it might happen, as for the preceding form, that x^s is the whole of $p'(u, v)$; for each of them, the function $p(y, x)$ would be replaced by unity.

The reduction for the third form can be made in another way, as follows. The equation can be written

$$\frac{dx}{dy} = \frac{p'(u, v)}{q(y, x)} e^{-G_1 (y, x)}.$$

On interchanging the variables in the statement of Weierstrass's theorem (p. 24) when there are two arguments of the functions there considered, we obtain a result which can be applied to this case. It shews that the equation can be put in the form

$$\frac{dx}{dy} = x^s \frac{p(y, x)}{q(y, x)} e^{-G (y, x)},$$

which is an equivalent of the third form.

The fourth form is

$$\frac{dy}{dx} = \frac{q''(u, v)}{p''(u, v)} e^{G_1 (y, x)},$$

where G_1 is regular in the vicinity of 0, 0, and the two functions q'', p'' have characteristic properties similar to those of q', p' in the preceding cases. The corresponding modifications and transformations, by means of Weierstrass's theorem, can be effected; and we have results such as

$$q''(u, v) = x^{s_1} q(y, x) e^{G_2 (y, x)},$$
$$p''(u, v) = x^{s_2} p(y, x) e^{G_3 (y, x)},$$

where s_1 and s_2 are positive integers, and the functions q and p are given by the equations

$$q(y, x) = y^n + q_1 y^{n-1} + q_2 y^{n-2} + \ldots + q_n,$$
$$p(y, x) = y^m + p_1 y^{m-1} + p_2 y^{m-2} + \ldots + p_m,$$

the coefficients q_r and p_r being functions of x which vanish at $x = 0$ and are regular in its vicinity. Hence writing

$$s_1 - s_2 = s,$$
$$G_1(y, x) + G_2(y, x) - G_3(y, x) = G(y, x),$$

so that G is a regular function of y, x in the vicinity of 0, 0, the equation becomes

$$\frac{dy}{dx} = x^s \frac{q(y, x)}{p(y, x)} e^{G (y, x)},$$

where s is an integer.

When $s = 0$, this equation is included in the first form; when s is positive, the equation is included in the second form; when s is negative, the equation is included in the third form.

Accordingly no new characteristic type of equation is introduced by the fourth form of § 20; and therefore summarising the results, it appears that, in the vicinity of an accidental singularity of the second kind of the equation

$$\frac{dw}{dz} = f(w, z)$$

at α, c, the equation can be expressed in the form

$$\frac{dy}{dx} = x^s \frac{q(y, x)}{p(y, x)} e^{G(y, x)},$$

where s is a finite integer, (zero, positive, or negative), and the functions q, p, G are of the assigned characters in the immediate vicinity of the singularity.

36.   The first result to be obtained is that integral (or class of integrals) of the equation which vanishes when $x = 0$. Such integrals may be regular or non-regular. Thus the integral of

$$\frac{dy}{dx} = \frac{y + ax + bx^2}{x},$$

which vanishes with x, is

$$y = ax \log x + bx^2 + Ax,$$

where A is still arbitrary; if a is zero, the integral is regular; if a is not zero, it is an integral with an unlimited number of branches.

But the equation in the form obtained may be propounded as an initial equation and not as a form representing a limiting case of another within the immediate vicinity of the origin; it will then be necessary to consider the integral of the equation for a range not so restricted as in the more special investigation for which the limiting form has been obtained.

37.   The determination of the forms has thus far been chiefly affected by the algebraical character of the occurrence of y in $q(y, x)$ and $p(y, x)$; but deviations from the apparent expressions can be caused by the mode of occurrence of x. Thus it might

happen that, in $q(y, x)$, the coefficients q_n, q_{n-1}, ..., q_{n-r} are evanescent, so that q contains a factor y^r, say

$$q(y, x) = y^r q_1(y, x),$$

where now $q_1(0, x)$ does not vanish for all values of x; the corresponding form of the equation is

$$\frac{dy}{dx} = x^s y^r \frac{q_1(y, x)}{p(y, x)} e^{G(y, x)}.$$

Or it might happen that, in $p(y, x)$, the coefficients p_m, p_{m-1}, ..., p_{m-r} are evanescent, so that p contains a factor y^r, say

$$p(y, x) = y^r p_1(y, x),$$

where now $p_1(0, x)$ does not vanish for all values of x; the corresponding form of the equation is

$$\frac{dy}{dx} = x^s y^{-r} \frac{q(y, x)}{p_1(y, x)} e^{G(y, x)}.$$

Accordingly, *the most general form of the equation in the vicinity of the singularity considered is*

$$\frac{dy}{dx} = x^s y^r \frac{q(y, x)}{p(y, x)} e^{G(y, x)},$$

where neither $q(0, x)$ nor $p(0, x)$ vanishes for all values of x; the integers s and r may be zero, positive, or negative, each independently of the other.

In very particular cases, it may happen that

   (i)   if s be a positive integer and $r \geqslant 0$, q may be a constant, though p must then be variable;

   (ii)   if r be a negative integer and $s \leqslant 0$, p may be a constant, though q must then be variable;

   (iii) if s be positive and r negative, p and q may each of them be constant.

Unless there is an explicit statement to the contrary, it will be assumed that $q(y, x)$, $p(y, x)$ are functions of y and x and are not constants; and then it is manifest, after the preceding explanations, that no generality is lost by further assuming that a term independent of y exists both in $q(y, x)$ and in $p(y, x)$.

Initial Form of Integral: Simplest Case.

38.   Taking the simplest of these forms, viz.,

$$\frac{dy}{dx} = \frac{q(y, x)}{p(y, x)} e^{G(y, x)},$$

we require primarily the character of the integral (or integrals) in the immediate vicinity of $x = 0$, which vanish at $x = 0$. Suppose that the order of y in powers of x is μ, the real part of which is positive: (it will appear that μ generally is real); then, in the vicinity of $x = 0$, y may be represented in the form

$$y = \rho x^\mu + \dots,$$

the indices in the remaining powers being greater than μ, and ρ being some constant.   Now

$$q(y, x) = y^n + y^{n-1} q_1 + \dots + q_n,$$

where the functions q_r are functions of x, which vanish at $x = 0$ and are regular in its vicinity.   Let

$$q_i = a_i x^{r_i} + \dots$$
$$= x^{r_i} Q_i,$$

where Q_i is a regular function of x which is equal to a_i when $x = 0$, and r_i is a positive integer ($r_0 = 0$); then the order of terms in q, other than those which occur in

$$y^n + a_1 x^{r_1} y^{n-1} + a_2 x^{r_2} y^{n-2} + \dots + a_n x^{r_n},$$

is greater than the order of the lowest term or set of lowest terms in that retained aggregate.   Similarly, let

$$p_i = b_i x^{s_i} + \dots$$
$$= x^{s_i} P_i,$$

where P_i is a regular function of x which is equal to b_i when $x = 0$, and s_i is a positive integer ($s_0 = 0$); then the order of terms in p, other than those which occur in

$$y^m + b_1 x^{s_1} y^{m-1} + b_2 x^{s_2} y^{m-2} + \dots + b_m x^{s_m},$$

is greater than the order of the lowest term or set of lowest terms in that retained aggregate.   Thus

$$\frac{dy}{dx} = \frac{y^n + \sum\limits_{l=1}^{n} (a_l x^{r_l} y^{n-l}) + \dots}{y^m + \sum\limits_{k=1}^{m} (b_k x^{s_k} y^{m-k}) + \dots} e^{G(y, x)}.$$

With the hypothesis adopted,

$$\frac{dy}{dx} = \mu\rho x^{\mu-1} + \ldots :$$

when the values of y and dy/dx are substituted, the differential equation is to be satisfied identically; and therefore the term of lowest order on the right-hand side must be of order $\mu - 1$. This term will be obtained by selecting the term (or the group of terms) in the numerator of lowest order in x, and the term (or the group of terms) in the denominator of lowest order in x.

For the value

$$y = \rho x^\mu + \ldots,$$

let $a_l x^{r_l} y^{n-l}$ be one of the set of terms of lowest order in the numerator; its leading term is

$$a_l \rho^{n-l} x^{r_l + \mu(n-l)},$$

and for all terms of this order in the numerator, the coefficient of $x^{r_l + \mu(n-l)}$ is

$$\Sigma a_l \rho^{n-l}.$$

For the same substitution, let $b_k x^{s_k} y^{m-k}$ be one of the set of terms of lowest order in the denominator; its leading term is

$$b_k \rho^{m-k} x^{s_k + \mu(m-k)},$$

and for all terms of this order in the denominator, the coefficient of $x^{s_k + \mu(m-k)}$ is

$$\Sigma b_k \rho^{m-k}.$$

Now the equation, after substitution of the value of y and multiplication of both sides by $p(y, x)$, may be satisfied identically in one of two ways; either (i) by having the terms of lowest order on the two sides equal or (ii) by having the terms of lowest order on one side lower in order than the corresponding terms on the other.

First, suppose that the terms of lowest order on the two sides are the same; by equating their orders, we have

$$\mu - 1 = r_l + \mu(n-l) - \{s_k + \mu(m-k)\},$$

and by equating their coefficients, we have

$$\rho\mu = C\frac{\Sigma a_l \rho^{n-l}}{\Sigma b_k \rho^{m-k}},$$

where C is the value of $e^{G(0,\,0)}$; this is the only part of $G(y, x)$ which occurs, because the index μ in the expression for y has its real part positive. The former gives a value of μ, viz.

$$\mu = \frac{r_l + 1 - s_k}{m + 1 - k - (n - l)};$$

a definite real value in general, when the appropriate terms have been selected.

APPLICATION OF PUISEUX'S DIAGRAM.

39.   To obtain the various possible values of μ, a method can be used which is similar to that adopted in connection with the branch-points of an algebraic equation*. We take two perpendicular axes $O\xi$, $O\eta$ in a plane; and we mark a first set of points $(m + 1 - k, s_k)$ for the various values of k, these being associated with the denominator; and a second set of points $(n - l, r_l + 1)$ for the various values of l, these being associated with the numerator. Now if the differential equation be

$$\frac{dy}{dx} = \frac{\Delta}{D}\, e^G,$$

where, for the present purpose, only the lowest terms in Δ and D are retained, the equation can be written

$$x\frac{dy}{dx}\, D = x\Delta e^G,$$

and the terms of lowest order on the two sides are then to be retained. A term on the left-hand side is of order

$$\mu + s_k + (m - k)\,\mu,$$

that is,

$$\mu (m + 1 - k) + s_k;$$

and a term on the right-hand side is of order

$$r_l + 1 + (n - l)\,\mu.$$

From the value of μ, these two are equal; and there are no terms of smaller index on either side. A straight line drawn in the plane, through the point $m + 1 - k$, s_k and making an angle $\tan^{-1}\mu$ with the negative direction of $O\xi$, is

$$\eta - s_k = -\mu\,[\xi - (m + 1 - k)],$$

* *Th. Fns.*, § 97.

so that the perpendicular from the origin on the line is

$$\{\mu\,(m+1-k)+s_k\}\,(1+\mu^2)^{-\frac{1}{2}}.$$

A parallel line through $n-l$, r_l+1 is

$$\eta-(r_l+1)=-\mu\,[\xi-(n-l)],$$

so that the perpendicular from the origin on the line is

$$\{\mu\,(n-l)+r_l+1\}\,(1+\mu^2)^{-\frac{1}{2}}\,;$$

the two perpendiculars being equal, the parallel lines must be one and the same: and the single line joins a point of one set to a point of the other set. Moreover, this perpendicular must be the least distance of all parallel lines through a point of the first set: it must also be the least distance of all parallel lines through a point of the second set.

On the basis of these properties and noting that no point of the first set is on the axis $O\eta$ and that no point of the second set is on the axis $O\xi$, while one point of the first set is on the axis $O\xi$ and one point of the second set is on the axis $O\eta$, we obtain possible values of μ as follows. Take the single point of the whole tableau of points which lies on $O\eta$; round this point, let a line turn from the position $O\eta$ in the counterclockwise sense until it meets one or more of the marked points; let it turn in the same sense round that one of these met points which lies nearest the axis $O\xi$, and continue turning until it meets one or more of the marked points; and so on, until it meets the point lying in $O\xi$. Then among the properties of this broken line, it is easy to note the following:—

(i)   all the values of μ are positive, real, and commensurable;

(ii)   when a number of the marked points lie on a portion of this line, the corresponding terms in the equation

$$x\,\frac{dy}{dx}\,D=x\Delta$$

are of order lower than all other terms for the substitution

$$y=\rho x^\mu\,;$$

(iii)   if the marked points on any portion of the line include at least one point from each set, then the corresponding value of μ gives rise to terms on both sides of the equation in the above form, the terms being of the proper lowest order;

(iv)   if the marked points on any portion of the line all belong to the first set only, the lowest terms that arise all occur on the left-hand side of the equation taken in the above form : while if all of them belong to the second set only, the lowest terms that arise all occur on the right-hand side of that equation.

As the line begins at a point of the second set and ends at a point of the first set, at least one part of it (and it may be all the parts of it) will contain marked points belonging to both sets.

40.   As a preliminary instance, consider

$$\frac{dy}{dx} = \frac{y^6 + a_1 xy^5 + a_2 x^5 y^4 + a_3 x^2 y^3 + a_4 xy^2 + a_5 x^5 y + a_6 x^6 + \dots}{y^4 + b_1 x^3 y^3 + b_2 xy^2 + b_3 x^2 y + b_4 x^3 + \dots} \, e^{G(y,\,x)},$$

the omitted parts referring to terms of higher order for all substitutions.   In the figure, the points A correspond to the terms in the numerator with the same suffix for a, A_0 corresponding to the first term there ; they are the second set of points.   The points B correspond to the terms in the denominator with the same suffix for b, B_0 corresponding to the first term there ; they are the first set of points.

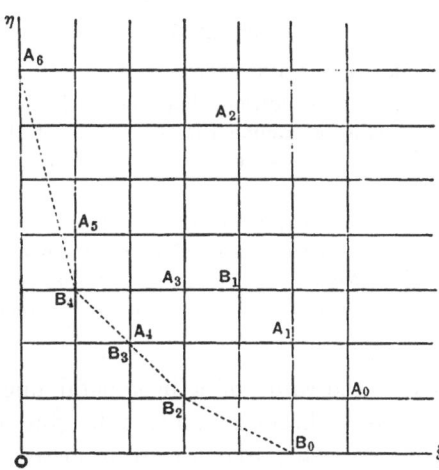

For the portion $A_6 B_4$, we have $\mu = 4$; if therefore

$$y = \rho x^4 + \dots,$$

we have

$$4\rho x^3 + \dots = C \frac{a_6 x^6 + \dots}{b_4 x^3 + \dots},$$

so that

$$\rho = \frac{1}{4} \frac{a_6}{b_4} C.$$

In this portion there is one point from each of the two sets.

For the portion $B_4 \dfrac{B_3}{A_4} B_2$, we have $\mu = 1$; if therefore

$$y = \rho x + \ldots,$$

we have

$$\rho + \ldots = C \, \frac{a_4 x^3 \rho^2 + \ldots}{(b_4 + b_3 \rho + b_2 \rho^2)\, x^3 + \ldots},$$

so that ρ is determined by a cubic equation

$$b_4 \rho^3 + b_3 \rho^2 + b_4 \rho = C a_4 \rho^2,$$

of which one root is zero : or neglecting the zero root, (which manifestly will occur in every case when, in the numerator on the right-hand side, the term free from y does not give rise to terms of lowest order), we have ρ determined by a quadratic

$$b_4 \rho^2 + (b_3 - C a_4)\, \rho + b_4 = 0.$$

For the portion $B_2 B_0$, we have $\mu = \tfrac{1}{2}$; if therefore

$$y = \rho x^{\frac{1}{2}} + \ldots,$$

we must have

$$\tfrac{1}{2} \rho x^{-\frac{1}{2}} + \ldots = C \, \frac{a_4 \rho^2 x^2 + \ldots}{(b_2 \rho^2 + \rho^4)\, x^2 + \ldots},$$

which could not possibly be satisfied unless, as a first condition, the aggregate of the terms in the denominator of lowest order vanish ; so that

$$b_2 \rho^2 + \rho^4 = 0 :$$

as a second condition, that the next aggregate of terms in the denominator be of order $\tfrac{5}{2}$: and as a third condition, that the coefficients of $x^{-\frac{1}{2}}$ be equal which, in effect, is a relation among the coefficients of the equation (those which are specified and others which are unspecified), ρ, and the coefficient of a succeeding term in y.

The statement that the points on each portion of the line correspond to the terms that give rise to quantities of lowest order after substitution is confirmed, it will be noticed, by the occurrence of the coefficients b and a of those terms that correspond to the points B and A respectively on each successive portion of the line.

41. As regards the value of μ in general, one remark should be made. If it arises through points belonging to the first set alone, its value is definite; likewise, if it arises through points belonging to the second set alone. When it arises through points, which belong some to one of the two sets and some to the other of them, then its value has the form

$$\frac{r_l + 1 - s_k}{m + 1 - k - (n - l)};$$

and this expression could be indefinite if

$$r_l + 1 = s_k, \quad n - l = m + 1 - k,$$

that is, if the point of the second set coincide with the point of the first. Moreover, this could occur for other pairs of coincident points.

If there be only one pair of coincident points, the portion of the line passing through that common point joins it to a point of either the first set or the second set. In each case, the portion of the line joins two points of one set; and by them the value of μ will be determined, though the coincident points do not determine its value.

If upon the portion of the line there be more than one pair of coincident points, the portion can be regarded as joining a point of one set (in one of the pairs) to another point of that same set (in another of the pairs); by means of them, its value is determined.

The alternative reduced forms for the value of dy/dx, adopted in § 37, made it possible to assume, for the case

$$\frac{dy}{dx} = \frac{q\,(y,\,x)}{p\,(y,\,x)}\, e^{G\,(y,\,x)}$$

under consideration, that $q\,(0,\,x)$ is not zero for all values of x and that $p\,(y,\,0)$ is not zero for all values of y; accordingly, there is always a point of the first set on the line $O\xi$ and there is always a point of the second set on the line $O\eta$. If all the other effective points occur in coincident pairs, the reduction is still possible by joining the one nearest $O\xi$ to the point on $O\xi$, and the one nearest $O\eta$ to the point on $O\eta$; so that, in the present case, the value or values of μ can always be determined.

42. In the initial derivation of μ, it was assumed (p. 90) that the terms of lowest order for the substitution adopted arose on both sides of the equation; but it was pointed out that terms of lowest order might arise on only one of the sides of the equation. In reality, so far as the determination of μ is concerned, the alternative has actually been taken into account. When terms of lowest order occur on both sides, the corresponding portion of the broken line contains points of both sets; when terms of lowest order occur on only one side of the equation

$$D\frac{dy}{dx} = \Delta e^{G},$$

then the corresponding portion of the broken line contains points of that set alone which belong to the side of the equation giving the lowest terms.

Accordingly, no further investigation for the value of μ is necessary at this stage: but discrimination must, of course, be made between the possible cases of the source of μ as arising from terms grouped in the various ways.

Construction of Critical Reduced Forms.

43. Consider, first, a portion of the line which contains at least one point from each set. Then since

$$\mu = \frac{r_l + 1 - s_k}{m + 1 - k - (n - l)},$$

and since μ is found to be a positive quantity by the preceding investigation, μ is a positive commensurable magnitude: let it be expressed as a fraction $\frac{p}{q}$ taken in its lowest terms. In particular cases, it may happen that q is unity. Then

$$p\{(m + 1 - k) - (n - l)\} = q(r_l + 1 - s_k),$$

and therefore

$$p(m + 1 - k) + qs_k = p(n - l) + q(r_l + 1)$$
$$= N,$$

say. Take

$$x = t^q, \quad y = ut^p,$$

so that, if the integral of order μ in x near the origin exist, u is to be different from zero when t vanishes. The equation for the determination of u is

$$\left\{ \Sigma b_k u^{m-k} t^{p(m-k)+qs_k} + \dots \right\} \left(pt^{p-1}u + t^p \frac{du}{dt} \right)$$
$$= Cqt^{q-1} \Sigma a_l u^{n-l} t^{p(n-l)+qr_l} + \dots,$$

where C is the value of $e^{G(0,\,0)}$; that is,

$$\left\{ \Sigma b_k u^{m-k} t^{p(m+1-k)+qs_k-1} + \dots \right\} \left(pu + t \frac{du}{dt} \right)$$
$$= Cq\Sigma a_l u^{n-l} t^{p(n-l)+q(r_l+1)-1} + \dots.$$

The terms of lowest order on the right-hand side in t contain it to the order $N - 1$: likewise for the terms in the coefficient of

$pu + t\dfrac{du}{dt}$ on the left-hand side, also to the order $N - 1$; for the substitution under consideration, all other indices of t are $\geqslant N$. Hence, dividing out by t^{N-1}, we have

$$\{\Sigma b_k u^{m-k} + tP_1(u, t)\}\left(pu + t\dfrac{du}{dt}\right) = Cq\Sigma a_l u^{n-l} + tP_2(u, t),$$

where P_1, P_2 are regular functions of u and t at and in the immediate vicinity of $0, 0$; in the summations on the two sides, only those terms occur which correspond to the marked points on the portion of the line determining the value of μ in the substitution adopted.

The variable u, if it exist, is required not to vanish with t; let ρ denote its value for $t = 0$. (When expressed as a function of t, it may be a regular function in the vicinity of $t = 0$, and then it would be expressible in a converging power-series containing a term ρ independent of t.) Since u satisfies the above equation, it follows that ρ is determined by the equation

$$F(\rho) = p\rho\Sigma b_k \rho^{m-k} - Cq\Sigma a_l \rho^{n-l} = 0.$$

44. Let ρ be a simple root of $F(\rho) = 0$, so that $F'(\rho)$ does not vanish for that value; and take

$$u = \rho + v,$$

so that v is (if possible) a function of t vanishing with t. Then the coefficient of $t\dfrac{dv}{dt}$ is

$$\Sigma b_k \rho^{m-k} + Q_1(v) + tR_1(v, t),$$

where Q_1 and R_1 denote regular functions, Q_1 vanishing with v. Let all the other terms be transferred to the right-hand side; their aggregate is

$$- vF'(\rho) - \tfrac{1}{2}v^2F''(\rho) - \ldots + tR_2(v, t),$$

where R_2 is a regular function. The equation thus becomes

$$t\dfrac{dv}{dt} = \dfrac{- vF'(\rho) - \tfrac{1}{2}v^2F''(\rho) - \ldots + tR_2(v, t)}{\Sigma b_k \rho^{m-k} + Q_1(v) + tR_1(v, t)}.$$

Suppose that

$$\Sigma b_k \rho^{m-k} \neq 0,$$

which also implies that

$$\Sigma a_l \rho^{n-l} \neq 0;$$

then the denominator does not vanish for the values $0, 0$; and therefore, expanding the whole fraction in the immediate vicinity of 0, 0 in a converging power-series, we have

$$t \frac{dv}{dt} = av + bt + \tfrac{1}{2}av^2 + \beta vt + \tfrac{1}{2}\gamma t^2 + \dots$$

$$= \phi_1(v, t) \dots\dots\dots\dots\dots\dots\dots\dots\dots \text{(I)},$$

where ϕ_1 is a regular function ; and the value of a is

$$a = \frac{-F'(\rho)}{\Sigma b_k \rho^{m-k}}.$$

This is the reduced form of the equation to be discussed ; it clearly is the case of widest generality, for the possibility of its derivation has depended only upon inequalities among the constants. Its relation to the original equation is given by the equations

$$w = \alpha + (\rho + v) t^p, \quad z = c + t^q ;$$

the quantity ρ is a simple root of the equation

$$p\rho \Sigma b_k \rho^{m-k} - Cq \Sigma a_l \rho^{n-l} = 0 ;$$

and $\dfrac{p}{q}$ is the expression of the equal fractions

$$\frac{r_l + 1 - s_k}{m + 1 - k - (n - l)},$$

when reduced to their lowest terms.

The form has been deduced on the supposition that ρ is a simple root of $F(\rho) = 0$. Suppose, in the next place, that it is a multiple root of $F(\rho) = 0$, of multiplicity κ, so that the first $\kappa - 1$ derivatives of F vanish for that value. It is easy to see that the equation becomes

$$t \frac{dv}{dt} = \frac{-v^\kappa \dfrac{F^{(\kappa)}(\rho)}{\kappa !} - \dots + tR_2(v, t)}{\Sigma b_k \rho^{m-k} + Q_1(v) + tR_1(v, t)}.$$

With the same assumption as before that

$$\Sigma b_k \rho^{m-k} \neq 0,$$

so that the denominator does not vanish at 0, 0, we can expand the whole fraction, which is a regular function of v and t, in a converging power-series, say

$$t \frac{dv}{dt} = gv^\kappa + bt + \beta vt + \gamma t^2 + \dots$$

$$= \phi_\kappa(v, t) \dots\dots\dots\dots\dots\dots\dots\dots \text{(II)},$$

where
$$g = -\frac{1}{\kappa !} \frac{F^{(\kappa)}(\rho)}{\Sigma b_k \rho^{m-k}},$$

and the function $\phi_\kappa (v, t)$ is such that $\phi_\kappa (v, 0)$, a regular function of v, has gv^κ for its lowest term. This is the corresponding reduced form.

45. In deducing these two forms, the assumption has been made that $\Sigma b_k \rho^{m-k}$ does not vanish ; but the assumption is not necessarily justified, for ρ would still be a root of $F(\rho) = 0$ if

$$\Sigma b_k \rho^{m-k} = 0, \quad \Sigma a_l \rho^{n-l} = 0,$$

simultaneously, the coexistence of these equations implying at least one relation between the coefficients a and b.

Suppose then that ρ is a simple root of

$$F_1 (\rho) = \Sigma b_k \rho^{m-k} = 0.$$

Then, as in the earlier investigation, the coefficient of $t\dfrac{dv}{dt}$ is

$$\Sigma b_k u^{m-k} + tP_1 (u, t) ;$$

this becomes

$$vF_1' (\rho) + \frac{v^2}{2} F_1'' (\rho) + \dots + tR_1 (v, t).$$

If ρ, being a simple root of $F_1 (\rho) = 0$ and also a root of $F(\rho) = 0$, is a simple root of the latter equation, then v is determined by

$$t\frac{dv}{dt} = \frac{\lambda'v + \mu't + \dots}{\lambda v + \mu t + \dots},$$

where the unexpressed terms are of the second and higher orders in v and t combined.

If however ρ, being a simple root of $F_1 (\rho) = 0$ and also a root of $F(\rho) = 0$, is a multiple root of the latter equation, then v is determined by

$$t\frac{dv}{dt} = \frac{\lambda'v^\kappa + \mu't + \dots}{\lambda v + \mu t + \dots},$$

where in the numerator the unexpressed terms independent of t have indices greater than κ, and the other terms are of the second and higher orders in v and t combined.

Suppose, next, that ρ is a multiple root of

$$F_1 (\rho) = \Sigma b_k \rho^{m-k} = 0$$

of multiplicity l. Then the coefficient of $t\dfrac{dv}{dt}$, being

$$\Sigma b_k u^{m-k} + tP_1 (u, t),$$

becomes

$$\frac{v^l}{l!} F_1^{(l)} (\rho) + \text{terms in higher powers of } v + tR_1 (v, t),$$

where $F_1^{(l)}(\rho)$ does not vanish. Now ρ is also a root of $F(\rho) = 0$; let it be of multiplicity l', so that $F^{(l')}(\rho)$ is the lowest derivative of F which does not vanish for the root; then the equation takes the form

$$t\,\frac{dv}{dt} = \frac{k'v^{l'} + \mu't + \ldots}{kv^l + \mu t + \ldots},$$

where, in the numerator, $v^{l'}$ is the lowest power of v in terms independent of t, and in the denominator, v^l is the lowest power of v in terms independent of t. The simplest case arises when $l' = 1$, so that the root ρ is a simple root of $F(\rho) = 0$.

46. Of all this set of forms, which can arise for particular equations when special relations among their constants are satisfied, some can be modified. When the numerator of the fraction on the right-hand side has a term of the first order in v, say as in the first of those (§ 45) which are obtained when $F_1(\rho) = 0$, we write

$$v + \frac{\mu'}{\lambda'}\,t = Vt;$$

then as v vanishes with t, V may (but need not necessarily) vanish with t. We have

$$\frac{dv}{dt} = -\frac{\mu'}{\lambda'} + V + t\,\frac{dV}{dt},$$

and therefore

$$t^2\frac{dV}{dt} = \frac{\mu'}{\lambda'}\,t - Vt + \frac{\lambda'Vt + \text{terms in squares of } t}{\dfrac{\mu\lambda' - \lambda\mu'}{\lambda'}\,t + \text{terms in squares of } t}$$

$$= a'V + b't + \ldots,$$

unless

$$\mu\lambda' - \lambda\mu' = 0.$$

But if $\mu\lambda' - \lambda\mu' = 0$, the new equation is

$$t^2\frac{dV}{dt} = \frac{\lambda'V + et + \ldots}{\lambda''V + e't + \ldots}.$$

A similar transformation, viz.

$$V + \frac{e}{\lambda'}\,t = V_1 t,$$

would then be made and would be found effective for the reduction unless

$$\lambda'e' - \lambda''e = 0:$$

and similarly for the alternative. The ultimate reduced form is seen to be

$$t^s \frac{dW}{dt} = aW + bt + \ldots$$

$$= \phi_1(W, t)\ldots\ldots\ldots\ldots\ldots\ldots(\text{III}),$$

with the condition that the positive integer s is not less than 2.

This also is a reduced form for

$$t \frac{dv}{dt} = \frac{\lambda' v + \mu' t + \ldots}{k v^l + \mu t + \ldots},$$

because it can be derived from the preceding case by taking $\lambda = 0$; the one condition, which then is necessary in order that (III) may be its reduced form, is that μ should not vanish.

Lastly, as regards the form

$$t \frac{dv}{dt} = \frac{k' v^{l'} + \mu' t + \ldots}{k v^l + \mu t + \ldots},$$

(which includes an earlier form when $l = 1$), we have $l' \geqslant 2$. This will be considered later, among the forms of the type

$$\frac{dy}{dx} = \frac{x^S}{y^R} \frac{q(y, x)}{p(y, x)} e^{G(y, x)},$$

where S is negative (§§ 54—56).

47. In the second place, consider a portion of the broken line in the tableau of points, which contains only points of the second set, *i.e.* points associated with the numerator in the expression for $\frac{dy}{dx}$. When the quantity μ for that portion of the line is in its lowest terms, let its value be $\frac{p}{q}$, where in particular cases it may happen that $q = 1$; as before, let

$$x = t^q, \quad y = u t^p;$$

so that, if the integral y exist of dimensions μ in x in the immediate vicinity of the origin, u must not be zero at $x = 0$. Then we have

$$\frac{dy}{dt} \left[\Sigma \left\{ b_k u^{m-k} t^{qs_k + p(m-k)} \right\} + \ldots \right]$$

$$= C \left[\Sigma \left\{ a_l u^{n-l} t^{qr_i + p(n-l)} \right\} + \ldots \right] \frac{dx}{dt}$$

$$= C q t^{q-1} \left[\Sigma a_l u^{n-l} t^{qr_i + p(n-l)} + \ldots \right].$$

The terms of lowest dimension on the right-hand side are those having

$$q(r_l + 1) + p(n - l) - 1$$

$= N$, say, for the index of t. The left-hand side of the equation is

$$\left(pu + t\frac{du}{dt}\right) t^{p-1} \left[\Sigma b_k u^{m-k} t^{qs_k + p(m-k)} + \ldots\right].$$

In this expression, all the quantities

$$qs_k + p(m + 1 - k) - 1$$

are greater than N, on the present hypothesis of the origin of μ; let the smallest of them be $N + \nu$, where $\nu \geqslant 1$, and let it arise from terms typically represented by that which has b_k in its coefficient. On the right-hand side, all the rest of these indices are $\geqslant N + 1$; on the left-hand side, they all are $\geqslant N + \nu + 1$; so that, dividing by t^N, we have

$$t^\nu \left(pu + t\frac{du}{dt}\right) \left[\Sigma b_k u^{m-k} + tP_1(u, t)\right] = qC\Sigma a_l u^{n-l} + tQ_1(u, t)$$

as the form of the equation, P_1 and Q_1 denoting regular functions.

It therefore follows that, if ρ be the value of u when $t = 0$, ρ being distinct from zero, then it must be a root of the equation

$$H(\rho) = \Sigma a_l \rho^{n-l} = 0,$$

and it may (or it may not) be such as to make

$$I(\rho), = \Sigma b_k \rho^{m-k},$$

vanish. Let $u = \rho + v$.

When ρ is a simple root of $H(\rho) = 0$ which does not make $I(\rho)$ vanish, then the reduced form of the equation is

$$t^{\nu+1}\frac{dv}{dt} = av + bt + \ldots$$

$$= \phi_1(v, t),$$

the unexpressed terms being of order higher than unity in v and t combined. Since $\nu \geqslant 1$, this form is included in the form (III).

When ρ is a multiple root of $H(\rho) = 0$ of multiplicity κ and does not make $I(\rho)$ vanish, then the reduced form of the equation is

$$t^{\nu+1}\frac{dv}{dt} = av^\kappa + bt + \ldots,$$

where av^κ is the lowest term on the right-hand side involving v alone, and the unexpressed terms are of order higher than unity in v and t combined.

When ρ is a simple root of $H(\rho) = 0$ and is also a simple root of $I(\rho) = 0$, the reduced form is, by analysis similar to that in the corresponding case, found to be

$$t^{\nu+s}\,\frac{dW}{dt} = a'W + b't + \dots,$$

where $s \geqslant 2$, and the unexpressed terms on the right-hand side are of order higher than unity in W and t combined. This form also is included in the form (III).

Lastly, if ρ is a multiple root of $H(\rho) = 0$ of multiplicity m' and if it is also a multiple root of $I(\rho) = 0$ of multiplicity m'' (where $m'' \geqslant 1$, taking account of preceding cases), then the reduced form of the equation is

$$t^{\nu+1}\,\frac{dv}{dt} = \frac{kv^{m'} + \mu t + \dots}{k'v^{m''} + \mu't + \dots},$$

with corresponding implications as regards the unexpressed terms alike in the numerator and in the denominator.

All these results are derivable exactly as in preceding investigations, and therefore they are merely stated without proof: the actual verification is simple.

48. In the third place, consider a portion of the broken line in the tableau of points, which contains only points of the first set, *i.e.* points associated with the denominator in the expression for $\dfrac{dy}{dx}$. When the quantity μ for that portion of the line is in its lowest terms, denote its value again by $\dfrac{p}{q}$, where for particular instances q may be unity; and, as before, let

$$x = t^q, \quad y = ut^p,$$

so that, if the integral y exist of dimensions μ in x in the immediate vicinity of $x = 0$, u must not be zero at $x = 0$. The equation is

$$t^{p-1}\left(pu + t\,\frac{du}{dt}\right)\left[\Sigma\left\{b_k u^{m-k} t^{qs_k + p(m-k)}\right\} + \dots\right]$$
$$= Cqt^{q-1}\left[\Sigma\left\{a_l u^{n-l} t^{qr + p(n-l)}\right\} + \dots\right].$$

The terms of lowest dimension on the left-hand side in the coefficient of $pu + t\dfrac{du}{dt}$ are those having

$$qs_k + p(m + 1 - k) - 1,$$

$= N$ say, for the index of t; all the others on that side have the index of $t \geqslant N + 1$. On the right-hand side, all the quantities

$$q(r_l + 1) + p(n - l) - 1$$

are greater than N, on the present hypothesis of the origin of μ; let the smallest be $N + \nu'$, where $\nu' \geqslant 1$, and let it arise from terms typically represented by that which has a_l in its coefficient. All the other indices of t on the right-hand side are $\geqslant N + \nu' + 1$; so that, on division by t^N, the equation becomes

$$\left(pu + t\frac{du}{dt}\right)\left[\Sigma b_k u^{m-k} + tP_1(u, t)\right]$$
$$= Ct^{\nu'}\left[\Sigma a_l u^{n-l} + tQ_1(u, t)\right],$$

where P_1 and Q_1 are regular functions. Now u is to be different from zero when $t = 0$; let its value there be ρ, and if it be assumed expansible in powers of t, let

$$u = \rho + \sigma t^\lambda + \dots,$$

so that, if the equation is identically satisfied, we must have

$$[p\rho + (p + \lambda)\sigma t^\lambda + \dots][I(\rho) + \text{powers of } t \text{ and } t^\lambda]$$
$$= Ct^{\nu'}[H(\rho) + \text{powers of } t \text{ and } t^\lambda]$$

identically satisfied. Because $\nu' \geqslant 1$, a first condition is given by taking

$$I(\rho) = 0,$$

which is the fundamental equation determining ρ. Clearly, the simplest cases arise when $\nu' = 1$, though this is not a necessary condition; but if $\nu' > 1$, some conditions would be satisfied by the constants in the differential equation. Taking then

$$u = \rho + v$$

in general, where ρ is a root of $I(\rho) = 0$, we have a reduced form

$$t\frac{dv}{dt} + pv + p\rho = t^{\nu'}\frac{\alpha + \dots}{av + bt + \dots},$$

if ρ is a simple root of $I(\rho) = 0$, and is not a root of $H(\rho) = 0$.

Without entering on the discussion of the various alternatives to these assumptions as to ρ, it may be noticed that, if the equation be taken in the form

$$\frac{dx}{dy} = \frac{p\,(y,\ x)}{q\,(y,\ x)}\ e^{-G(y,\ x)},$$

then the particular substitution giving y as of dimensions μ arises solely from terms in the numerator and thus can be included in the last case (§ 47). Since

$$y \propto x^\mu$$

so far as concerns its most important term, so also

$$x \propto y^{\frac{1}{\mu}}$$

$$\propto y^{\frac{q}{p}}.$$

We should then introduce new variables such that

$$y = \tau^q, \quad x = \xi\tau^p,$$

where ξ does not vanish when $\tau = 0$; and the reductions for the respective possibilities would be similar to those in the preceding discussion.

THE REMAINING CASES OF § 37.

49. It still remains to consider the rest of the forms of § 37, which can be represented generally by

$$\frac{dy}{dx} = \frac{x^S}{y^R} \frac{q\,(y,\ x)}{p\,(y,\ x)}\ e^{G(y,\ x)},$$

where neither $q\,(0,\ x)$ nor $p\,(0,\ x)$ vanishes for all values of x: the integers R and S may be positive, zero, or negative, though it is now unnecessary to consider simultaneous zero values. Moreover, if S be positive, then q may be a constant; if R be positive, p may be a constant; if both R and S be positive, then both p and q may be constants. This last case is relatively simple; we then have

$$y^R \frac{dy}{dx} = x^S\,(C + ay + bx + \dots).$$

If $S > R$, and b be not zero, then in the immediate vicinity of the origin, the two most important terms of y are

$$y = \left(C\,\frac{R+1}{S+1}\right)^{\frac{1}{R+1}} x^{\frac{S+1}{R+1}} + \gamma x^{\frac{S+1}{R+1}+1};$$

while if $S < R$ and a be not zero, then in the immediate vicinity of the origin, the two most important terms of y are

$$y = \left(C \frac{R+1}{S+1} \right)^{\frac{1}{R+1}} x^{\frac{S+1}{R+1}} + \gamma' x^{2\frac{S+1}{R+1}} ;$$

if $S = R$, these pairs of terms coincide.

It is clear that, without discussing this very special case further, the origin (being an accidental singularity of the second kind) is in general an algebraical branch-point of the integral of the original equation; the character of the branch-point, and the number of branches of the function, depend upon the expression for $\frac{S+1}{R+1}$ in its lowest terms.

The assumption has been made that the form of the equation has arisen as a particular form of an earlier equation for an exceptional combination of values constituting an accidental singularity of the second kind: so that the limiting condition on the integral y is that it is to vanish with x. If however, as is possible, the given equation be propounded as an initial form, and be not subjected to this condition, then the conditions actually imposed may form either an ordinary combination for the equation or an accidental singularity of the first kind; in each case, the properties of the integral can be regarded as known, after the earlier investigations.

Accordingly, we shall now consider that the case when both q and p are constants, S and R being positive integers, has been dealt with, so that it may be excluded from further discussion.

There are, in fact, four distinct general cases: but they can be effectively considered in three forms.

First, it may happen that neither R nor S is negative; the equation then has the form

$$\frac{dy}{dx} = \frac{x^S}{y^R} \frac{q(y, x)}{p(y, x)} \cdot e^{G(y, x)}$$

Secondly, it may happen that R is negative and S is not negative; changing the signification of R, we shall take

$$\frac{dy}{dx} = x^S y^R \frac{q(y, x)}{p(y, x)} e^{G(y, x)},$$

with R and S not negative, as the representative form.

Thirdly, it may happen that S is negative and that R is not negative ; changing the signification of S, we take

$$\frac{dx}{dy} = y^R x^S \frac{p(y, x)}{q(y, x)} e^{-G(y, x)}$$

as the representative form, which manifestly is the same as in the last case when the variables are interchanged.

Lastly, it may happen that both R and S are negative; changing the signification of both R and S, we take

$$\frac{dy}{dx} = \frac{y^R}{x^S} \frac{q(y, x)}{p(y, x)} e^{G(y, x)}$$

as the representative form, where now neither R nor S is negative.

Case I.

50. Taking the equation in the form

$$\frac{dy}{dx} = \frac{x^S}{y^R} \frac{q(y, x)}{p(y, x)} e^{G(y, x)},$$

where in the first place it will be assumed that S and R are positive integers either of which (though not both) may be zero, what is sought is an integral (if any) which shall vanish with x. The functions q and p have the same general functional expression as before ; and therefore, for the present purpose, we may take

$$q(y, x) = y^n + a_1 x^{r_1} y^{n-1} + a_2 x^{r_2} y^{n-2} + \ldots\ldots + a_n x^{r_n} + \ldots,$$

$$p(y, x) = y^m + b_1 x^{s_1} y^{m-1} + b_2 x^{s_2} y^{m-2} + \ldots\ldots + b_m x^{s_m} + \ldots,$$

where a_n and b_m are different from zero.

Suppose that the order in powers of x of the integral y, if it exist in the vicinity of the origin, is μ; so that it may there be represented by

$$y = \rho x^\mu + \ldots,$$

ρ being a constant that does not vanish, and the indices in the remaining powers being greater than μ.

Now in $q(y, x)$, let

$$a_l x^{r_l} y^{n-l}$$

be one of the set of terms which are the lowest in order for the substitution $y = \rho x^\mu$; so that in $q(y, x)$ the leading term of lowest index is

$$x^{r_l + \mu(n-l)} \Sigma a_l \rho^{n-l}.$$

In $p(y, x)$, let

$$b_k x^{s_k} y^{m-k}$$

be one of the set of terms which are the lowest in order for the substitution $y = \rho x^{\mu}$; so that in $p(y, x)$ the leading term of lowest index is

$$x^{s_k + \mu (m-k)} \, \Sigma b_k \rho^{m-k}.$$

If the equation is to be identically satisfied by the postulated value of y, the lowest term in

$$y^R \frac{dy}{dx} p(y, x)$$

and the lowest term in

$$x^S q(y, x) \, e^{G(y, x)}$$

must have equal indices and equal coefficients. That the indices may be equal, we must have

$$\mu R + \mu - 1 + s_k + \mu (m - k) = S + r_l + \mu (n - l);$$

that the coefficients may be equal, we must have

$$F(\rho) = \rho^{R+1} \mu \Sigma b_k \rho^{m-k} - C \Sigma a_l \rho^{n-l} = 0,$$

where C is the value of $e^{G(0, 0)}$. The former gives a value of μ, viz.

$$\mu = \frac{r_l + 1 + S - s_k}{m + R + 1 - k - (n - l)},$$

a definite value in general, when the appropriate terms have been selected.

51.   To determine the values of μ, we adopt the same method as in the case when $R = 0$, $S = 0$. Drawing two perpendicular axes $O\xi$, $O\eta$ in a plane, we mark a first set of points

$$(m + R + 1 - k, \; s_k^i)$$

for the various values of k, these being associated with the denominator in the equation

$$\frac{dy}{dx} = \frac{x^S}{y^R} \frac{q(y, x)}{p(y, x)} e^{G(y, x)};$$

and we mark a second set of points $(n - l, \; r_l + 1 + S)$ for various values of l, these being associated with the numerator in that equation. Writing the equation in the form

$$x y^R p(y, x) \frac{dy}{dx} = x^{1+S} q(y, x) \, e^{G(y, x)},$$

and on the two sides keeping terms of order lowest for the substitution $y = \rho x^\mu + \ldots$, a term on the left-hand side is of order

$$\mu (R + 1) + s_k + (m - k) \mu,$$

that is,

$$\mu (m + R + 1 - k) + s_k;$$

and a term on the right-hand side is of order

$$1 + S + r_l + \mu (n - l);$$

these two orders are equal on account of the value of μ. There are no terms of smaller index on either side. Drawing through a point $(m + R + 1 - k, s_k)$ a line making an angle $\tan^{-1} \mu$ with the negative direction of $O\xi$, so that its equation is

$$\eta - s_k = - \mu \left\{ \xi - (m + R + 1 - k) \right\},$$

the perpendicular from the origin on the line is

$$\left\{ \mu (m + R + 1 - k) + s_k \right\} (1 + \mu^2)^{-\frac{1}{2}};$$

and drawing a parallel line through a point $(n - l, r_l + 1 + S)$, the perpendicular from the origin upon it is

$$\left\{ r_l + 1 + S + \mu (n - l) \right\} (1 + \mu^2)^{-\frac{1}{2}}.$$

The two perpendiculars are equal: and the two lines are therefore one and the same, which thus is a line joining a point of one set to a point of the other set. Moreover, this perpendicular is the least distance from the origin of all parallel lines through a point of either set.

Accordingly, we can construct the broken line in the tableau of points, corresponding to that in § 39 : for there is one point of the first set (but none of the second set) on $O\xi$, and there is one point of the second set (but none of the first set) on $O\eta$. We take a line coincident with $O\eta$, make it turn in the counterclockwise sense round the single marked point on $O\eta$ until it meets some point or points in the tableau : then make it turn about the last of them until it meets others : and so on, until it passes through the single marked point on $O\xi$.

For all equations having integers R and S that are not negative, the properties of the positions of the broken line are similar, detail by detail, to those before enumerated in § 39 : they need not here be repeated. There are corresponding explanations as regards the determination of μ for any part of the broken line, if it should happen that some expression for μ in that portion becomes indeterminate (§ 41).

52. Consider, first, a portion of the line containing at least one point from each set. Then

$$\mu, = \frac{r_l + 1 + S - s_k}{m + R + 1 - k - (n - l)},$$

is a real, positive, commensurable magnitude; let it be expressed in its lowest terms, say $\dfrac{p}{q}$, so that

$$p\{(m + R + 1 - k) - (n - l)\} = q\{(r_l + 1 + S) - s_k\},$$

and therefore

$$p(m + R + 1 - k) + qs_k = p(n - l) + q(r_l + 1 + S)$$
$$= N,$$

say. Take

$$x = t^q, \quad y = ut^p$$

as before, so that, as y is to be of order $\dfrac{p}{q}$ in x in the immediate vicinity of the origin, u must be equal to ρ, that is, different from zero, when $t = 0$. The equation for the determination of u becomes

$$\left(pu + t\frac{du}{dt}\right)[\Sigma b_k u^{m-k+R} t^{p(m+1+R-k)+qs_k-1} + \ldots]$$
$$= Cq[\Sigma a_l u^{n-l} t^{p(n-l)+Sq+q(r_l+1)-1} + \ldots].$$

The terms, containing the lowest powers of t on the right-hand side, contain it to a power $N - 1$; all other indices of t on that side are $\geqslant N$. Also, in the coefficient of $pu + t\dfrac{du}{dt}$ on the left-hand side, the lowest power of t that occurs is that given in the expressed terms: its index is $N - 1$, and all the other indices are $\geqslant N$. Hence, dividing out by t^{N-1}, we have

$$\Sigma\{b_k u^{m-k+R} + tP_1(u, t)\}\left(pu + t\frac{du}{dt}\right)$$
$$= Cq\,\Sigma a_l u^{n-l} + tP_2(u, t),$$

where P_1 and P_2 are regular functions of u and t in the immediate vicinity; and in the summations on the two sides, those terms are taken that correspond to the points on the portion of the line which determines the value of μ adopted.

The further reduction to typical forms is obviously similar to that for the earlier case. Let

$$F(\rho) = p\rho\,\Sigma b_k \rho^{m-k+R} - Cq\,\Sigma a_l \rho^{n-l}.$$

If ρ is a simple root of $F(\rho) = 0$, we take

$$u = \rho + v\,;$$

then, unless ρ is such as to make

$$F_1(\rho), \ = \Sigma b_k \rho^{m-k+R},$$

vanish, the equation for v in the immediate vicinity of $t = 0$ is

$$t\,\frac{dv}{dt} = av + bt + \tfrac{1}{2}av^2 + \beta vt + \tfrac{1}{2}\gamma t^2 + \dots$$
$$= \phi_1(v,\,t),$$

where ϕ_1 is a regular function of v, t vanishing with $v = 0$, $t = 0$; and the value of a is

$$a = \frac{-F'(\rho)}{F_1(\rho)}.$$

The relation to the integral of the original equation is

$$w = \alpha + (\rho + v)\,t^p, \quad z = c + t^q\,;$$

and $\dfrac{p}{q}$ is the common value of the equal fractions

$$\frac{r_l + 1 + S - s_k}{m + R + 1 - k - (n - l)}\,,$$

when expressed in their lowest terms.

If ρ is a multiple root of $F(\rho) = 0$ of multiplicity m_1 and is not a root of $F_1(\rho) = 0$, then the equation becomes

$$t\,\frac{dv}{dt} = gv^{m_1} + bt + \beta vt + \gamma t^2 + \dots$$
$$= \phi_{m_1}(v,\,t),$$

where ϕ_{m_1} is regular in the vicinity of $0, 0$; the lowest term in v in $\phi_{m_1}(v, 0)$ is gv^{m_1}, and the value of g is

$$g = -\frac{1}{m_1!}\,\frac{F^{m_1}(\rho)}{F_1(\rho)}.$$

If ρ is a simple root of $F(\rho) = 0$ and is also a simple root of $F_1(\rho) = 0$, the equation can be reduced to the form

$$t\,\frac{dv}{dt} = \frac{\lambda' v + \mu' t + \dots}{\lambda v + \mu t + \dots}\,;$$

by suitable transformations, this can be changed to the form

$$t^m\,\frac{dW}{dt} = aW + bt + \dots,$$

where the integer $m \geqslant 2$.

If ρ is a multiple root of $F(\rho) = 0$ of multiplicity l' and at the same time is a multiple root of $F_1(\rho) = 0$ of multiplicity l, then the equation can be reduced to the form

$$t \frac{dv}{dt} = \frac{k'v'^l + \mu't + \ldots}{kv^l + \mu t + \ldots},$$

which can be transformed as was the other.

In fact, with the appropriate changes in the form of the equation

$$F(\rho) = 0$$

and of the function $F_1(\rho)$, all the cases that occur, when either R or S or both are different from zero, are of the same types respectively as when both R and S vanish. No new forms therefore arise for consideration.

It is not necessary to advert, except with the utmost brevity after the preceding discussion, to the other two possibilities that can occur in connection with portions of the broken line.

For a portion which contains points of the second set alone, being the points associated with the numerator in the expression for $\frac{dy}{dx}$, the typical forms that arise are respectively similar to those where $R = 0$, $S = 0$; we substitute for the former functions $H(\rho)$ and $I(\rho)$, functions defined by

$$H(\rho) = \Sigma a_l \rho^{n-l},$$

$$I(\rho) = \rho^R \Sigma b_k \rho^{m-k};$$

and the reduced forms belong to one or other of the types

$$t^{\nu+1} \frac{dv}{dt} = av + bt + \ldots,$$

$$t^{\nu+1} \frac{dv}{dt} = gv^m + bt + \ldots,$$

$$t^{\nu+s} \frac{dW}{dt} = a'W + b't + \ldots, \qquad\qquad (s \geqslant 2),$$

$$t^{\nu+1} \frac{dv}{dt} = \frac{kv^m + \mu t + \ldots}{k'v^{m'} + \mu't + \ldots},$$

where $\nu \geqslant 1$, and in the last form m and m' are not unity together.

For a portion of the broken line which contains points of the first set alone, being the points associated with the denominator in the expression for $\dfrac{dy}{dx}$, the simplest plan of discussion is to interchange the variables (as suggested in § 35, p. 86, for the corresponding case when $S = 0$, $R = 0$), so that the equation is

$$\frac{dx}{dy} = \frac{y^R}{x^S} \frac{p\,(y,\,x)}{q\,(y,\,x)}\, e^{-G\,(y,\,x)}\,;$$

the discussion then resolves itself into a repetition (with the appropriate modifications) of the discussion of which the summarised results have just been given.

In fact, it appears that, when R and S are positive, the equation provides no canonical types other than those found for the case $R = 0$, $S = 0$. The discussion has been given at some length : as will be seen immediately, some of the analysis needed to establish this result will be found useful in the discussion of the remaining instances for negative values of R, or of S, or of both.

Note. The discussion of the integral of the equation

$$\frac{dy}{dx} = \frac{x^S}{y^R} \frac{q\,(y,\,x)}{p\,(y,\,x)}\, e^{G\,(y,\,x)}$$

has been limited by the condition that the integral y is to vanish with x.

If however the equation be propounded as an initial equation, not subject to this particular condition, then the simultaneous initial values that are imposed upon y and x form either an ordinary combination for the right-hand side or they form an accidental singularity of the first kind. The mode of obtaining the characteristics of the integral in either case has been indicated in preceding sections.

53. Further, it is proper to notice the two cases which were pointed out (§ 37) as special cases, viz. that, when $R > 0$, p may be a constant; and, when $S > 0$, q may be a constant.

The same method for the determination of μ can be adopted as before.

If $R > 0$, the suggested form of the equation becomes

$$y^R \frac{dy}{dx} = C x^S q\,(y,\,x)\, e^{G\,(y,\,x)}.$$

The corresponding equation for the determination of μ is

$$\mu(R+1) - 1 = S + r_l + \mu(n-l),$$

so that

$$\mu = \frac{S+1+r_l}{R+1-(n-l)}.$$

We mark the set of points $(n-l, S+1+r_l)$ for the various values of l; the first of them (for the highest value of l) lies on the line $O\eta$; and we mark the point $R+1, 0$, which is a point on the line $O\xi$. The critical line for the value (or values) of μ can be constructed as before; the simplest instance will be when it consists of a single piece joining the point $(0, S+1+r_n)$ to the point $(R+1, 0)$, and then the value of μ is

$$\mu = \frac{S+1+r_n}{R+1}.$$

Whether the line consist of one portion only or of several portions, it is easy to see that no new typical forms arise other than those already included in the typical reduced forms : those which would actually be obtained are only special cases of those which have been retained.

If $S > 0$ and q is a constant, then the equation may be taken in the form

$$x^S \frac{dx}{dy} = C'p(y, x) e^{-G(y, x)},$$

which, with the interchange of variables, is merely the preceding case: the results of that case, when inverted as between the variables, apply to the present.

Case II.

54. We now consider the equation when one (but not both) of the integers R and S is negative: and as has been seen, we may take the equation in the form

$$\frac{dy}{dx} = x^S y^R \frac{q(y, x)}{p(y, x)} e^{G(y, x)},$$

which covers the two possibilities (§ 49). The functions q and

p have the same functional expressions as before, so that for our present purpose we may take

$$q\,(y,\,x) = y^n + a_1 x^{r_1} y^{n-1} + a_2 x^{r_2} y^{n-2} + \dots + a_n x^{r_n} + \dots,$$

$$p\,(y,\,x) = y^m + b_1 x^{s_1} y^{m-1} + b_2 x^{s_2} y^{m-2} + \dots + b_m x^{s_m} + \dots,$$

and a_n, b_m are different from zero.

Suppose that the order of the integral y, if it exist in the vicinity of the origin, is μ in powers of x; so that it may there be represented by

$$y = \rho x^\mu + \dots,$$

ρ being a constant that does not vanish, and the indices of the remaining terms being greater than μ.

As before, let

$$a_l x^{r_l} y^{n-l}$$

be one of the set of terms in $q\,(y,\,x)$ which are the lowest in order for the substitution $y = \rho x^\mu$, so that, in $q\,(y,\,x)$, the leading term of lowest order is

$$x^{r_l + \mu\,(n-l)} \Sigma a_l \rho^{n-l}.$$

Similarly, let

$$b_k x^{s_k} y^{m-k}$$

be one of the set of terms which are the lowest in order for the same substitution, so that, in $p\,(y,\,x)$, the leading term of lowest order is

$$x^{s_k + \mu\,(m-k)} \Sigma b_k \rho^{m-k}.$$

When the equation is taken in the form

$$p\,(y,\,x) \frac{dy}{dx} = x^S y^R q\,(y,\,x)\, e^{G\,(y,\,x)},$$

it may be satisfied by having the lowest terms the same on both sides, as regards coefficients and indices, or by having a vanishing set of terms on one side only—with, of course, equality of subsequent terms. The latter may (as was seen to be the fact in § 42) be regarded as a limiting case of the former; and so, generally, we take the condition in the form that the lowest terms on the two sides must be the same. To secure the equality of the indices, we have

$$\mu - 1 + s_k + \mu\,(m-k) = S + \mu R + r_l + \mu\,(n-l);$$

to secure the equality of the coefficients, we must have

$$F\,(\rho) = \rho\mu\Sigma b_k \rho^{m-k} - C\rho^R \Sigma a_l \rho^{n-l} = 0,$$

where C is the value of $e^{G(0,\,0)}$. The former gives a value of μ, viz.

$$\mu = \frac{r_l + 1 + S - s_k}{m - R + 1 - k - (n - l)},$$

a definite value in general, when the appropriate terms have been selected.

55. To determine values of μ, we adopt the same method as before. Drawing two perpendicular axes $O\xi$, $O\eta$ in a plane, we mark a first set of points $(m - R + 1 - k,\ s_k)$ for the values of k, these being associated with the denominator $p\,(y,\,x)$ in the original form of the equation; and we mark a second set of points $(n - l,\ r_l + 1 + S)$ for the values of l, these being associated with the numerator $q\,(y,\,x)$ in that original form.

Each portion of the broken line, drawn as before to join the points determined by some values k of the first set and some values l of the second set, gives rise to the terms of the lowest order for the substitution $y = \rho x^\mu$, the value of μ being the tangent of its inclination to the negative direction of $O\xi$.

But to draw the broken line, we do not necessarily begin with the point $(0,\ r_n + 1 + S)$ on the axis $O\eta$, as the first point round which the line turns. In that case, it would be necessary to begin with a direction, not coincident with $O\eta$ but parallel to $O\xi$, and make the line turn in the counterclockwise direction. For if R be greater than 1, some of the points in the first set are on the negative side of the axis $O\eta$; and then, if

$$s_m > r_n + 1 + S,$$

there would arise a negative value of μ; if

$$s_m = r_n + 1 + S,$$

there would arise a zero value of μ; and if

$$s_m < r_n + 1 + S,$$

there would arise a positive value of μ.

In the first of these instances, there is no corresponding fitting solution for the equation; we have assumed that $y = \rho x^\mu + \dots$, and are seeking the integral that vanishes with x and, to obtain it, higher powers of y (among other combinations) have been neglected: it is clear that a negative value of μ does not satisfy the postulated conditions and the implicit assumptions. Moreover, it would make $G(y, x)$, in the exponential, an infinite quantity. (If the equation is given as an original form, and not as a typical form, then we should no longer be restricted to the particular type of integral indicated, and no longer have the variables restricted to the immediate vicinity of the origin.)

In the second case and in the third case, the value of μ not being negative, the portion of line can be considered as beginning in the point $-(R-1), s_m$. If no other of the quantities

$$r_l + 1 + S$$

is less than $r_n + 1 + S$, then in the second case the value $\mu = 0$ will give the terms of lowest order; but this is not so if any of the quantities $r_l + 1 + S$ be less than $r_n + 1 + S$. And in the third case, the value of μ will give the terms of lowest order if no one of the quantities $r_l + 1 + S$ be less than $r_n + 1 + S$; should however this condition not be satisfied, then the value of μ so obtained may or may not be an appropriate value giving rise to terms of lowest order.

Similarly, if R be equal to 1; there then is a point $0, s_m$, in the first set, and there is a point $0, r_n + 1 + S$, in the second set. Taking a broken line as before, clearly the first part of it lies along the axis $O\eta$: the corresponding value of μ is

$$\mu = \infty.$$

For all other points in the line, the value of μ is positive.

It thus appears that, in the present typical form of differential equation, the method adopted for obtaining the lowest terms gives rise to three forms of μ that were not obtained for the earlier typical forms. Negative values are to be rejected, for reasons already stated: they do not give rise to any integral of the required character. If there be infinite values of μ, then as regards the first term, we have

$$y = \rho x^\mu,$$

and so

$$|y| = |\rho| |x|^\mu;$$

so that, as μ is infinitely large, $|x|^\mu$ becomes infinitesimally small for all values of x within the circle $|x| = 1$, that is, $|y|$ is steadily zero for such values of x or, in other words, there is an integral of the equation which is a constant zero while x varies in a finite range; and then $w = \alpha$ would be a solution of the original equation for a finite region round the point $z = c$. If there be zero values of μ, then as regards the first term we have

$$y \propto x^\mu,$$

and so

$$x \propto y^{\frac{1}{\mu}},$$

that is, when x is regarded as a function of y, it is of an infinite order in powers of y, near the origin. Manifestly, the argument of the preceding case applies now: we infer the meaning of the result to be that x is steadily zero while y can vary through a finite range, in other words, that y is not a function of x, or that w is not a function of z in reference to the original equation.

These results have been deduced on the initial assumption that the integral can be expressed in the form

$$y = \rho x^\mu + \dots$$

in the vicinity of the origin. This is a form of function which is regular if μ be an integer, and which has a limited number of branches if μ be a fraction, μ being positive in each case ; the point is a definite point for the function when the conditions are actually satisfied. But if the assumption is distinctly contravened, or cease to be given in a definite form (as for $\mu = \infty$, or for $\mu = 0$), the expression for the function is no longer necessarily admissible ; the point may be a point of indeterminateness for the integral, and the proper tests must be applied to determine whether this is the fact or is not.

56. Excluding, therefore, portions of the broken line which might give rise to values of μ that could be negative, or zero, or infinite, we draw the broken line as follows. When R is greater than unity, there are points with negative abscissæ; we choose as the initial point that which has the greatest negative abscissa, viz. $-(R-1)$, s_m. When R is equal to unity, there is a point on $O\eta$ belonging to the first set, viz. 0, s_m: and there is a point on $O\eta$ belonging to the second set, viz. 0, $r_n + 1 + S$; we choose as the initial point that one of the pair which has the smaller ordinate. Through the initial point take a line parallel to $O\eta$, and make it turn in the counterclockwise sense until it meets one or more of the marked points in the tableau ; then make it turn about the last of these met points, that is. the one most distant from the initial point, until it meets one or more of the remainder ; and so on. Since s_m in the first case, and either s_m or $r_n + 1 + S$ in the

second case, is greater than zero, the initial point is off the axis $O\xi$; since s_0 is 0, there is a point $m - R + 1$, 0, of the first set on the axis $O\xi$; and therefore the last part of the broken line will end at a point on the line $O\xi$.

Clearly the value of μ, for each portion of the broken line thus constructed, is a positive quantity; its typical value is

$$\frac{r_l + 1 + S - s_k}{m - R + 1 - k - (n - l)},$$

when the portion of the line, which determines it, contains points from each of the two sets.

The further reduction to typical forms is effected in the same manner as before, with the appropriate changes of the significant functions.

When the portion of the broken line contains points of both sets, the function $F(\rho)$ is

$$F(\rho) = p\rho\Sigma b_k \rho^{m-k} - Cq\Sigma a_l \rho^{n-l+R}.$$

If ρ is a simple root of $F(\rho) = 0$, then we take

$$x = t^q, \quad y = ut^p, \quad u = \rho + v,$$

where $\dfrac{p}{q}$ is the value of μ reduced to its lowest terms as a proper fraction, and $u = \rho$ when $t = 0$; then unless ρ is a root also of $F_1(\rho) = 0$, where

$$F_1(\rho) = \Sigma b_k \rho^{m-k},$$

the equation for v in the immediate vicinity of $t = 0$ is found to be

$$t\frac{dv}{dt} = av + bt + \tfrac{1}{2}\alpha v^2 + \beta vt + \tfrac{1}{2}\gamma t^2 + \dots$$

$$= \phi_1(v, t),$$

where ϕ_1 is a regular function of v, t in the vicinity of 0, 0 : and the value of a is

$$-\frac{F'(\rho)}{F_1(\rho)}.$$

If ρ is a multiple root of $F(\rho) = 0$ and not a root of $F_1(\rho) = 0$: or if it is a simple root of $F(\rho) = 0$ and a root of $F_1(\rho) = 0$; or if it is a multiple root of $F(\rho) = 0$ and a multiple root of $F_1(\rho) = 0$: in every case, we obtain one of the types before considered, and there is no new type thus deduced.

When the portion of the broken line contains points of only one set, then the terms of lowest order in the equation

$$p\,(y,\,x)\,\frac{dy}{dx} = x^s\,y^r\,q\,(y,\,x)\,e^{G\,(y,\,x)}$$

arise through terms on only one side of the equation. With these we proceed as in earlier cases: it appears that various relations among constants must be satisfied so that all terms on that side, which are of order lower than the lowest on the other side, may disappear: and then subsequent terms of the same order on the two sides must have the same coefficients, a relation which serves to determine the remaining part of the leading term.

It thus appears that, in this case, positive values of μ lead in general to typical forms: but for values of μ, that arise out of a particular group of terms on one side of the equation, conditions among the coefficients must be satisfied identically.

Case III.

57. We now consider the equation when both of the integers R and S of the original form are negative; and changing the significance of the integers, we consider the equation in the form

$$\frac{dy}{dx} = \frac{y^R}{x^S}\,\frac{q\,(y,\,x)}{p\,(y,\,x)}\,e^{G\,(y,\,x)},$$

where $R,\,S$ now are positive. As before, we construct a tableau of points. In the present case, the points of the first set—the set connected with the denominator—are given by

$$m - R + 1 - k,\ s_k,$$

for the values of k; and the points of the second set—the set connected with the numerator—are given by

$$n - l,\ r_l + 1 - S,$$

for the values of l.

When negative values, or infinite values, or zero values, of μ occur, they are put on one side as in the preceding case and for similar reasons; but the origin may be a point of indeterminateness for the integral, and this question requires separate investigation.

Of the first set of points, some have negative abscissæ when $R > 1$, and, when $R = 1$, one is on the axis $O\eta$: all have positive ordinates except for $k = 0$. Of the second set of points, all have positive abscissæ, except for $l = n$, when there is a point on the axis $O\eta$. If $S = 1$, all the points (except for $l = 0$) have positive ordinates; but if $S > 1$, one of the points (for $l = 0$) certainly has a negative ordinate, and others may have negative ordinates.

To construct the appropriate broken line, we begin with the point of the first set which has the largest negative abscissa; through this draw a line parallel to $O\eta$, and make it turn about the point in the counterclockwise sense until it meets one or more of the points in the tableau; choose the point of those now on the line which is most distant from the initial point, and about it make the line turn in the counterclockwise sense until it meets other points: and so on. The last portion of the line may be part of the axis $O\xi$: in that case $\mu = 0$, and there is no corresponding integral to be retained: or it may be a part which lies below the axis $O\xi$, this case corresponding to a value $S > 1$.

The value of μ being determined, the reduction to the typical forms is by the same process as before: and no new types of final reduced forms are obtained.

58.   The simplest, and perhaps the most interesting, instance of the present form occurs when $R = 1$, $S = 1$, the equation then having the form

$$\frac{dy}{dx} = \frac{y}{x} \frac{q\,(y,\,x)}{p\,(y,\,x)}\, e^{G\,(y,\,x)}.$$

The points in the tableau are

$$m - k,\, s_k, \qquad\qquad (k = 0, 1, \ldots, m),$$

being the first set; and

$$n - l,\, r_l, \qquad\qquad (l = 0, 1, \ldots, n),$$

being the second set. Each set has a point on the axis $O\eta$; the corresponding value of μ is infinite. Each set has a point on $O\xi$; the corresponding value of μ is zero. All other parts of the broken line give positive values of μ; for each of them, there is the corresponding reduction.

The values $\mu = \infty$, $\mu = 0$ are put on one side, for the same reason as before; the first gives merely a constant zero for y, and a constant α for w; the second gives no function at all for w. But the origin may be a point of indeterminateness: the decision as to whether this is so or not, requires separate investigation.

59.   Summarising the discussion as regards the various forms of the equation in §§ 49—58, we have the following results.

When the integers R and S of the original equation

$$\frac{dy}{dx} = \frac{x^S}{y^R} \frac{q(y, x)}{p(y, x)} e^{G(y, x)},$$

are (either or both) zero or positive, the origin stands in the same general relation to the integral as it does when both S and R are zero. There are various values of μ, all positive, such that, in the immediate vicinity, $y \propto x^\mu$; all these values of μ are real commensurable quantities.

When S is zero or positive, and R is negative, the origin may be a point of indeterminateness for some integral or integrals; but, in general, some value or values of μ exist in the form of real commensurable quantities, such that in the immediate vicinity there are integrals for which $y \propto x^\mu$.

When R is zero or positive and S is negative, again the origin may be a point of indeterminateness for some integral or integrals; but, in general, some value or values of μ' exist in the form of real commensurable quantities, such that in the immediate vicinity there are integrals for which $x \propto y^{\mu'}$, and therefore

$$y \propto x^{\frac{1}{\mu'}}.$$

Lastly, when both R and S are negative, again the origin may be a point of indeterminateness for some integral or integrals; but, in general, some value or values of μ exist in the form of real commensurable quantities, such that in the immediate vicinity of the origin there exist integrals for which $y \propto x^\mu$.

For any such value of μ, arising in any of the cases indicated, let p/q denote its expression when reduced to its lowest terms: the corresponding integral is obtained as follows. Let

$$x = t^q, \quad y = (\rho + v) t^p;$$

then ρ is determined by an algebraical equation

$$F(\rho) = 0,$$

and v is determined by a differential equation which, in the immediate vicinity of $t = 0$, has one or other of the forms

$$t \frac{dv}{dt} = av + bt + \tfrac{1}{2}\alpha v^2 + \beta vt + \tfrac{1}{2}\gamma t^2 + \ldots$$
$$= \phi_1(v, t),$$

where ϕ_1 is a regular function of v, t, and

$$\phi_1(v,\ 0) = vP(v),$$

where $P(v)$ is a regular function of v not vanishing with v; or

$$t\frac{dv}{dt} = gv^m + bt + \ldots$$
$$= \phi_m(v,\ t),$$

where ϕ_m is of the same character as ϕ_1, and

$$\phi_m(v,\ 0) = v^m P(v),$$

where $P(v)$ is a regular function of v not vanishing with v; or

$$t^\kappa \frac{dv}{dt} = \phi_1(v,\ t),\ \text{ or }\ \phi_m(v,\ t),$$

where ϕ_1 and ϕ_m are as before, and the positive integer $\kappa \geqslant 2$; or

$$t^\kappa \frac{dv}{dt} = \frac{k'v^{l'} + \mu't + \ldots}{kv^l + \mu t + \ldots},$$

where the positive integer $l' \geqslant 2$, and $\kappa \geqslant 1$.

The character of each integral thus obtained is determined by the quantity v which satisfies one or other of these equations.

For each distinct non-zero root of $F(\rho) = 0$, there are as many integrals as can arise out of the respective reduced differential equations for the determination of the quantity v. The aggregate of all such integrals must be taken: they represent the aggregate of integrals of the original equation in the immediate vicinity of the accidental singularity of the second kind, which can be made to depend upon algebraical transformations of the variables.

In addition to this aggregate, and depending upon the form of the equation, there may be integrals for which the singularity in question is a point of indeterminateness.

In every instance, the discussion has been approached as though the differential equation were a limiting form of the original differential equation, for values of the variables in the immediate vicinity of some definite combination of values; and on this basis, integrals y are required which vanish with x. The various forms may, however, be propounded as initial forms; in that case, the origin still has its characteristic property as regards the equation, but the variables are no longer restricted, as regards their variation, to the immediate vicinity of the origin. Other singularities possessed by the functions q and p belong to the classes already considered : and, in particular, it may be necessary to take the vicinities of infinite values of y or of x or of both, in order to render the discussion complete.

Before discussing the typical forms of the equations for v which have been obtained, some illustrations of the preceding general theory may now conveniently be considered.

EXAMPLES.

60. *Ex.* 1. The simplest example of all is that in which both q and p are of the first degree in y, such that $q(0, x)$, $p(0, x)$ both begin with a term in the first power of x : and $G(y, x)$ is a constant. We then have

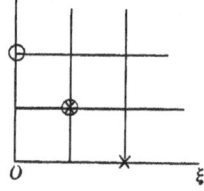

$$\frac{dy}{dx} = C \frac{y + a_1 x + \beta_1 x^2 + \dots}{y + a_0 x + \beta_0 x^2 + \dots},$$

where the remaining terms in the numerator and the denominator are higher powers of x alone.

Adopting the preceding method*, we have the points as in the diagram. Connected with integrals of the equation that vanish with x, there manifestly is a single value of μ, viz. $\mu = 1$. To determine the leading term, we write

$$y = \rho x + \dots,$$

no transformation to a new independent variable t being necessary, because μ is an integer ; and then we have

$$\rho = C \frac{\rho + a_1}{\rho + a_0},$$

that is, ρ is determined by the equation

$$\rho^2 + (a_0 - C)\rho - a_1 C = 0.$$

When the roots are unequal, denote them by ρ_1, ρ_2 ; if equal, denote the common value by ρ'.

We now write

$$y = x(\rho + v),$$

so that

$$\rho + x \frac{dv}{dx} + v = C \frac{\rho + v + a_1 + x P_1(x)}{\rho + v + a_0 + x P_0(x)},$$

where P_1, P_0 are regular functions of x. Thus

$$x \frac{dv}{dx} = C \left\{ \frac{a_1 + \rho + v + x P_1(x)}{a_0 + \rho + v + x P_0(x)} \right\} - \rho - v$$

$$= \frac{v(C - 2\rho - a_0) - v^2 + Cx P_1(x) - \rho x P_0(x) - vx P_0(x)}{a_0 + \rho + v + x P_0(x)}.$$

* The points associated with terms in the numerator of the fraction for $\frac{dy}{dx}$ are throughout marked by $\times$; those associated with its denominator, by o.

When the roots of the quadratic are unequal, $C - 2\rho - a_0$ is not zero ; and $\rho + a_0$ is not zero, when a_1 and a_0 are different from one another. In this case, we can regard the whole function on the right-hand side as a regular function of v and x ; when expanded in the immediate vicinity of 0, 0, we have

$$x \frac{dv}{dx} = av + bx + \dots,$$

where

$$a = \frac{C - 2\rho - a_0}{a_0 + \rho}.$$

This result holds for each of the roots of the quadratic equation : so that in connection with each of them, there is a differential equation for v. When the character of the quantity v, thus determined, has been obtained, the integral of the original equation y is known for each of the values of ρ.

It may happen that a_0 and a_1 are the same ; in that case, one root of the quadratic is C, the other is $-a_0$. Still assuming that the roots are unequal, so that for the present instance $C + a_0$ does not vanish, we see at once that the preceding reduction is valid for the root $\rho = C$; the value of a is -1, and the equation for v is

$$x \frac{dv}{dx} = -v + bx + \dots.$$

But for the root $\rho = -a_0$, it is not a satisfactory reduction ; we then have

$$x \frac{dv}{dx} = \frac{v(C + a_0) + Cx P_1(x) + a_0 x P_0(x) - v^2 - vx P_0(x)}{v + x P_0(x)}.$$

Here

$$P_1(x) = \beta_1 + \gamma_1 x + \delta_1 x^2 + \dots,$$
$$P_0(x) = \beta_0 + \gamma_0 x + \delta_0 x^2 + \dots ;$$

we take

$$v + \frac{C\beta_1 + a_0\beta_0}{C + a_0} x = x V,$$

and then

$$x^2 \frac{dV}{dx} = -xV + x \frac{C\beta_1 + a_0\beta_0}{C + a_0} + \frac{(C + a_0) x V + x^2 R(V, x)}{\frac{C(\beta_0 - \beta_1)}{C + a_0} x + x V + x^2 Q(x)},$$

where $R(V, x)$ is linear in V and regular in x. If β_0 is distinct from β_1, this becomes

$$x^2 \frac{dV}{dx} = -xV + x \frac{C\beta_1 + a_0\beta_0}{C + a_0} + \frac{(C + a_0) V + x R(V, x)}{\frac{C(\beta_0 - \beta_1)}{C + a_0} + V + x Q(x)}$$

$$= \frac{(C + a_0)^2}{C(\beta_0 - \beta_1)} V + \kappa x + \dots ;$$

while, if $\beta_0 = \beta_1$, we obtain for $x^2 \dfrac{dV}{dx}$ a form similar to that for $x \dfrac{dv}{dx}$. As P_0 and P_1 are not identical, there will be a limit to the number of necessary transformations ; and the final form would be

$$x^s \frac{dW}{dx} = \theta W + \theta' x + \dots,$$

where $s \geqslant 2$.

When the roots of the quadratic are equal, and their common value ρ' is not $-a_0$ so that $C+a_0$ is not zero, then $a_0+\rho'$ is not zero ; the equation is

$$x\frac{dv}{dx} = \frac{-v^2 + CxP_1(x) - \rho'xP_0(x) - vxP_0(x)}{a_0 + \rho' + v + xP_0(x)}$$

$$= -\frac{1}{a_0+\rho'}\, v^2 + cx + \dots,$$

on expanding the right-hand side in the vicinity of 0, 0 ; the remaining terms in v alone are of order higher than 2, those in x alone are of order higher than 1, and there are terms in x and v combined.

When the roots of the quadratic are equal and their common value ρ' is $-a_0$, so that $C+a_0$ is zero and $a_1=a_0$, then $a_0+\rho'$ is zero. The original equation is then

$$\frac{dy}{dx} = -a_0\frac{y + a_0 x + \beta_1 x^2 + \dots}{y + a_0 x + \beta_0 x^2 + \dots},$$

so that

$$\frac{d}{dx}(y + a_0 x) = -a_0\frac{(\beta_1 - \beta_0)x^2 + (\gamma_1 - \gamma_0)x^3 + \dots}{y + a_0 x + \beta_0 x^2 + \dots}.$$

When β_1 is not equal to β_0, a diagram, corresponding to that in the general theory, shews that

$$y + a_0 x \propto x^{\frac{3}{2}}$$

in the vicinity of $x=0$; accordingly, we write

$$x = \xi^2, \qquad y + a_0 x = u\xi^3,$$

and then

$$\tfrac{1}{2}\left(\xi^2\frac{du}{d\xi} + 3u\xi\right) = -\frac{a_0(\beta_1 - \beta_0)\xi + \text{higher powers of } \xi}{u + \beta_0\xi + \dots}.$$

Hence if

$$\theta_0^2 = -\tfrac{2}{3}(\beta_1 - \beta_0)a_0,$$

we have a value of u beginning with $u=\theta_0$; and it appears that the value of u is of the form

$$u = \theta_0 + \theta_1\xi + \dots,$$

a regular function of ξ for each of the two values of θ_0. Similarly, when β_1 is equal to β_0 ; and so on in succession for the various alternatives that may occur.

Another method for dealing with the equation under consideration has been used*, wherein the variables x and y are regarded as the coordinates of a point in the plane, and thus are real quantities ; the differential equation then represents a curve. A subsidiary variable t is introduced in the form

$$\frac{dy}{a'y + b'x + \dots} = \frac{dx}{ay + bx + \dots} = \frac{dt}{t}.$$

* Poincaré, *Sur les courbes définies par les équations différentielles*, Liouville, 3me Sér., t. vii (1881), pp. 375—422, *ib.*, 3me Sér., t. viii (1882), pp. 251—296, *ib.*, 4me Sér., t. i (1885), pp. 167—244. See also Picard, *Cours d'Analyse*, t. iii, ch. ii, ch. ix, where the subject is resumed with independent treatment ; and Bendixson, *Sur les points singuliers d'une équation différentielle linéaire*, Öfversigt af Kongl. Vet.-Akad. Förh., Årg. 52, (1895), pp. 81—99.

Let new constants λ and μ be chosen, so that

$$a\lambda + a'\mu = \xi\mu, \quad b\lambda + b'\mu = \xi\lambda \ ;$$

then ξ is a root of the quadratic equation

$$(\xi - a')(\xi - b) - ab' = 0.$$

In general, there are two distinct roots of the quadratic, say ξ_1 and ξ_2 : and then there are two sets of constants λ and μ. Now

$$\lambda_1(ay + bx) + \mu_1(a'y + b'x) = \xi_1(\lambda_1 x + \mu_1 y) \ ;$$

so that, if $X = \lambda_1 x + \mu_1 y$, we have

$$t\frac{dX}{dt} = \xi_1 X + \text{terms of the second and higher orders.}$$

Similarly, if $Y = \lambda_2 x + \mu_2 y$, so that Y is a variable connected with the value ξ_2, then similarly

$$t\frac{dY}{dt} = \xi_2 Y + \text{terms of the second and higher orders.}$$

Hence if t, X, Y are kept so that their moduli are small, (that is, x and y are in the immediate vicinity of their respective origins), and ξ_1 and ξ_2 have their real parts positive, then approximately

$$X = at^{\xi_1}, \quad Y = \beta t^{\xi_2} \ ;$$

or introducing another variable τ, where $\tau = t^{\xi_1}$, the integrals there are

$$X = a\tau, \quad Y = \beta\tau^{\frac{\xi_2}{\xi_1}},$$

that is,

$$Y = \gamma X^{\frac{\xi_2}{\xi_1}}.$$

At present, it is not so much the integral of the equation which is being considered as its inclusion (by reduction) in a typical form. The equations used can be connected with the quadratic in the preceding investigation. Manifestly, in the immediate vicinity of the origin, the significant term in y can be obtained by taking $X = 0$, or by taking $Y = 0$. These give

$$y = \rho x,$$

where

$$\rho = -\frac{\lambda}{\mu} = \frac{a' - \xi}{a}.$$

But

$$(\xi - a')(\xi - b) = ab' \ ;$$

and therefore

$$(a\rho + b)\rho = a'\rho + b',$$

equivalent to the former quadratic equation.

Ex. 2. Consider the equation

$$\frac{dy}{dx} = \frac{ay^2 + 2bxy + cx^2}{ay^3 + 3\beta y^2 x + 3\gamma yx^2 + \delta x^3} = \frac{(a, b, c\ \chi y, x)^2}{(a, \beta, \gamma, \delta\ \chi y, x)^3}.$$

From the diagram (where the points o are the points $3+1-k$, s_k, of the denominator, and the points × are the points $2-l$, r_l+1, of the numerator), it appears that two values of μ must be considered, viz.

$$\mu=1, \quad \mu=\tfrac{1}{2}.$$

First, let $\mu=1$: as the value is derived from the numerator terms alone, at least the first term in the expression for y will be settled by means of them. Take

$$y=\rho x+ux^2,$$

where ρ is a constant ; then

$$\rho+x^2\frac{du}{dx}+2ux=\frac{x^2(a,\,b,\,c\chi\rho,\,1)^2+2x^3u(a,\,b\chi\rho,\,1)+x^4u^2a}{x^3(a,\,\beta,\,\gamma,\,\delta\chi\rho,\,1)^3+3x^4u(a,\,\beta,\,\gamma\chi\rho,\,1)^2+3x^5u^2(a,\,\beta\chi\rho,\,1)+x^6u^3a}.$$

Choose ρ, so that

$$(a,\,b,\,c\chi\rho,\,1)^2=0\;;$$

then unless $(a,\,\beta,\,\gamma,\,\delta\chi\rho,\,1)^3$ for one or other of the two values of ρ is zero—a result that can occur only if the resultant of the binary form $(a,\,b,\,c\chi\varkappa)^2$ and the cubic form $(a,\,\beta,\,\gamma,\,\delta\chi\varkappa)^3$ vanishes, and that will be assumed as not occurring, for then a reduction in the expression for dy/dx could be effected— we can remove a factor x^3 from the numerator and the denominator of the fraction. We then have

$$\rho+x^2\frac{du}{dx}+2ux=\frac{2u(a,\,b\chi\rho,\,1)+xu^2a}{(a,\,\beta,\,\gamma,\,\delta\chi\rho,\,1)^3+3xu(a,\,\beta,\,\gamma\chi\rho,\,1)^2+3x^2u^2(a,\,\beta\chi\rho,\,1)+x^3u^3a}.$$

It is clear that there are functions u which do not vanish with x ; let the value, when $x=0$, be σ, so that σ is determined by the equation

$$\sigma=\frac{\rho(a,\,\beta,\,\gamma,\,\delta\chi\rho,\,1)^3}{2(a\rho+b)}.$$

Then writing

$$u=\sigma+v,$$

it is not difficult to prove that the equation determining v is

$$x^2\frac{dv}{dx}=v\frac{\rho}{\sigma}-\kappa x+\text{terms of second and higher orders,}$$

where

$$\kappa=\frac{3a\rho\sigma+3\rho^2(a,\,\beta,\,\gamma\chi\rho,\,1)^2+4b\sigma}{2(a\rho+b)}.$$

This is the typical form for the reduction when $\mu=1$; the relation to the integral is given by

$$y=\rho x+x^2(\sigma+v),$$

where ρ is a root of the quadratic

$$(a,\,b,\,c\chi\rho,\,1)^2=0,$$

and σ has the above value.

Secondly, let $\mu = \frac{1}{2}$; the value is derived from a term in the numerator and a term in the denominator. In accordance with the general theory, we take

$$x = t^2, \quad y = ut = (\rho + v)\, t,$$

where v is to be zero when $t = 0$. We have

$$\rho + t\frac{dv}{dt} + v = 2t\, \frac{at^2\,(\rho+v)^2 + 2bt^3\,(\rho+v) + ct^4}{at^3\,(\rho+v)^3 + 3\beta t^4\,(\rho+v)^2 + 3\gamma t^5\,(\rho+v) + \delta t^6}$$

$$= 2\, \frac{a\,(\rho+v)^2 + 2bt\,(\rho+v) + ct^2}{a\,(\rho+v)^3 + 3\beta t\,(\rho+v)^2 + 3\gamma t^2\,(\rho+v) + \delta t^3}$$

$$= 2\, \frac{a\rho^2 + 2a\rho v + 2b\rho t + \ldots}{a\rho^3 + 3a\rho^2 v + 3\beta\rho^2 t + \ldots}.$$

Choose a non-vanishing value of ρ so that

$$\rho = 2\frac{a\rho^2}{a\rho^3},$$

that is,

$$\rho^2 a = 2a.$$

Then after some comparatively simple reduction, we find

$$t\frac{dv}{dt} = -2v + t\left(2\frac{b}{a} - 3\frac{\beta}{a}\right) + \ldots,$$

as the reduced form.

Ex. 3.　Discuss, in a similar manner, the equation

$$\frac{dy}{dx} = \frac{(a,\ \beta,\ \boldsymbol{\gamma},\ \delta\mathfrak{X}y,\ x)^3}{(a,\ b,\ c\mathfrak{X}y,\ x)^2}.$$

Ex. 4.　Consider the equation

$$\frac{dy}{dx} = \frac{y^2}{x^2}\, \frac{(a,\ b,\ c\mathfrak{X}y,\ x)^2}{(a,\ \beta,\ \boldsymbol{\gamma},\ \delta\mathfrak{X}y,\ x)^3}\ ;$$

here $S = -2,\ R = -2$.

From the diagram (where the points o are the points $3 - 1 - k,\ s_k$ of the denominator, and the points × are the points $2 - l,\ r_l - 1$ of the numerator), it appears that two values of μ arise for consideration, viz. $\mu = 2,\ \mu = 1$.

First, take $\mu = 2$. The leading term in y is ρx^2, where

$$2\rho x = \rho^2 x^2\, \frac{cx^2}{\delta x^3},$$

that is,

$$\rho = 2\frac{\delta}{c}.$$

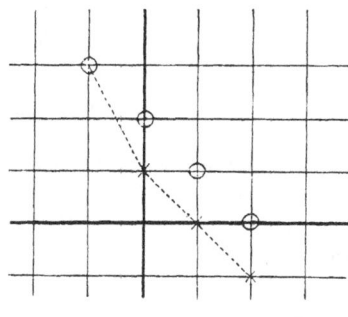

We then take

$$y = x^2 \left(2 \frac{\delta}{c} + v \right),$$

so that

$$4 \frac{\delta}{c} x + x^2 \frac{dv}{dx} + 2vx = x^2 4 \frac{\delta}{c} \left(\frac{\delta}{c} + v + \ldots \right) \frac{cx^2 + 2bx^3 \cdot 2 \frac{\delta}{c} + \ldots}{\delta x^3 + 3\gamma x^4 \cdot 2 \frac{\delta}{c} + \ldots},$$

and therefore

$$x \frac{dv}{dx} + 2v = 4 \frac{\delta}{c} \left\{ -1 + \left(\frac{\delta}{c} + v + \ldots \right) \frac{c + 4 \frac{b\delta}{c} x + \ldots}{\delta + 6 \frac{\gamma\delta}{c} x + \ldots} \right\}$$

$$= 4v + 8 \frac{\delta}{c^2} \left(2 \frac{b\delta}{c} - 3\gamma \right) x + \ldots,$$

that is,

$$x \frac{dv}{dx} = 2v + 8 \frac{\delta}{c^2} \left(2 \frac{b\delta}{c} - 3\gamma \right) x + \ldots,$$

on reduction and expansion. This is the typical reduced form for the value $\mu = 2$.

Secondly, take $\mu = 1$; then the leading term in y is $y = \kappa x$. We write

$$y = x (\kappa + v),$$

and substitute; thus

$$\kappa + x \frac{dv}{dx} + v = \frac{(\kappa + v)^2}{x} \frac{(a, b, c \chi \kappa + v, 1)^2}{(a, \beta, \gamma, \delta \chi \kappa + v, 1)^3}.$$

In order that this may be satisfied when $v = 0$, we must have

$$(a, b, c \chi \kappa, 1)^2 = 0$$

as a first condition; and the leading term in v must be a multiple of x, say $v = xV$, where V is not zero when $x = 0$. The original transformation thus is

$$y = x\kappa + x^2 V ;$$

when this is substituted, we have

$$\kappa + x^2 \frac{dV}{dx} + 2xV = \frac{(\kappa + xV)^2}{x} \frac{2 (a\kappa + b) xV + ax^2 V^2}{(a, \beta, \gamma, \delta \chi \kappa + xV, 1)^3}.$$

Let

$$(a, \beta, \gamma, \delta \chi \kappa, 1)^3 = f_3,$$

$$3 (a, \beta, \gamma \chi \kappa, 1)^2 = \frac{\partial f_3}{\partial \kappa} = f_2 ;$$

then

$$x^2 \frac{dV}{dx} + 2xV = -\kappa + (\kappa + xV)^2 \frac{2 (a\kappa + b) V + ax V^2}{f_3 + f_2 x V}.$$

If ρ denote the value of V when $x = 0$, we clearly have

$$0 = -\kappa + \kappa^2 \frac{2 (a\kappa + b) \rho}{f_3},$$

that is,

$$\rho = \frac{f_3}{2\kappa (a\kappa + b)}.$$

Now substitute
$$V = \rho + W,$$
so that W vanishes with x; the value of y is
$$y = x\kappa + x^2\rho + x^2 W,$$
and the equation for W is, after some straightforward reductions, found to be
$$x^2 \frac{dW}{dx} = \frac{\kappa}{\rho} W + \frac{x}{f_3}(a\kappa^2\rho^2 - \kappa f_2\rho) + \ldots,$$
which is the reduced typical form.

Ex. 5. Consider the equation
$$x \frac{dy}{dx} = \lambda \frac{y^l + \mu x}{y^{l'} + \mu' x},$$
which is one of the unreduced typical forms (§ 46): here $S = -1$, $R = 0$.

The numerator points (marked $\times$) are $l-j$, r_j, though the only terms entering are those for which $j = l$, $j = 0$; the denominator terms (marked o) are $1 + l' - k$, s_k, though the only terms entering are those for which $k = l'$, $k = 0$. The diagram can easily be constructed; the points o are $(1 + l', 0)$, $(1, 1)$; and the points $\times$ are $(l, 0)$, $(0, 1)$. There are three cases, according as $1 + l' > l$, $= l$, $< l$.

(i) Let $1 + l' > l$; there is only one value of μ, and it is $\frac{1}{l}$. Write
$$x = t^l, \quad y = ut;$$
and let $1 + l' = l + s$, where s is a positive integer $\geqslant 1$. Then
$$t \frac{du}{dt} + u = \frac{l\lambda}{t} \frac{u^l t^l + \mu t^l}{\mu' t^l + t^{l'} u^{l'}}$$
$$= \frac{l\lambda}{t} \frac{\mu + u^l}{\mu' + t^{s-1} u^{l'}}.$$

Now u is to be different from 0 when $t = 0$; let ρ be its value, and take
$$u = \rho + vt,$$
where v may or may not be zero when $t = 0$. We have
$$t^2 \frac{dv}{dt} + 2vt + \rho = \frac{l\lambda}{t} \frac{\mu + \rho^l + l\rho^{l-1} vt + \frac{1}{2}l(l-1)\rho^{l-2}v^2t^2 + \ldots}{\mu' + t^{s-1}(\rho^{l'} + l'\rho^{l'-1}vt + \ldots)}.$$

There are two sub-cases, according as $s > 1$ or $s = 1$. First, let $s > 1$; take
$$\mu + \rho^l = 0,$$
$$\rho = \frac{l\lambda}{\mu'} l\rho^{l-1}a;$$
so that
$$t^2 \frac{dv}{dt} + 2vt + \rho = l\lambda \frac{l\rho^{l-1}v + \frac{1}{2}l(l-1)\rho^{l-2}v^2t + \ldots}{\mu' + t^{s-1}\rho^{l'} + \ldots},$$

and a is the value of v when $t=0$. Take $v=a+W$, where $W=0$ when $t=0$, so that

$$y = \rho t + at^2 + t^2 W;$$

then

$$t^2 \frac{dW}{dt} + 2at + 2Wt = -\rho + l\lambda \frac{l\rho^{l-1}a + l\rho^{l-1}W + \frac{1}{2}l(l-1)\rho^{l-2}a^2t + \dots}{\mu' + t^{s-1}\rho^{l'} + \dots}$$

$$= \frac{l\lambda}{\mu'}\{l\rho^{l-1}W + \frac{1}{2}l(l-1)\rho^{l-2}a^2t - \rho^{l'+1}t^{s-1}\} + \dots,$$

and therefore

$$t^2 \frac{dW}{dt} = \frac{l^2\lambda\rho^{l-1}}{\mu'} W + \left\{\frac{1}{2}\frac{l^2(l-1)}{\mu'}\lambda\rho^{l-2}a^2 - 2a\right\}t - \frac{l\lambda}{\mu'}\rho^{l'+1}t^{s-1} + \dots,$$

where if $s=2$, the third term combines with the second, and if $s>2$, it combines with a later term.

Secondly, let $s=1$, so that $l'=l$; take

$$\mu + \rho^l = 0,$$

$$\rho = \frac{l\lambda}{\mu' + \rho^l} l\rho^{l-1}a;$$

so that

$$t^2 \frac{dv}{dt} + 2vt + \rho = l\lambda \frac{l\rho^{l-1}v + \frac{1}{2}l(l-1)\rho^{l-2}v^2t + \dots}{\mu' + \rho^l + l\rho^{l'-1}vt + \dots},$$

and a is the value of v when $t=0$. Take $v=a+W$, so that $W=0$ when $t=0$, and then

$$y = \rho t + at^2 + t^2 W;$$

we have

$$t^2 \frac{dW}{dt} + 2at + 2Wt = -\rho + l\lambda \frac{l\rho^{l-1}a + l\rho^{l-1}W + \frac{1}{2}l(l-1)\rho^{l-2}a^2t + \dots}{\mu' + \rho^l + l\rho^{l-1}at + \dots}$$

$$= \frac{l\lambda}{\mu' + \rho^l}[l\rho^{l-1}W + t\{\frac{1}{2}l(l-1)\rho^{l-2}a^2 - l\rho^l a\} + \dots],$$

and therefore

$$t^2 \frac{dW}{dt} = \frac{l^2\lambda\rho^l}{\mu' + \rho^l} W + \left[\frac{l\lambda\{\frac{1}{2}l(l-1)\rho^{l-2}a^2 - l\rho^l a\}}{\mu' + \rho^l} - 2a\right]t + \dots.$$

These are the respective reduced forms for the case $l' \geqslant l$.

(ii)   Let $1 + l' = l$. There is a single value of μ, and it is $\frac{1}{l}$. Again, write

$$x = t^l, \quad y = ut;$$

then

$$t\frac{du}{dt} + u = \frac{l\lambda}{t} \frac{u^l t^l + \mu t^l}{\mu' t^l + u^{l-1}t^{l-1}}$$

$$= \frac{l\lambda(\mu + u^l)}{u^{l-1} + \mu' t}.$$

Now when $t=0$, u must be different from zero, say ρ; so that

$$\rho = \frac{l\lambda(\mu + \rho^l)}{\rho^{l-1}},$$

and therefore

$$\rho^l = \frac{l\lambda\mu}{1 - l\lambda}.$$

Let $u = \rho + v$, so that
$$y = (\rho + v)\, t\, ;$$
then the equation for v is
$$t\frac{dv}{dt} + v = -\rho + \frac{l\lambda\,(\mu + \rho^l + l\rho^{l-1}v + \ldots)}{\mu' t + \rho^{l-1} + (l-1)\rho^{l-2}v + \ldots}\, ,$$
leading to
$$t\frac{dv}{dt} = \left(l^2\lambda - \frac{l-1}{\rho} - 1\right)v - \frac{\mu'}{\rho^{l-2}}\, t + \ldots,$$
the reduced typical form.

(iii)   Let $1 + l' < l,\ = l - \sigma$, where the integer $\sigma \geqslant 1$. There is a single value of μ, and it is $\dfrac{1}{l - \sigma}$.   Write
$$x = t^{l-\sigma}, \quad y = ut = (\rho + v)\, t\, ;$$
then taking
$$\rho^{l-\sigma} = \lambda\mu\,(l - \sigma),$$
the equation for v is found to be
$$t\frac{dv}{dt} = -(l - \sigma)\, v - \frac{\mu'}{\rho^{l-\sigma-2}}\, t + \frac{\rho^{l+1}}{\mu}\, t^{\sigma} + \ldots,$$
the two terms in t that are explicitly given coalescing when $\sigma = 1$. This is the reduced typical form.

Ex. 6.   Consider the equation
$$\frac{dy}{dx} = \kappa x^2 y^5\, \frac{y^3 + ax^4 y^2 + bxy + cx^{14}}{y^{15} + fx^7 y^2 + gx^9 y + hx^{15}}\, ;$$
here $S = 2,\ R = -5$.

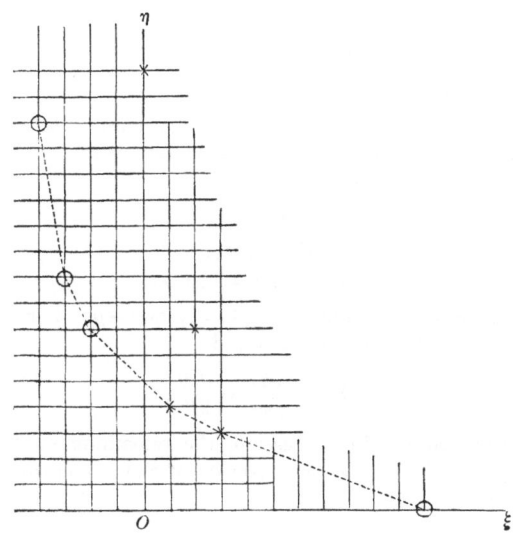

From the diagram (where the points o are the points $15-4-k$, s_k of the denominator, and the points × are the points $3-l$, r_l+3 of the numerator), it appears that possible values of μ are 6, 2, 1, $\frac{1}{2}$, $\frac{3}{8}$.

First, take $\mu=6$; then when we substitute

$$y = x^6 \left(\rho + \sigma_1 x^{a_1} + \sigma_2 x^{a_2} + \sigma_3 x^{a_3} + \sigma_4 x^{a_4} + \sigma_5 x^{a_5} + \dots\right)$$

and retain the lowest terms, we find that, in the numerator on the right-hand side (including the factor $\kappa x^2 y^5$), the term of lowest order is $\kappa b \rho^6 x^{39}$; consequently, in order that the equation may be satisfied identically, the terms in the denominator of order lower than x^{34} must disappear. To secure this, we must have

$$a_1 = 4, \quad a_2 = 8, \quad a_3 = 12, \quad a_4 = 16, \quad a_5 = 19 ;$$

and then the necessary relations among the coefficients are

$$\begin{cases} 0 = h + g\rho \\ 0 = g\sigma_1 + f\rho^2 \\ 0 = g\sigma_2 + 2f\rho\sigma_1 \\ 0 = g\sigma_3 + f(2\rho\sigma_2 + \sigma_1^2) \\ 0 = g\sigma_4 + f(2\rho\sigma_3 + 2\sigma_1\sigma_2), \end{cases}$$

which determine ρ, σ_1, σ_2, σ_3, σ_4. That the coefficient of x^5 may be the same on the two sides, we must have

$$6\rho = \frac{\kappa b \rho^6}{g\sigma_5},$$

or

$$\sigma_5 = \frac{\kappa b}{6g}\rho^5 = -\frac{\kappa b h^5}{6g^6} ;$$

and so for the other powers in succession, the remaining coefficients of powers in the expansion of y involving the constants κ, a, b, c of the numerator in the expression for $\frac{dy}{dx}$.

Next, take $\mu = 2$; then when we substitute

$$y = x^2 \left(\rho + \lambda_1 x + \lambda_2 x^2 + \lambda_3 x^3 + \dots\right),$$

the lowest power in the numerator is that which occurs in $\kappa b \rho^6 x^{15}$; accordingly, all terms in the denominator of order lower than x^{14} must disappear. That this may be the case, we must have

$$0 = g\rho + f\rho^2,$$
$$0 = g\lambda_1 + 2f\rho\lambda_1,$$
$$0 = g\lambda_2 + f(\lambda_1^2 + 2\rho\lambda_2),$$

which can be satisfied by taking $\lambda_1 = 0$, $\lambda_2 = 0$, $f\rho = -g$. For the next term, we equate powers on the two sides and find

$$\lambda_3 = -\frac{\kappa b \rho^5}{2g} = \frac{\kappa b g^4}{2f^5} ;$$

and so for other powers in succession, the remaining coefficients in the expansion of y again involving the constants of the numerator in the expression for $\frac{dy}{dx}$.

Next, take $\mu = 1$.   Substituting $y = \rho x$, we find

$$\rho = \kappa \rho^5 \frac{b\rho}{f\rho^2},$$

that is,

$$\rho^3 = \frac{f}{\kappa b}.$$

Now, let

$$y = x(\rho + v) ;$$

on substitution and reduction, we find that the equation for v is

$$x \frac{dv}{dx} = 3v + x\left(\frac{\rho}{b} - \frac{g}{f}\right) + \dots,$$

the typical reduced form.

Next, take $\mu = \frac{1}{2}$, and change the independent variable to t, where

$$x = t^2 ;$$

then in the expansion of y, the leading term is ρt.   A similar analytical process shews that ρ is determined solely by the terms in the numerator ; and if

$$y = \rho t + \gamma t^2 + \sigma t^3 + \dots,$$

then

$$\rho^2 + b = 0,$$

$$\gamma = 0,$$

$$\sigma = \frac{1}{4}\frac{\rho b^4}{\kappa},$$

and similarly for succeeding powers of t.   To obtain the typical reduced form, take

$$y = (\rho + v)\, t = \rho t + (\sigma + V)\, t^3 ;$$

then after reduction, we find

$$t^3 \frac{dV}{dt} = 4\frac{\kappa}{b^4}\, V - \frac{f}{b^6}\, t + \dots.$$

Lastly, take $\mu = \frac{3}{8}$; we change the independent variable from x to t, where

$$x = t^8 ;$$

then in the expansion of y, the leading term is ρt^3, where

$$\rho^8 = \tfrac{8}{3}\kappa.$$

According to the general theory, we take

$$y = t^3(\rho + v);$$

and after reduction, we find

$$t \frac{dv}{dt} = -24v + 3\frac{b}{\rho}\, t^2 + \dots,$$

the typical reduced form.

Note.   The object of the examples is to indicate how the typical reduced forms arise in particular cases.   It can, however, be inferred as a suggestion from this and from other examples that the simplest typical reduced forms arise in connection with those values of μ which are determined by a portion of the broken line containing points from both sets in the tableau ; the character of the corresponding integral or integrals is determined by the character of the integral of the typical reduced form. In other cases, where the values of μ arise from a portion of the broken line containing points of only one set, there can be a branch (or a set of algebraical branches) of the integral vanishing with x and expansible in regular form, either in powers of x or in powers of another variable, the existence of the branch depending upon the convergence of the series ; for the determination of the branch or branches, no typical reduced form is necessary. It is manifest that the latter class of integrals (when they exist) is highly special ; the initial terms in the expansion are obtained without reference to the differential coefficient, for they are obtained by making terms, in the numerator or the denominator as the case may be, of lowest orders vanish.

It may, however, be the case that such special integrals are only particular portions of some more general integral, which would be determined in connection with a typical form. In the present instance, the simpler plan would be to proceed from the form

$$\kappa x^2 y^5 \frac{dx}{dy} = \frac{y^{15} + fx^7 y^2 + gx^9 y + hx^{15}}{y^3 + ax^4 y^2 + bxy + cx^{14}} \; ;$$

the two cases that lead to typical reduced forms, which were not obtained in the preceding analysis, are

$$x = \rho y^{\frac{3}{5}} + \dots, \quad x = \rho y^{\frac{1}{2}} + \dots \; ;$$

they would be treated as was the case $\mu = \frac{1}{2}$.

Ex. 7.   Consider the equation

$$\frac{dy}{dx} = \frac{y}{x} \frac{ay + bx}{a'y + b'x} \; ;$$

here $R = -1$, $S = -1$.

The tableau consists of two coincident pairs of points at 1, 0 and 0, 1 ; the only value of μ is unity. If $y = \rho x$, then

$$\rho = \rho \frac{a\rho + b}{a'\rho + b'} ,$$

so that

$$\rho = \frac{b' - b}{a - a'} .$$

If now, in accordance with the general theory, we substitute

$$y = x \, (\rho + v),$$

we easily find

$$x \frac{dv}{dx} = \frac{(a - a') \, (b' - b)}{ab' - a'b} \, v Q \, (v),$$

where $Q\,(v)$ is a regular function of v in the vicinity of $v = 0$, and $Q\,(0) = 1$.

This form is satisfactory unless $ab' - a'b = 0$. If, however, $ab' - a'b = 0$, then the original equation degenerates to

$$\frac{dy}{dx} = \frac{a}{a'}\frac{y}{x},$$

the integral of which is obvious.

The integral of the original equation can be obtained by quadratures. Let

$$y = ux,$$

so that

$$x\frac{du}{dx} + u = u\frac{au+b}{a'u+b'};$$

hence

$$\frac{dx}{x} = \frac{du}{u}\frac{a'u+b'}{(a-a')u+b-b'}$$

$$= du\left\{\frac{\theta}{u} + \frac{\phi}{(a-a')u+b-b'}\right\},$$

where

$$\theta = \frac{b'}{b-b'}, \qquad \phi = \frac{a'b-ab'}{b-b'}.$$

Consequently

$$x = Au^{\frac{b'}{b-b'}}\{(a-a')u+(b-b')\}^{\frac{a'b-ab'}{(a-a')(b-b')}},$$

and therefore

$$x^{\frac{a}{a-a'}} = Ay^{\frac{b'}{b-b'}}\{(a-a')y+(b-b')x\}^{\frac{a'b-ab'}{(a-a')(b-b')}},$$

which is the integral.

Taking

$$u = \rho + v,$$

with the earlier value of ρ, and absorbing a constant into A, we have

$$x = A'(\rho+v)^{\frac{b'}{b-b'}}v^{\frac{a'b-ab'}{(a-a')(b-b')}}.$$

Accordingly, if

$$\lambda = \frac{a'b-ab'}{(a-a')(b-b')}$$

is positive, v tends to zero with x. When λ is a positive commensurable quantity, other than the reciprocal of a positive integer, then there are a finite number of branches for v as a function of x, all vanishing with x and circulating round the origin. If λ be the reciprocal of a positive integer, then v is a uniform function of x, vanishing with x.

Implicit assumptions have been made that a and a' are unequal, also that b and b' are unequal: the discussion of the alternatives is simple.

Ex. 8. Consider the equation

$$\frac{dy}{dx} = \frac{y}{x}\frac{y^4+a_1x^2y^3+a_2x^4y^2+a_3x^3y+a_4x^8}{y^5+b_1x^2y^4+b_2x^3y^3+b_3x^2y^2+b_4x^5y+b_5x^6};$$

here $S = -1$, $R = -1$.

From the diagram (where the points o are the points $5-k$, s_k of the denominator, and the points × are the points $4-l$, r_l of the numerator), it appears that the possible values of μ are ∞, 3, 1, 0. Of these we put the infinite value on one side, for it corresponds to a constant zero value of y while x varies ; and we put the zero value on one side, for it corresponds to the case when y varies without variation of x, that is, y is not then a function of x.

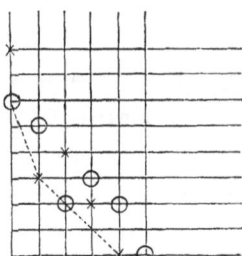

First, take $\mu=3$ and substitute $y=\rho x^3$, retaining only the lowest powers of x ; the equation for ρ is

$$3\rho = \rho\,\frac{a_3\rho}{b_5},$$

so that

$$\rho = 3\,\frac{b_5}{a_3}.$$

Now write

$$y = x^3\left(3\,\frac{b_5}{a_3}+v\right);$$

then substituting and reducing, we find that the equation for v is

$$x\,\frac{dv}{dx} = 3v + 3x^2\left(\frac{a_4}{a_3} - 9\,\frac{b_4 b_5}{a_3^2} - 27\,\frac{b_3 b_5^2}{a_3^3}\right) + \dots .$$

Secondly, take $\mu=1$ and substitute $y=\rho x$. Proceeding as before, we find the equation for ρ to be

$$\rho = \rho\,\frac{\rho^4 + a_3\rho}{b_3\rho^2},$$

that is, ρ is a root of the cubic

$$\rho^3 - b_3\rho + a_3 = 0,$$

the roots of which are distinct from one another unless $4b_3^3 = 27a_3^2$, a condition which we shall assume is not satisfied. Then taking

$$y = x\,(\rho + v),$$

we find that the equation for v is

$$x\,\frac{dv}{dx} = \left(3\,\frac{\rho^2}{b_3} - 1\right)v + \frac{a_1\rho^2 - \rho^4}{b_3}\,x + \dots ,$$

where the coefficient of v is distinct from zero, because the cubic for ρ does not possess equal roots.

Ex. 9. Consider the equation

$$\frac{dy}{dx} = \frac{y}{x^7}\,(y + 2a_4 x^8).$$

Here $R=-1$, $S=-7$; and the function $p\,(y,\,x)$ is unity. The diagram at once shews that the possible values for μ are ∞, 6.

The former gives a constant zero value for y.

For the latter, we take $y = \rho x^6$ and find

$$6\rho = \rho^2,$$

that is, $\rho = 6$. Hence substituting

$$y = x^6 (6 + v),$$

we find

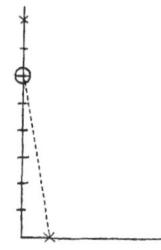

$$x \frac{dv}{dx} = 6v + 12a_4 x^2 + v^2 + 2a_4 v x^2$$

$$= 6v + 12a_4 x^2 + \dots,$$

which is the reduced equation for v.

But as a matter of fact, the integral of the equation can be obtained by quadratures. We have

$$\frac{1}{y^2} \frac{dy}{dx} - \frac{2a_4 x}{y} = \frac{1}{x^7},$$

and therefore

$$\frac{1}{y} e^{a_4 x^2} = A - \int \frac{1}{x^7} e^{a_4 x^2} dx$$

$$= A + \frac{1}{6x^6} + \frac{1}{4} \frac{a_4}{x^4} + \frac{1}{4} \frac{a_4^2}{x^2} + \frac{1}{6} a_4^3 \log x + P(x),$$

where $P(x)$ is a regular function of x. This corresponds with the above form

$$y = x^6 (6 + v) \ ;$$

and clearly $x = 0$ is a point of indeterminateness for y, since v contains terms in $\log x$.

Ex. 10.   Discuss the following equations :—

(i) $\dfrac{dy}{dx} = \dfrac{ay^4 + bx^2 y^2 + cx^5}{ay^5 + \beta x^2 y^2 + \gamma x^5}$;

(ii) $x \dfrac{dy}{dx} = \dfrac{ay^4 + bx^2 y^3 + cx^6}{\lambda y^6 + \lambda' y x^3}$;

(iii) $x^2 \dfrac{dy}{dx} = \dfrac{ay^4 + bxy^3 + cx^4}{y^4 + ax^2 y^3 + \beta x^3 y + \gamma x^4}$;

(iv) $\dfrac{dy}{dx} = \dfrac{\lambda}{x} \dfrac{y^3 + \mu x^3}{y^2 + \kappa x}$;

(v) $x^\theta \dfrac{dy}{dx} = a \dfrac{y^l + \beta x}{y^m + \gamma x}$, where $\theta \geqslant 2$;

(vi) $\dfrac{dy}{dx} = \kappa x^3 y^5 \dfrac{y^3 + a_1 x^2 y^2 + a_2 xy + a_3 x^4}{y^3 + b_1 x^6 y^2 + b_2 x^7 y + b_3 x^8}$;

(vii) $\dfrac{dy}{dx} = \kappa x^2 y \dfrac{y^3 + ax^4 y^2 + bxy + cx^{14}}{y^4 + fx^7 y^2 + gx^9 y + hx^{15}}$;

(viii) $\dfrac{dy}{dx} = \kappa x^2 y^3 \dfrac{y^8 + a_1 x^2 y^7 + a_2 xy^6 + a_3 x^4 y^5 + a_4 x^2 y^4 + a_5 x^9 y^3 + a_6 x^4 y^2 + a_7 x^2 y + a_8 x}{y^7 + b_1 x^2 y^6 + b_2 x^4 y^5 + b_3 x^3 y^4 + b_4 x^3 y^3 + b_5 x^6 y^2 + b_6 x^5 y + b_7 x^8}$;

(ix) $\dfrac{dy}{dx} = \kappa x^2 y^2 \dfrac{y^4 + a_1 xy^3 + a_2 x^2 y^2 + a_3 xy + a_4 x^3}{y^7 + b_1 x^2 y^6 + b_2 x^4 y^5 + b_3 x^3 y^4 + b_4 x^3 y^3 + b_5 x^6 y^2 + b_6 x^5 y + b_7 x^8}$;

(x) $\dfrac{dy}{dx} = \dfrac{1}{x^2 y^2} \dfrac{ay + bx}{ay + \beta x}$.

CHAPTER VI.

The Integrals of the various Reduced Forms of the Original Differential Equation in the Vicinity of the Accidental Singularity of the Second Kind*.

61. It has appeared that the determination of the integral of the original differential equation in the immediate vicinity of an accidental singularity of the second kind depends upon the character of the integral of a subsidiary differential equation which belongs to one or other of a number of reduced typical forms. Moreover, the algebraical substitutions, adopted in §§ 43, 47, 52, tacitly assumed that, subject to this consideration, the integral of the original equation is of a regular form or is changeable into a regular form. It may be the case with some of the equations that the portions of the integrals thus obtained constitute, when taken in the aggregate, the complete elements of the solution in the vicinity of the singularity. But for some of the equations these portions, if any, represent only part of the solution: in particular, it may be necessary to consider independently whether the singularity is or is not a point of indeterminateness for the general integral.

We proceed, in the first place, to the consideration of the reduced typical forms in succession.

---

* Many references are given in the course of the chapter. In addition to those quoted, there may be mentioned, Picard, *Cours d'Analyse*, t. III, ch. II, IX; Jordan, *Cours d'Analyse*, t. III, pp. 112—122; and the papers of Poincaré in the footnote to Ex. 1, § 60.

Neither the method originally given by Briot and Bouquet, nor that given by Poincaré and to some extent followed by Picard, has been adopted: it will be seen that Jordan's method is more closely followed.

The forms which have to be considered are

$$t^\kappa \frac{dv}{dt} = av + bt + \ldots = \phi_1(v, t),$$

where κ is an integer $\geqslant 1$;

$$t^\kappa \frac{dv}{dt} = gv^m + bt + \ldots = \phi_m(v, t),$$

where κ is an integer $\geqslant 1$, and m is an integer $\geqslant 2$;

$$t^\kappa \frac{dv}{dt} = \frac{gv^n + bt + \ldots}{hv^m + ct + \ldots}$$

$$= \frac{\phi_n(v, t)}{\phi_m(v, t)},$$

where κ is an integer $\geqslant 1$, m and n are integers $\geqslant 1$; but if they are both equal to 1, the form can (by transformation) be included in the first of the set.

Of all these forms, the most important, as regards the natural occurrence, is the first when $\kappa = 1$. It will be seen that the characteristics of v, as determined by

$$t \frac{dv}{dt} = \phi_1(v, t),$$

are settled for all the varieties of this form that can occur; but it will be seen that the same claim cannot be made for the other forms, and they offer opportunities for further investigation.

THE FIRST TYPICAL REDUCED FORM.

62. The first of the typical reduced forms is

$$t \frac{dv}{dt} = \phi_1(v, t)$$

$$= av + bt + \tfrac{1}{2}\alpha v^2 + \beta vt + \tfrac{1}{2}\gamma t^2 + \ldots,$$

where ϕ_1 is a regular function of v and t in the immediate vicinity of the origin. The quantity v thus defined is required to vanish when $t = 0$.

It will appear that the general character of v is to some extent determined by the nature of the constant a. When a is not a positive integer, transformations of the dependent variable can be effected, so that the new equation is the same in form as before but with a new coefficient a which has its real part zero or negative; and when a is a positive integer, the corresponding transformations lead to a new equation the same in form but having unity instead of the coefficient a. The transformations indicated are of the type given by

$$v = \left(-\frac{b}{a-1} + v'\right) t;$$

after substitution and division by t, we find

$$t \frac{dv'}{dt} = (a-1) v' + b't + \dots.$$

This transformation is effective, if a be not unity. If a be not a positive integer, say $\alpha + i\alpha'$, where α may be integral or fractional when α' is not zero, and where α is fractional when α' is zero, then it can be applied any number of times in succession: after a definite number of operations equal to or greater than the integer in α, the real part of the coefficient which emerges in place of a is either zero or negative. If a be a positive integer, the transformation can be effected until the coefficient which emerges in place of a is unity, that is, it can be effected $a-1$ times.

63. Since $\phi_1(v, t)$ is a regular function of v and t in the vicinity of 0, 0, expressible in a converging power-series, let the radius of convergence in the plane of t be r and the radius of convergence in the plane of v be ρ; within this region of existence, let the maximum value of $|\phi_1(v, t)|$ be M.

If a regular integral of the equation exists which vanishes with t, it can be expanded in a form

$$v = \alpha_1 t + \alpha_2 t^2 + \alpha_3 t^3 + \dots;$$

when this is substituted in

$$t \frac{dv}{dt} = \phi_1(v, t)$$
$$= av + bt + \tfrac{1}{2}(\alpha v^2 + 2\beta vt + \gamma t^2) + \dots,$$

the new result must be an identity. Accordingly, from the coefficients of t^n, we have the relation

$$(n - a) \alpha_n = \text{coefficient of } t^n \text{ in } bt + \tfrac{1}{2}(\alpha v^2 + 2\beta vt + \gamma t^2) + \dots$$
$$= \text{integral function of } \alpha_1, \dots, \alpha_{n-1}, \text{ which is linear in}$$
$$\text{the coefficients in } \phi_1;$$

thus

$$(1 - a) \alpha_1 = b,$$
$$(2 - a) \alpha_2 = \tfrac{1}{2}(\gamma + 2\beta\alpha_1 + \alpha\alpha_1^2),$$

and so on. When the values of α_1, α_2, ..., α_{n-1} are obtained in succession and are substituted in the equation for α_n, the final expression for α_n is an integral function of the coefficients in ϕ_1, divided by a quantity, which is a product of factors $1 - a$, $2 - a$, ..., $n - a$, some of them being repeated in that product.

Further, it may be noted that

$$v = \frac{b}{1-a} t + \alpha_2 t^2 + \alpha_3 t^3 + \ldots,$$

so that

$$v' = \alpha_2 t + \alpha_3 t^2 + \ldots;$$

and therefore the quantity v', which occurs in the equation after a single transformation has been effected as above, is also a quantity that vanishes with t when v is a regular quantity. And similarly after any number of transformations. Hence we may assume, *ab initio*, that the transformations are effected; and therefore we have two cases to consider, viz.

(i)   when the real part of a is zero or negative, the imaginary part of a not vanishing when the real part is zero;

(ii) when $a = 1$.

The former assumption will now be made: the latter case is reserved for later consideration (§ 68).

Case I : $a \neq$ positive integer.

64.   Now as the real part of a either is negative, or is zero and then the imaginary part does not vanish, the value of $|n - a|$ for any integer n is certainly greater than unity. For $n = 1$, the quantity $|n - a|$ is less than for all other values of n; let this be θ, so that $\theta > 1$. Hence when the modulus of the final expression for α_n is taken, and in the denominator each factor $|m - a|$, for $m = 1, \ldots, n$, is replaced by θ, the modified quantity will be greater than the proper value of $|\alpha_n|$. Further, the modulus of the numerator is increased when every term (with its proper sign) is replaced by its modulus. Moreover, it is known* that the modulus of the coefficient of $v^m t^{m'}$ in ϕ_1 is less than $M \div \rho^m r^{m'}$; consequently if, in the expression obtained as being already greater than $|\alpha_n|$, the quantities $M \div \rho^m r^{m'}$ replace the moduli of

* *Th. Fns.*, § 22.

all the coefficients of $v^m t^{m'}$, which occur in the numerator of the expression (and occur there only), a still higher value will be obtained for the newly modified expression than the former. When these cumulative appreciations in value are effected upon the modulus of the expression for α_n, let the new quantity be A_n; thus

$$|\alpha_n| < A_n.$$

This quantity A_n is the value of α_n when each of the numbers $1-a$, $2-a$, $3-a$, ... is replaced by θ, when the coefficients of $v^m t^{m'}$ which occur in ϕ_1 are replaced by $M \div \rho^m r^{m'}$, for the respective combinations of m and m', and when all the terms are made positive. Let V denote $A_1 t + A_2 t^2 + A_3 t^3 + \dots$. The effect of the changes, which lead from $|\alpha_n|$ to A_n, is to replace $t \dfrac{dv}{dt}$ by θV and $\phi_1(v, t)$ by

$$\frac{M}{r} t + \frac{M}{r^2} t^2 + \frac{M}{r\rho} tV + \frac{M}{\rho^2} V^2 + \dots;$$

hence V satisfies the equation

$$\theta V = \frac{M}{r} t + \frac{M}{r^2} t^2 + \frac{M}{r\rho} tV + \frac{M}{\rho^2} V^2 + \dots$$

$$= M \left\{ \frac{1}{\left(1 - \dfrac{t}{r}\right)\left(1 - \dfrac{V}{\rho}\right)} - 1 - \frac{V}{\rho} \right\}.$$

This, being a quadratic equation, has two roots: one of them vanishes with t, and the other is different from zero when $t = 0$. The former clearly is the proper value $A_1 t + A_2 t^2 + A_3 t^3 + \dots$ of V; it is easily found to be given by

$$\frac{2(\theta\rho + M)}{\rho^2} V = \theta - \left\{ \theta^2 - \frac{4Mt}{r-t} \frac{\theta\rho + M}{\rho^2} \right\}^{\frac{1}{2}},$$

where that branch of the (two-valued) radical is to be taken which becomes equal to θ when $t = 0$. Now this branch remains regular and can be expanded in a series of ascending powers of t, converging for all values of t such that

$$\left| \frac{t}{r-t} \right| < \frac{\theta^2 \rho^2}{4M(M + \theta\rho)}$$

$$< \kappa,$$

say; and the values of t are such that $|t| \leqslant r$. Now trace the curve

$$|t| = \kappa |r - t|.$$

It is a circle; its diameter lies between the two points which divide, internally and externally in the ratio of $\kappa : 1$, the radius of the circle of radius r drawn from the centre in the positive direction of the axis of real quantities. Since κ is not an infinitesimal quantity, the part of the area of this circle which lies within the circle $|t| = r$ is finite, not infinitesimal; every point lying within this part and not on its boundary satisfies all the required conditions. Accordingly, there is a region of values of t which is finite in extent and throughout the whole of which the series

$$A_1 t + A_2 t^2 + A_3 t^3 + \dots$$

converges absolutely. But we have seen that

$$|a_n| < A_n;$$

consequently, throughout the same finite region, the series

$$v = \alpha_1 t + \alpha_2 t^2 + \alpha_3 t^3 + \dots$$

converges absolutely.

Now remove the restriction that the real part of a is zero or negative; and replace it by the condition that a is not a positive integer. When the equation

$$t \frac{dv}{dt} = \phi_1 (v,\, t)$$

is in this form, it is known that a finite number of operations is sufficient to transform it to the preceding case, these operations being transformations of the type

$$v = \left(- \frac{b}{a-1} + v_1 \right) t,$$

$$v_1 = \left(- \frac{b'}{a-2} + v_2 \right) t,$$

$$\vdots$$

The last of the dependent variables v thus arising is that which occurs in the preceding investigation; it has been expressed as an absolutely converging power-series. When the successive substitutions are carried out, so as to give the initial dependent variable, the result manifestly is to give a regular function of t, that vanishes with t and, owing to its construction, satisfies the differential equation. Hence we have the theorem :—

The differential equation

$$t \frac{dv}{dt} = av + bt + \dots$$

$$= \phi_1(v, t),$$

where the coefficient a is not a positive integer, possesses a regular integral which vanishes when $t = 0$ and exists over a finite part of the region of existence of the function $\phi_1(v, t)$.

The argument of § 12 may be applied, or a proof similar to that in § 13 may be constructed, to shew that the regular integral thus obtained is the only regular integral which vanishes with t and satisfies the equation.

65. Though the integral thus obtained is unique as a regular integral vanishing with t, there may be other integrals, non-regular, which vanish with t; to determine whether such integrals exist or not, we proceed as follows. Let u denote the regular integral which has been obtained; and take

$$v = u + U,$$

so that U is a quantity vanishing with t; but unless the integral u already obtained is the only integral that satisfies the condition, U is not zero everywhere. (It is not a regular function of t; but it might be a regular function of some other variable, such as $t \log t$, which itself is not regular in t.) We have

$$t \frac{dU}{dt} = \phi_1(u + U, t) - \phi_1(u, t)$$

$$= U \frac{\partial \phi_1}{\partial u} + \frac{1}{2!} U^2 \frac{\partial^2 \phi_1}{\partial u^2} + \dots ;$$

since $\phi_1(u + U, t)$ and $\phi_1(u, t)$ are regular functions of their arguments, the right-hand side of the expression for $t \dfrac{dU}{dt}$ is an absolutely converging power-series. Now

$$\phi_1(u, t) = au + bt + \tfrac{1}{2}(\alpha u^2 + 2\beta ut + \gamma t^2) + \dots ,$$

so that

$$\frac{\partial \phi_1}{\partial u} = a + \alpha u + \beta t + \text{terms of higher orders,}$$

$$\frac{\partial^2 \phi_1}{\partial u^2} = \alpha + \delta u + \epsilon t + \dots\dots\dots\dots\dots ,$$

and so on. In each of these quantities, let the value of u be substituted; then we have

$$\frac{\partial \phi_1}{\partial u} = a + b_1 t + b_2 t^2 + \dots,$$

$$\frac{\partial^2 \phi_1}{\partial u^2} = \alpha + b_1' t + b_2' t^2 + \dots,$$

and so on, so that the equation for U is

$$t \frac{dU}{dt} = aU \{1 + g(U)\} + tUh(U, t),$$

where $g(U)$ is a regular function of U vanishing with U, and $h(U, t)$ is a regular function of U and t that does not vanish with U and t, unless b_1 vanishes, that is, unless

$$\beta + \alpha \frac{b}{1 - a}$$

vanishes: which, in general, is not the case. Hence

$$\frac{dU}{U\{1 + g(U)\}} = a \frac{dt}{t} + \frac{h(U, t)}{1 + g(U)} dt,$$

and therefore

$$\frac{dU}{U} (1 - \gamma_0 U - \gamma_1 U^2 + \dots) = a \frac{dt}{t} + \frac{b_1 + \dots}{1 + \gamma_0 U + \dots} dt,$$

that is,

$$\frac{dU}{U} - a \frac{dt}{t} = (\gamma_0 + \gamma_1 U + \dots) dU + \frac{b_1 + \dots}{1 + \gamma_0 U + \dots} dt.$$

In the vicinity of $t = 0$ take a point t_0; and suppose that, if possible, U_0 is a value of U at that point, U_0 being finite and different from zero. Join the point t_0 to the origin by any curve such that U tends from U_0 to a zero value as t moves from t_0 along the curve to the origin; and assume first, that the curve is finite in length and that, if desirable, it can pass an infinite number of times round the origin, (for an infinite number of revolutions round the origin is possible with a curve of finite length such as an equiangular spiral); and secondly, that the origin is not a point of indeterminateness for U. Let integration be effected along this curve between t and t_0; the equation gives

$$\log \frac{U}{U_0} - a \log \frac{t}{t_0} = \int_{V_0}^{U} (\gamma_0 + \gamma_1 U + \dots) dU + \int_{t_0}^{t} \frac{b_1 + \dots}{1 + \gamma_0 U + \dots} dt,$$

at any point of the curve.

All the series on the right-hand side converge absolutely. Hence in the first integral, which is

$$\gamma_0 (U - U_0) + \tfrac{1}{2}\gamma_1 (U^2 - U_0^2) + \dots,$$

as U tends to zero with t, the sum of the series tends to a definite and finite limit. Similarly, as t tends to zero, the second integral also tends to a definite and finite limit. The right-hand side in these circumstances tends to a definite and finite limit: let its value be $C + \tau$, where τ, depending on t, vanishes with t. Thus

$$\frac{U}{t^a} = \frac{U_0}{t_0^a} e^{C+\tau},$$

so that, in the limit as t tends to vanish,

$$U = At^a,$$

where A is a finite constant, which is arbitrary so far as concerns this equation : it depends upon the arbitrary quantities t_0, U_0.

The curve of variation has been limited by the condition that U should tend to vanish with t; the expression for U shews that the curve must be such as to make t^a vanish with t.

Let $a = \alpha' + \beta' i$, where (by the initial hypothesis) α' is not a positive integer when β' vanishes ; and let $t = r e^{\theta i}$; thus

$$| t^a | = e^{\alpha' \log r - \beta' \theta},$$

and therefore, as t tends towards zero, the quantity $\alpha' \log r - \beta' \theta$ must tend to a value $- \infty$.

(i)   If β' be zero, the condition is satisfied when α' is positive : but it cannot be satisfied if α' then be zero or negative. Hence when a is real and not an integer, there exists an integral U vanishing with t if a be positive ; and there is no such integral if a be negative.

(ii)   If β' be not zero and α' be positive, the condition is satisfied provided that, as t tends towards zero, θ either remains finite or, becoming infinite, has the same sign as β'. The latter case gives a spiral with an unlimited number of turns in a prescribed sense ; and the integral exists for such variations.

(iii)   If β' be not zero and α' be negative, the condition can be satisfied, only if θ tend to an infinite limit having the same sign as β' and such that $\alpha' \log r - \beta'\theta$ tends to the limit $-\infty$. Again the curve is a spiral with an unlimited number of turns in a prescribed sense; and the integral exists for such variations.

In the respective cases, these are the curves along which the variable t must tend to its origin.

It therefore follows that, in certain circumstances, a quantity U exists satisfying the differential equation

$$t\frac{dU}{dt} = \phi_1(u + U, t) - \phi_1(u, t),$$

vanishing when $t = 0$ and distinct from zero in an infinitesimal region round the origin; consequently, for that region it follows that the solution u is not the unique solution of the original equation. Moreover, the expression of U, when it exists, is

$$U = At^a$$

in the immediate vicinity of $t = 0$, where A is a finite constant that is arbitrary so far as the equation is concerned.

66. This result, valid over the infinitesimal region specified, suggests a form of result for the equation when the region of variation of t is not restricted in that manner. Returning, therefore, to the original differential equation

$$t\frac{dv}{dt} = av + bt + \dots$$
$$= \phi_1(v, t),$$

the integrals that vanish with t are indicated by the following theorem* :—

The differential equation possesses an infinitude of integrals that vanish with t, when a is not a positive integer but has its real part positive; and they have the form

$$v = \sum_{n=1}^{\infty} a_n t^n + t^a \xi,$$

where $\sum_{n=1}^{\infty} a_n t^n$ is the regular integral of the equation the existence of which has already been established, and ξ can be expressed as a doubly-infinite series of powers of t and t^a, which converges in a finite region round $t = 0$ and acquires a value A at $t = 0$, where A is an arbitrarily assigned constant.

---

* It was enunciated in part by Briot and Bouquet, *Journal de l'Éc. Polytech.*, t. xxi (1856), p. 172; and was rendered complete by Poincaré, *ib.*, t. xxviii (1878), p. 14.

It is evident that, for extremely small values of $|t|$, the value of $t^a \xi$ is the same as that of U in the preceding investigation, so that ξ tends to the value A as t tends to the limit zero. To establish the theorem, it must be shewn that a doubly-infinite series of the specified kind can be obtained which, when substituted for U, will render the equation for U identically satisfied. That equation is

$$t\frac{dU}{dt} = U\frac{\partial \phi_1}{\partial u} + \frac{1}{2!} U^2 \frac{\partial^2 \phi_1}{\partial u^2} + \ldots$$

$$= aU(1 + \gamma_0 U + \ldots) + tU(b_1 + \text{powers of } U \text{ and } t),$$

where

$$b_1 = \beta + a\frac{b}{1-a}, \quad \gamma_0 = \tfrac{1}{2}\frac{\alpha}{a};$$

substituting $t^a \xi$ for U, the equation for ξ is

$$t\frac{d\xi}{dt} = \xi(b_1 t + a\gamma_0 t^a \xi + \ldots)$$

$$= \xi\psi(t, t^a \xi),$$

where $\psi(t, t^a \xi)$ is a regular function of t, t^a (in some vicinity of $t = 0$), and of ξ (for values of ξ that tend to acquire an arbitrarily assigned finite value A as t tends to the value zero), the powers of ξ always occurring through powers of $t^a \xi$; and ψ vanishes when $t = 0$. The equation thus obtained is the differential equation for ξ; the establishment of the existence of a solution ξ, subject to the assigned condition, can be effected in a manner similar to that adopted for the regular integral of the original equation (§ 64).

Let N denote the greatest value of $|\psi(t, t^a \xi)|$ for values of the variables within the region of convergence of ψ, say $|t| \leqslant r$, $|\xi| \leqslant \rho$, $|t^a| \leqslant \sigma$; so that, as is known*, the modulus of the coefficient of $t^{m+na}\xi^n$ in the expansion of ψ is less than

$$\frac{N}{r^m \sigma^n \rho^n},$$

for all positive values of the integers m and n. Now as ξ is to be a function of t and t^a, take

$$\theta = t^a,$$

so that

$$t\frac{d\xi}{dt} = t\frac{\partial \xi}{\partial t} + a\theta\frac{\partial \xi}{\partial \theta},$$

* *Th. Fns.*, § 22.

when ξ is expressed as a function of t and θ; the equation to determine ξ then is

$$t \frac{\partial \xi}{\partial t} + a\theta \frac{\partial \xi}{\partial \theta} = \xi \psi\,(t,\,\theta\xi).$$

Now if ξ is to be a regular function of t and θ, which is finite when t and θ vanish, it must be expressible as a converging power-series, say of the form

$$\xi = \sum_{m=0} \sum_{n=0} c_{m,\,n} t^m \theta^n,$$

where $c_{0,\,0}$ is not zero ; when this series is substituted for ξ, the equation must be an identity. Thus*, for values of m and n that are not simultaneous zeros,

$$(m + an)\, c_{m,\,n} = \text{coefficient of } t^m \theta^n \text{ in } \xi \psi\,(t,\,\theta\xi)$$
$$= C_{m,\,n},$$

where $C_{m,\,n}$ is a rational integral function of the coefficients in ψ and of the coefficients $c_{m',\,n'}$ in ξ, such that $m' \leqslant m$, $n' \leqslant n$, and $m' + n' < m + n$. But there is no constant term either on the left-hand side or on the right-hand side of the equation after substitution : so that no limitation on the quantity $c_{0,\,0}$ is introduced by the equation. When the series for ξ converges, its value for $t = 0$ is $c_{0,\,0}$: the preceding investigation shews that, as t tends to the value zero, ξ tends to a limit A which is an arbitrary constant: and therefore, if the series converges, we have

$$c_{0,\,0} = A.$$

For the other coefficients $c_{m,\,n}$, the equations

$$(m + an)\, c_{m,n} = C_{m,n},$$

solved in succession for increasing values of $m + n$, lead to results of the form

$$c_{m,\,n} = \gamma_{m,\,n},$$

---

* A possible difficulty might be suggested, as regards equating the coefficients, in the case where a is a real positive commensurable quantity, say $= \frac{1}{2}$: for then the even powers of θ would be integral powers of t and the identification in the form adopted would not be possible.

The difficulty ceases to be valid when, as is the actual fact, the equation in ξ is regarded as a partial differential equation in two independent variables t and θ. The integral of the partial differential equation leads to the integral of the original ordinary equation by making $\theta = t^a$; but it can lead to the integrals of other ordinary equations by making other substitutions for θ in terms of t.

where $\gamma_{m,n}$ is an integral function of the coefficients that occur in ψ, this function being divided by a product of factors of the form $m + an$.

Now as the real part of a is positive, $|m + an|$ increases with increasing values of m and n; it is therefore at least equal to 1, or to $|a|$ if $|a| < 1$, for m and n are not simultaneously zero; and except for the lowest values of m and n, it is greater than unity. Hence when the modulus of the expression for $\gamma_{m,n}$ is constructed, its value will be increased if each of the factors $|m + an|$ in its denominator be replaced by ϵ, where ϵ is the smallest among them all. The value of the expression, thus modified, will be further increased if the modulus of the numerator be replaced by the sum of the moduli of its various terms; and in this later form, its value will be still further increased, if the modulus of the coefficient of $t^p (\theta \xi)^q$ in the expansion of ψ be replaced by

$$\frac{N}{r^p \sigma^q \rho^q},$$

for all values of p and q. Hence, if $\Gamma_{m,n}$ is the result of making all these appreciations in the value of the expression for $|\gamma_{m,n}|$, we have

$$|c_{m,n}| < \Gamma_{m,n}.$$

We take $\Gamma_{0,0} = |c_{0,0}|$. The series obtained from substituting $\Gamma_{m,n}$ for $c_{m,n}$ gives a quantity η, such that

$$\eta = \sum_{m=0} \sum_{n=0} \Gamma_{m,n} t^m \theta^n.$$

From the law of formation of $\Gamma_{m,n}$, it is easy to see, as on p. 144, that η satisfies the equation

$$\epsilon \eta = \epsilon A + \eta \left(\frac{N}{r} t + \frac{N}{\sigma \rho} \theta \eta + \frac{N}{r \sigma \rho} t \theta \eta + \frac{N}{r^2} t^2 + \dots \right)$$

$$= \epsilon A + N \eta \left\{ \frac{1}{\left(1 - \dfrac{t}{r}\right)\left(1 - \dfrac{\theta \eta}{\sigma \rho}\right)} - 1 \right\}$$

Also, η must be that root of the quadratic equation which becomes A when t and θ vanish. The solution of this quadratic and the choice of the appropriate root lead to the result

$$\eta \frac{2(\epsilon + N) \dfrac{\theta}{\sigma \rho}}{\epsilon + \dfrac{\epsilon A \theta}{\sigma \rho} - \dfrac{Nt}{r-t}} = 1 - \left\{ 1 - \frac{4 \epsilon A (\epsilon + N) \dfrac{\theta}{\sigma \rho}}{\left(\epsilon + \dfrac{\epsilon A \theta}{\sigma \rho} - \dfrac{Nt}{r-t} \right)^2} \right\}^{\frac{1}{2}},$$

where that branch of the radical is chosen which is equal to 1 when $\theta = 0$. The branch of the radical remains regular, and tends to the value 1 when $\theta = 0$, for a region of variation of the variables represented by

$$\left| \epsilon + \frac{\epsilon A}{\sigma \rho} t^a - N \frac{t}{r - t} \right|^2 > 4 \left| \frac{\epsilon A}{\sigma \rho} (\epsilon + N) t^a \right|$$

Since ϵ is a finite and not an infinitesimal quantity, this region is finite in extent; clearly $t = 0$ is included within the region, and therefore the area common to the region and the circle $|t| = r$—but not necessarily the boundary of this area—is a region of existence for the function η; and it satisfies the conditions precedently laid down.

Within this region, the quantity η can be expanded in an absolutely converging power-series in t and θ; the constant term in the expansion is at once seen to be A, from the expression for the appropriate root of the quadratic; consequently the series

$$\sum_{m=0} \sum_{n=0} \Gamma_{m,n} t^m \theta^n$$

converges absolutely for the range indicated. But

$$|c_{m,n}| < |\Gamma_{m,n}| ;$$

and therefore the series

$$\sum_{m=0} \sum_{n=0} c_{m,n} t^m \theta^n$$

converges absolutely. It is a solution of the equation

$$t \frac{\partial \xi}{\partial t} + a\theta \frac{\partial \xi}{\partial \theta} = \xi \psi (t, \theta \xi),$$

and its value is A when $t = 0$.

When these various results are combined, the theorem is established as enunciated.

67. It thus appears that an integral of the equation can be associated with an arbitrarily assigned constant A, provided A be different from zero. (If A be zero, we merely fall back upon the regular integral of the original equation.) It may be noted that, when a is a real commensurable positive quantity (and therefore a fraction, for integer values are excluded by the theorem), each integral associated with an assigned constant A is composed of a finite number of different branches, which have the origin for their common branch-point and circulate round it in one cycle. When a is not real, the integral, associated with an assigned constant A,

is a regular function of t and θ in the vicinity of $t = 0$, $\theta = 0$; regarded as a function of t alone, there is no limit to the number of its branches for each constant A,—the origin is, in fact, a point of indeterminateness.

Next, when a is a real negative quantity (whether an integer or not), or when a, being complex, has its real part negative, there is no integral of the equation, other than the regular integral, which vanishes when $t = 0$.

Lastly, when these integrals other than the regular integral are known to exist, that is, when the real part of a is positive (a not being itself an integer), the earlier stages in the theorem just proved give a practical method of obtaining their analytical expression: as a matter of fact, the coefficients $c_{m, n}$ in the double-series for ξ are explicitly deduced in a succession from the first, which is the arbitrary constant of the integral.

Case II: $a = positive\ integer$.

68.   It now is necessary to consider the sole case omitted from the preceding investigation, viz. that in which a is a positive integer.

The particular form that arises when a is zero can be omitted for the present: the implication then is that the term involving the first power of v alone is absent from $\phi_1(v, t)$. When this happens, then, if $\phi_1(v, 0)$ is zero for all values of v, $\phi_1(v, t)$ has a factor t, which could be removed from both sides of the equation and would leave it in the form

$$\frac{dv}{dt} = \psi_1(v, t),$$

where ψ_1 is a regular function of v, t; this form has already been considered.   If $\phi_1(v, 0)$ is not zero for all values of v, let gv^m be the term with lowest index which it contains: on the hypothesis adopted, $m \geqslant 2$; the equation is then

$$t \frac{dv}{dt} = gv^m + bt + \ldots,$$

where $m \geqslant 2$.   This is one of the forms reserved for later consideration, being a typical reduced form (§ 44).

Accordingly, we may assume that the positive integer a is $\geqslant 1$.   When $a > 1$, it has been seen (§ 62) that transformations of the dependent variable can be made in succession which at each stage diminish the coefficient of the new dependent variable by 1, while leaving the general form of the equation unaltered : and that this transformation can be effected so long as the coefficient is greater than 1, but that it cannot be effected when the coefficient is unity. We shall therefore assume that, when $a > 1$, these transformations have been effected as often as is possible ; and consequently the form of the equation to be discussed is

$$t \frac{dv}{dt} = v + bt + \tfrac{1}{2}(\alpha v^2 + 2\beta vt + \gamma t^2) + \ldots$$
$$= \theta_1 (v, t),$$

say.

Some hints as to possible forms of the theorems that apply to this equation may be obtained by taking the theorems for the former case, say when a is real, positive, and not an integer ; regarding a as parametric, we make it pass to the limit when it becomes unity. For this purpose, it will suffice to take

$$a = 1 + \delta,$$

where δ will be made zero in the limit.

As regards the regular integral in the former case, it was

$$v = \alpha_1 t + \alpha_2 t^2 + \alpha_3 t^3 + \ldots,$$
where

$$\alpha_1 = \frac{b}{1 - a},$$

and the succeeding coefficients $\alpha_2, \alpha_3, \ldots$ depend upon α_1 in part, when it occurs. Now with the adopted value of a, we have

$$\alpha_1 = \frac{b}{-\delta},$$

so that, if b be not zero, α_1 becomes infinite in the limit contemplated ; and the remaining coefficients α then become infinite. The regular integral would then cease to be significant ; hence it may be expected that $b = 0$ is a primary condition for the possession of a regular integral by the equation

$$t \frac{dv}{dt} = \theta_1 (v, t).$$

If, however, b be zero, then in the former instance α_1 is 0; the value of α_2 then is

$$\alpha_2 = \frac{\frac{1}{2}\gamma}{2 - a},$$

and so for the other coefficients in succession. Passing to the limit in the present case, we have

$$\alpha_2 = \tfrac{1}{2}\gamma,$$

and so for the other coefficients; it therefore appears as if, when $b = 0$, the regular integral survives. But this suggested inference cannot be regarded as thereby established.

As regards the non-regular integrals of the former case, it was proved that they could be expressed as regular functions of t and t^a. Now

$$t^a = t^{1+\delta}$$

$$= t\{1 + \delta \log t + \ldots\},$$

so that, if we replace t^a by this value, a function of t and t^a becomes changed into a function of t and $t \log t$. Hence it appears possible that there may be non-regular integrals of the equation

$$t \frac{dv}{dt} = \theta_1(v, t),$$

which can be expressed as regular functions of t and $t \log t$; but again, this suggested inference cannot be regarded as thereby established.

We consider first, the possibility of regular integrals: secondly, the possibility of non-regular integrals.

69. The existence of regular integrals is defined by the following theorem, due to Briot and Bouquet:—

The differential equation

$$t \frac{dv}{dt} = v + bt + \ldots = \theta_1(v, t)$$

possesses no regular integral vanishing with t, if the coefficient b be different from zero; but if the coefficient b vanish, then the equation possesses a regular integral, which vanishes with t and involves an arbitrary constant.

Any regular function of t, which vanishes when $t = 0$, can be expanded in a converging power-series

$$v = \alpha_n t^n + \alpha_{n+1} t^{n+1} + \dots ,$$

where $n \geqslant 1$. If this can be an integral of the equation

$$t \frac{dv}{dt} = \theta_1 (v, t)$$

$$= v + bt + \tfrac{1}{2} (\alpha v^2 + 2\beta vt + \gamma t^2) + \dots ,$$

the equation must be satisfied identically when the value of v is substituted in it. We ought then to have

$$n\alpha_n t^n + (n + 1) \alpha_{n+1} t^{n+1} + \dots$$

$$= \alpha_n t^n + \alpha_{n+1} t^{n+1} + \dots$$

$$+ bt + \tfrac{1}{2}\gamma t^2 + \beta t (\alpha_n t^n + \dots) + \tfrac{1}{2}\alpha (\alpha_n t^n + \dots)^2 + \dots ,$$

satisfied as an identity. If $n > 1$, the only term in the equation involving the first power of t, is bt: this cannot disappear if b is not zero. If then b is not zero, n cannot be greater than 1; it must be equal to 1. In order that the first power of t should then disappear, we must have

$$\alpha_n = \alpha_n + b,$$

which cannot be satisfied by any finite value of α_n if b be different from zero.

The series accordingly is not possible; and therefore there is no regular integral of the equation when the coefficient b does not vanish.

Next, suppose that $b = 0$, so that the equation is

$$t \frac{dv}{dt} = v + \tfrac{1}{2}(\alpha v^2 + 2\beta vt + \gamma t^2) + \dots .$$

If this equation possesses a regular integral that vanishes when $t = 0$, we may write $v = t\xi$, and the sole condition attaching to ξ is that it is to be a regular function of t in the vicinity of $t = 0$: but it need not vanish there, in order to secure a zero value for v when $t = 0$. Substituting, we have

$$\frac{d\xi}{dt} = \tfrac{1}{2}(\gamma + 2\beta\xi + \alpha\xi^2)$$

$$+ \tfrac{1}{6}t (\delta', \gamma', \beta', \alpha' \rangle\!\langle 1, \xi)^3 + \dots ,$$

where the function on the right-hand side is a regular function of t and ξ; let the region of its existence be defined by

$$|t| \leqslant r, \quad |\xi| \leqslant \rho.$$

To this equation for ξ, Cauchy's original existence-theorem can be applied. If λ be any arbitrary quantity such that

$$|\lambda| < \rho,$$

then an integral of the equation for ξ exists, which is a regular function of t in the vicinity of $t = 0$ and assumes the value λ when $t = 0$.

Denote this integral by ξ_λ; then

$$v = t\xi_\lambda$$

is a regular integral of the original equation, which vanishes when $t = 0$; and it contains an arbitrary constant λ.

Moreover, ξ_λ is the only regular integral of the equation for ξ, determined by the condition that it shall assume the value λ when $t = 0$; hence $t\xi_\lambda$ is the only regular integral of the equation

$$t\frac{dv}{dt} = v + \tfrac{1}{2}(\alpha v^2 + 2\beta vt + \gamma t^2) + \ldots,$$

determined by the conditions that v vanishes when $t = 0$ and $\dfrac{dv}{dt}$ assumes a value λ when $t = 0$.

If the limiting condition be solely that v shall vanish when $t = 0$, then as λ is arbitrary, subject solely to the restriction $|\lambda| < \rho$, it follows that the equation in v, when $b = 0$, has an infinitude of regular integrals which vanish with t.

70.   The existence of non-regular integrals is defined by the following theorems * :—

An infinitude of integrals of the equation

$$t\frac{dv}{dt} = v + bt + \ldots = \theta_1(v, t)$$

exist, which vanish with t and, within a region round $t = 0$ that is not infinitesimal, are regular functions of t and t Log t, where Log t

---

* The first was enunciated definitely by Poincaré, *Journ. de l'Éc. Polytech.*, t. xxviii (1878), p. 26; Briot and Bouquet had previously (*l.c.*) shewn that the equation possesses an infinitude of non-regular integrals.

is that branch of the function $\log t$, *whose argument lies between* 0 *and* 2π.

Further, *in the immediate vicinity of the origin, these integrals take the form*

$$bt \operatorname{Log} t + V,$$

where $|V|$ *can be made smaller than any finite quantity on sufficiently diminishing* $|t|$, *and is infinitesimally small compared with* $|t \operatorname{Log} t|$.

The second of these theorems will first be established, in order to follow as closely as possible the line of investigation adopted for the case when a is not a positive integer. We take the equation in the form

$$t \frac{dv}{dt} = v\{1 + vg(v)\} + bt + vt\{\beta + vh(v)\} + t^2 R(v, t),$$

where $g(v)$, $h(v)$, $R(v, t)$ are regular functions which may (but do not necessarily) vanish for zero values of their arguments; and rearranging it, we construct expressions to which quadratures can be applied.

We have, for any function $F(v)$,

$$\frac{d}{dt}\left[\frac{F(v)}{t} e^{-\sigma t + tG(v)}\right] = e^{-\sigma t + tG(v)} U,$$

where

$$U = \frac{F'(v)}{t} \frac{dv}{dt} - \frac{F(v)}{t^2} + \frac{F(v)}{t}\{-\sigma + G(v)\} + \frac{F(v)}{t} G'(v) t \frac{dv}{dt},$$

σ being any constant. Let $F(v)$ be determined by the condition

$$\frac{F(v)}{F'(v)} = v\{1 + vg(v)\},$$

so that

$$\frac{F'(v)}{F(v)} = \frac{1}{v\{1 + vg(v)\}}$$

$$= \frac{1}{v} - \tfrac{1}{2}\alpha + \text{positive powers of } v$$

$$= \frac{1}{v} + \frac{dV_1}{dv},$$

where V_1 is a regular function of v, vanishing with v; thus we may take

$$F(v) = ve^{V_1},$$

arbitrary constants not being important for this purpose. And then

$$F'(v) = 1 - \alpha v + \text{higher powers of } v.$$

We now have

$$U = F'(v) \left[\frac{1}{t} \frac{dv}{dt} - \frac{v}{t^2} \{1 + vg(v)\} \right] + J$$

$$= F'(v) \left[R(v, t) + \frac{b}{t} + \frac{v}{t} \{\beta + vh(v)\} \right] + J,$$

where

$$J = \frac{F(v)}{t} \{-\sigma + G(v)\} + \frac{F(v)}{t} G'(v) t \frac{dv}{dt}$$

$$= \frac{F(v)}{t} \{-\sigma + G(v)\} + \frac{F(v)}{t} G'(v) [v \{1 + vg(v)\} + t\Phi]$$

$$= \frac{F(v)}{t} \{-\sigma + G(v)\} + \frac{F(v) G'(v)}{t} v \{1 + vg(v)\} + \Phi F(v) G'(v);$$

here Φ denotes

$$b + v \{\beta + vh(v)\} + tR(v, t),$$

so that the term $\Phi F(v) G'(v)$ in J is a regular function of v and t. Also $F'(v) R(v, t)$ in U is a regular function. Now choose the function $G(v)$ so that

$$[b + v \{\beta + vh(v)\}] F'(v)$$
$$+ F(v) [-\sigma + G(v) + vG'(v) \{1 + vg(v)\}] = b;$$

then, if this choice be possible, the term in U in $\frac{1}{t}$ has b for its coefficient. We must have

$$ve^{V_1} [-\sigma + G(v) + vG'(v) \{1 + vg(v)\}]$$
$$= b - [b + v \{\beta + vh(v)\}] F'(v)$$
$$= b(\alpha v - \dots) - v \{\beta + vh(v)\} (1 - \alpha v + \dots)$$
$$= (b\alpha - \beta) v + v^2 \gamma_1 + \dots.$$

The constant σ is, as yet, unspecified; let

$$\sigma = \beta - b\alpha,$$

so that

$$G(v) + vG'(v) \{1 + vg(v)\}$$
$$= \{-(\beta - b\alpha) + v\gamma_1 + \dots\} e^{-V_1} + \beta - b\alpha$$
$$= 2v\delta_1 + \text{higher powers of } v$$
$$= vP(v), \text{ say};$$

thus

$$G(v) + G'(v) \frac{F(v)}{F'(v)} = vP(v),$$

and therefore

$$G(v)\,F(v) = \int v P(v)\,F'(v)\,dv$$
$$= \delta_1 v^2 + \text{higher powers of } v$$
$$= v^2 e^{V_2};$$

hence

$$G(v) = v e^{V_2 - V_1}$$
$$= v e^{V_2},$$

so that $G(v)$ is a regular function of v, vanishing when $v = 0$. (The object is merely to obtain some regular function vanishing with v, and therefore constants of integration are neglected.)

With these values, we have

$$U = \frac{b}{t} + \Phi F(v)\,G'(v) + F'(v)\,R(v,t)$$
$$= \frac{b}{t} + \Theta,$$

where Θ is a regular function of v and t, in the immediate vicinity of $v = 0$, $t = 0$.

Thus

$$\frac{d}{dt}\left[\frac{F(v)}{t}\,e^{-\sigma t + t G(v)}\right] = \left(\frac{b}{t} + \Theta\right) e^{-\sigma t + t G(v)}$$
$$= \frac{b}{t} + \Theta_1,$$

where Θ_1 is a regular function of t and v in the immediate vicinity of $t = 0$, $v = 0$.

Now the required integral v is to vanish with t; take a point t_0 near $t = 0$, and let the value of v there be v_0. Join t_0 to the origin by a curve, which is of finite length and does not make an infinite number of circuits round the origin; and integrate along this curve from t_0 to t. We have

$$\frac{v e^{V_1}}{t}\,e^{-\sigma t + t G(v)} - \frac{v_0}{t_0} e^{V_1(v_0)} e^{-\sigma t_0 + t_0 G(v_0)} = b \log \frac{t}{t_0} + \int_{t_0}^{t} \Theta_1\,dt.$$

As Θ_1 is a regular function of t and v, then, if the variable v be such as to have only definite finite values along the curve of integration, the integral

$$\int_{t_0}^{t} \Theta_1\,dt$$

is a finite quantity, say E; and therefore

$$\frac{v}{t}\,e^{-\sigma t + V_1 + t G(v)} = b \log t + A + E,$$

where A is a constant, depending upon v_0 and t_0.   Hence finally

$$v = bt \log t \, e^{\sigma t - V_1 - tG(v)} + (At + Et)\, e^{\sigma t - V_1 - tG(v)}$$

$$= bt \log t + (At + Et)\, e^{\sigma t - V_1 - tG(v)} + bt \log t \, \{ e^{\sigma t - V_1 - tG(v)} - 1 \}.$$

Now when $|t|$ is a small quantity, so also is $|t \log t|$; for if $\mathrm{Log}\, t$ denote that branch of the logarithmic function the argument of which lies between 0 and 2π, then

$$| t \log t | = | t \,\mathrm{Log}\, t + t\, 2k\pi i |,$$

where k is a finite integer on account of the properties of the curve drawn through t from 0 to t_0.   Now

$$| t\, 2k\pi i |$$

is small compared with $| t \,\mathrm{Log}\, t |$ when t is small.   Also

$$| t \,\mathrm{Log}\, t | = r\, \{ (\mathrm{Log}\, r)^2 + \theta^2 \}^{\frac{1}{2}},$$

where $t = r e^{\theta i}$; that is, when r is very small, $| t \,\mathrm{Log}\, t |$ differs from $r \,\mathrm{Log}\, r$ by a quantity infinitesimal compared with its value; moreover, this value can be made smaller than any finite quantity, on sufficiently diminishing r.

From the foregoing equation, it therefore follows that

$$v = bt \log t + \ldots,$$

where the unexpressed parts are powers of t, of $t \log t$, and of combinations of their powers; the term retained is the most important term, for

$$| t | < | t \log t |,$$

$$| t^m | \, | t \log t |^n < | t \log t |.$$

In particular, if we choose the branch $\mathrm{Log}\, t$ of the logarithmic function, we have

$$v = bt \,\mathrm{Log}\, t + \ldots$$

$$= bt \,\mathrm{Log}\, t + V,$$

where $| V |$ can be made less than any assignable finite quantity on sufficiently diminishing $| t |$, and is at the same time infinitesimally small compared with $| t \,\mathrm{Log}\, t |$.

71.   To establish the first theorem in § 70, it would be possible to pass from the equation

$$\frac{d}{dt} \left[\frac{F(v)}{t}\, e^{-\sigma t + tG(v)} \right] = \frac{b}{t} + \Theta_1 (v, t),$$

where $\Theta_1 (v, t)$ is a regular function of v and t, to the more extended theorem dealing with the general character of v in a finite region round the origin; but the argument is difficult. We can make a definite use of the argument for the case when a is not a positive integer, by proceeding (of course, with proper strictness) to the limit when a assumes a value unity*. The equation, when a is not a positive integer, is

$$t\frac{dv}{dt} = av + a_{01}t + \sum_{m=0}\sum_{n=0} a_{m,n}v^m t^n,$$

where $m + n \geqslant 2$ in the double summation. To pass from this form to the required case, we proceed as follows.

Let $A_{m,n} = |a_{m,n}|$, and consider the equation

$$t\frac{dV}{dt} - aV + A_{01}t = \Sigma\Sigma A_{m,n}V^m t^n, \qquad (m+n \geqslant 2),$$

choosing a to be a real positive fraction less than unity but tending to the value 1. Let a new variable θ be defined by the relation

$$\frac{t^a - t}{a - 1} = -\theta,$$

and assign to t^a its principal value $e^{a\,\mathrm{Log}\,t}$; then when a becomes 1 in the limit, θ is $-t\,\mathrm{Log}\,t$.

By the theorem of § 66, it is known that there is an integral of the differential equation which vanishes with t and which can be expressed as a series in powers of t and t^a, converging absolutely for sufficiently small values of $|t|$. When t^a is replaced by $t + (1-a)\theta$, the converging series becomes a converging series of powers of t and θ, and it satisfies the preceding equation for V. In order to obtain its expression in this modified form, let

$$V = \Sigma\Sigma C_{m,n}\theta^m t^n,$$

where C_{00} is zero; then as

$$t\frac{d\theta}{dt} = a\theta - t,$$

it follows that

$$t\frac{dV}{dt} = \Sigma\Sigma C_{m,n}\{n\theta^m t^n + m\theta^{m-1}t^n(a\theta - t)\}$$
$$= \Sigma\Sigma C_{m,n}\{(n + am)\theta^m t^n - m\theta^{m-1}t^{n+1}\};$$

hence the coefficients C are given by

$$\Sigma\Sigma C_{m,n}\{(n+am)\theta^m t^n - m\theta^{m-1}t^{n+1}\} - a\Sigma\Sigma C_{m,n}\theta^m t^n + A_{01}t$$
$$= \Sigma\Sigma A_{\rho,\sigma}t^\sigma(\Sigma\Sigma C_{p,q}\theta^p t^q)^\rho, \qquad (\rho + \sigma \geqslant 2).$$

* This is the method used by Jordan, Cours d'Analyse, t. III, pp. 118 et seq.; his proof has been adopted in the text.

The terms involving θ give no relation; in order that the terms in t may disappear, the equation

$$C_{10} = (1 - a) C_{01} + A_{01}$$

must be satisfied. But there is no equation to determine C_{01}: let it be assumed positive. The condition, that terms in $\theta^m t^n$ shall disappear, is

$$(n + am - a) C_{m,n} - (m + 1) C_{m+1,\,n-1} = B_{m,n},$$

where $B_{m,n}$ is a sum of terms of the form $BA_{\rho,\sigma} C_{\mu_1 \nu_1} C_{\mu_2 \nu_2} \dots C_{\mu_\rho \nu_\rho}$, such that

$$\left.\begin{array}{c} \rho + \sigma \geqslant 2 \\ \mu_1 + \mu_2 + \dots + \mu_\rho = m \\ \sigma + \nu_1 + \nu_2 + \dots + \nu_\rho = n \end{array}\right\} \;;$$

in particular, when $m = 0$, the term A_{0n} must be included in $B_{m,n}$; and the numerical constant B is the number of those simultaneous solutions of the equations

$$\mu_1 + \dots + \mu_\rho = m, \quad \nu_1 + \dots + \nu_\rho = n - \sigma,$$

which do not allow μ_p and ν_p, for $p = 1, \dots, \rho$, to be zero together.

These equations can be solved in succession, by taking first the three equations for which $m + n = 2$, next the four equations for which $m + n = 3$, and so on. The result is to give

$$C_{m,n} = \Gamma_{m,n},$$

where $\Gamma_{m,n}$ is the sum of a number of positive terms; each of these terms is the quotient, by a number of factors of the form $n + am - a$, of a quantity which is integral in the coefficients $A_{\rho,\sigma}$, contains a power of C_{01} of index not higher than n, a power of C_{10} of index not higher than m, and a numerical factor that (being a combination of the constants B) is otherwise independent of a.

The form of the divisor of a term in $\Gamma_{m,n}$ is limited by the property that the number of factors $\leqslant m + 2n - 1$, a result which can be established by induction. Let it be supposed valid for all expressions $\Gamma_{m,n}$, which have their second suffix less than n and for all those which, having their second suffix equal to n, have their first suffix less than m. Then when the equation

$$(n + am - a) C_{m,n} - (m + 1) C_{m+1,\,n-1} = B_{m,n}$$

is solved for $C_{m,n}$, the terms in $C_{m,n}$ arising from $C_{m+1,\,n-1}$ have the number of factors in the denominator

$= 1$, on account of $n + am - a$

$+$ an integer $\leqslant m + 1 + 2\,(n-1) - 1$, on account of $C_{m+1,\,n-1}$

$\leqslant m + 2n - 1$.

And the terms in $C_{m,n}$ arising from $BA_{\rho\sigma}C_{\mu_1\nu_1} \ldots C_{\mu_\rho\nu_\rho}$ have the number of factors in their denominator

$= 1$, on account of $n + am - a$

$+$ a sum of integers $\leqslant \Sigma\,(\mu_1 + 2\nu_1 - 1)$.

Now $\qquad \Sigma\,(\mu_1 + 2\nu_1 - 1) = \Sigma\mu_1 + 2\Sigma\nu_1 - \rho$

$$= m + 2\,(n - \sigma) - \rho$$

$$\leqslant m + 2n - \sigma - 2,$$

because $\rho + \sigma \geqslant 2$; and therefore the number of factors

$$\leqslant 1 + m + 2n - \sigma - 2$$

$$\leqslant m + 2n - 1 - \sigma$$

$$\leqslant m + 2n - 1.$$

Hence, on the hypothesis adopted, the result is valid also for $C_{m,n}$.

The first set of equations, viz. those for which $m + n = 2$, are

$$aC_{20} = A_{20}C_{10}{}^2,$$

$$C_{11} - 2C_{20} = A_{11}C_{10} + 2A_{20}C_{10}C_{01},$$

$$(2 - a)\,C_{02} - C_{11} = A_{02} + A_{11}C_{01} + A_{20}C_{01}{}^2;$$

so that the integer in question is 1 for C_{20}, 1 for C_{11}, 2 for C_{02}. The result is consequently valid for C_{30}, for C_{21}, for C_{12}, for C_{03}; and so on, it is valid in general.

And on account of the existence of the integral, established by the previous theorem in § 66, it is known that the series

$$\Sigma\Sigma C_{m,n}\,\theta^m t^n$$

converges absolutely.

Now proceed to the limit in which a increases to the value unity. The effect on the equation is to change it to the form

$$t\,\frac{dV}{dt} - V + A_{01}t = \Sigma\Sigma A_{m,n}V^m t^n\,;$$

and the effect on the integral is to change it to

$$\Sigma\Sigma C'_{m,n}\,(-\,t\,\mathrm{Log}\,t)^m t^n,$$

where $C'_{m,n}$ is the value of $C_{m,n}$ when $a = 1$.

The quantity $|\Gamma_{m,n}|$ is the sum of a series of positive terms. Let T be any one of these terms, and T' the value of the term T when $a = 1$. The changes thus made are, first, $|C_{10}|$, which is $(1-a)C_{01} + A_{01}$, is replaced by A_{01}, which is smaller than the modulus of C_{10}: and secondly, the change in the numerical denominator. Each factor $n + am - a$ is replaced by $n + m - 1$, that is, by a greater factor when m does not vanish; while if m vanishes, $n \geqslant 2$, so that

$$\frac{n - a}{n - 1} \leqslant 2 - a.$$

As the number of factors is not greater than $m + 2n - 1$, it follows that their product, say Π, is such that

$$\Pi \left(\frac{n + am - a}{n + m - 1} \right) \leqslant (2 - a)^{m+2n-1}$$
$$\leqslant (2 - a)^{2m+2n}.$$

Consequently, as the changes made have depreciated the numerator of T, it follows that

$$\frac{T'}{T} < \Pi \frac{n + am - a}{n + m - 1}$$
$$< (2 - a)^{2m+2n} ;$$

and therefore

$$\frac{|C'_{m,n}|}{|\Gamma_{m,n}|} < (2 - a)^{2m+2n}.$$

If the series

$$\Sigma\Sigma\Gamma_{m,n}\theta^m t^n$$

converge for values of t given by $|t| \leqslant r$, and $|\theta| < |s|$, and if M be the maximum value of the modulus of the series for the range of variation considered, then

$$\Gamma_{m,n} < \frac{M}{s^m r^n},$$

and therefore

$$C'_{m,n} < \frac{M}{\left\{ \dfrac{s}{(2-a)^2} \right\}^m \left\{ \dfrac{r}{(2-a)^2} \right\}^n}.$$

Consequently, the series

$$\Sigma\Sigma C'_{m,n} (t \operatorname{Log} t)^m t^n$$

converges absolutely for values of t such that $|t| < r$.

The existence of the integral, as stated (§ 70) in the theorem, can now be established for the original equation

$$t\frac{dv}{dt} = v + bt + \Sigma\Sigma a_{m,n} v^m t^n.$$

Should an integral of this equation exist, vanishing with t and expansible in a series of powers of t and $t\log t$, it must have the form

$$v = \Sigma\Sigma c_{m,n}(t\log t)^m t^n.$$

Choose c_{01} so that $|c_{01}|$ is the quantity C_{01} of the preceding case. When this value of v is substituted, the differential equation must be satisfied identically. When the relations, necessary to ensure this result, are solved to give the coefficients $c_{m,n}$, it is easy to see that $c_{m,n}$ can be deduced from $C'_{m,n}$ on replacing A_{01}, $A_{\rho\sigma}$ by $-b$, $a_{\rho\sigma}$ (for all values of ρ and σ) in the expression for $C'_{m,n}$. Consequently, when the modulus of the expression for $c_{m,n}$ is formed, a superior limit is obtained by writing, for $-b$, $a_{\rho\sigma}$, their moduli A_{01}, $A_{\rho\sigma}$ respectively; and therefore

$$|c_{m,n}| < C'_{m,n}.$$

But the series

$$\Sigma\Sigma C'_{m,n}(t\operatorname{Log} t)^m t^n$$

converges absolutely for all values of t such that $|t| < r$; and therefore the series

$$\Sigma\Sigma c_{m,n}(t\operatorname{Log} t)^m t^n$$

also converges absolutely for the same range of values.

The quantity C_{01} in the case when a is not a positive integer was left undetermined: it is arbitrary. So also, in the present case, the coefficient c_{01} in this expansion is arbitrary, and the coefficients $c_{m,n}$, such that $m + n \geqslant 2$, involve c_{01}; the expression for v thus is

$$v = bt\log t + c_{01}t + \underset{m\,n}{\Sigma\Sigma} c_{m,n}(t\log t)^m t^n,$$

where, in the double summation, $m + n \geqslant 2$, and the coefficient c_{01} is arbitrary.

72. Having now established the existence of the infinitude of integrals, their actual expression in any particular case can be obtained as follows. Let the equation be

$$t\frac{dv}{dt} = v + bt + \Sigma\Sigma a_{m,n} v^m t^n;$$

write $\theta = t \operatorname{Log} t$, so that $t\dfrac{d\theta}{dt} = \theta + t$; then $t\dfrac{dv}{dt} = t\dfrac{\partial v}{\partial t} + \theta\dfrac{\partial v}{\partial \theta} + t\dfrac{\partial v}{\partial \theta}$.
It is known, from the existence-theorem, that an integral exists, given by

$$v = b\theta + At + \Sigma\Sigma c_{m,n}\theta^m t^n.$$

Take together those terms which are of dimensions p in θ and t combined, and denote them by v_p, so that

$$v_p = \sum_{n=0}^{p} c_{n,p-n}\theta^n t^{p-n};$$

then

$$v = b\theta + At + v_2 + v_3 + \ldots = v_1 + v_2 + v_3 + \ldots \text{ say.}$$

Now

$$t\frac{\partial v}{\partial t} + \theta\frac{\partial v}{\partial \theta} = b\theta + At + 2v_2 + 3v_3 + \ldots;$$

so that our equation is

$$bt + b\theta + At + 2v_2 + 3v_3 + \ldots + t\frac{\partial v_2}{\partial \theta} + t\frac{\partial v_3}{\partial \theta} + \ldots$$

$$= bt + b\theta + At + v_2 + v_3 + \ldots + \Sigma\Sigma a_{m,n}v^m t^n,$$

where $m + n \geqslant 2$ on the right-hand side.   Hence

$$\left(v_2 + t\frac{\partial v_2}{\partial \theta}\right) + \left(2v_3 + t\frac{\partial v_3}{\partial \theta}\right) + \left(3v_4 + t\frac{\partial v_4}{\partial \theta}\right)$$

$$= \Sigma\Sigma a_{m,n}v^m t^n, \qquad\qquad (m + n \geqslant 2).$$

All that now is necessary is to arrange the right-hand side after substitution for v, in groups of terms that are homogeneous in t and θ of succeeding degrees; and to equate the respective groups.   Thus

$$v_2 + t\frac{\partial v_2}{\partial \theta} = a_{20}v_1^2 + a_{11}v_1 t + a_{02}t^2,$$

which determines the coefficients in v_2;

$$2v_3 + t\frac{\partial v_3}{\partial \theta} = 2a_{20}v_1 v_2 + a_{11}v_2 t + (a_{30}, a_{21}, a_{12}, a_{03} \mathbb{X} v_1, t)^3,$$

which determines the coefficients in v_3; and so on.

COROLLARY.   It is easy to prove that, when $b = 0$, then $c_{m,n} = 0$ for all values of $m > 0$.   The integral then, in fact, becomes the regular integral, possessed by the equation and containing an arbitrary constant (§ 69).

73. The explicit form of the integral of the equation, when a is a positive integer, is easily inferred from the foregoing result. Let

$$v = \frac{b}{1-a}\, t + tv_1$$

be substituted in the equation

$$t\,\frac{dv}{dt} = av + bt + f(v,\, t);$$

and

$$v_1 = \frac{b'}{2-a}\, t + tv_2,$$

$$\dotfill$$

$$v_{a-2} = \frac{b^{(a-2)}}{-1}\, t + tv_{a-1},$$

in the transformed equations in succession; the last of them is

$$t\,\frac{dv_{a-1}}{dt} = v_{a-1} + Bt + \Im\,(v_{a-1},\, t).$$

The integral of this is known to be

$$v_{a-1} = Bt \log t + Kt + G\,(t \log t,\, t),$$

where K is an arbitrary constant, and G is the regular function indicated by the preceding investigation. Hence, writing θ for $t \log t$, we have

$$v = \alpha_1 t + \alpha_2 t^2 + \ldots + \alpha_{a-1} t^{a-1} + Kt^a + Bt^{a-1}\theta + t^{a-1}\,G\,(\theta,\, t),$$

where the coefficients $\alpha_1, \ldots,\ \alpha_{a-1}$, B are definite, the coefficient K is arbitrary, and the function G contains terms of no dimension lower than 2 in t and θ combined.

Actual substitution of this value in

$$t\,\frac{dv}{dt} = av + bt + f(v,\, t)$$

leads to the equations determining the coefficients.

As an example, consider the equation

$$t\frac{dv}{dt} = 3v + bt + (a_2,\, b_2,\, c_2 \mathord{\)\!\(} v,\, t)^2 + (a_3,\, b_3,\, c_3,\, d_3 \mathord{\)\!\(} v,\, t)^3 + \ldots.$$

Then

$$v = a_1 t + a_2 t^2 + Kt^3 + Bt^2\theta + t^2 G\,(\theta,\, t),$$

where

$$G\,(\theta,\, t) = (a_2',\, b_2',\, c_2' \mathord{\)\!\(} \theta,\, t)^2 + (a_3',\, b_3',\, c_3',\, d_3' \mathord{\)\!\(} \theta,\, t)^3 + \ldots;$$

say

$$v = v_1 + v_2 + v_3 + v_4 + \ldots,$$

where v_m is the aggregate of terms in v which are of dimensions m in t and θ combined. Now

$$t\frac{dv}{dt} = t\frac{\partial v}{\partial t} + \theta\frac{\partial v}{\partial \theta} + t\frac{\partial v}{\partial \theta}$$

$$= v_1 + 2v_2 + 3v_3 + 4v_4 + 5v_5 + \dots$$

$$+ t^3B + t\frac{\partial v_4}{\partial \theta} + t\frac{\partial v_5}{\partial \theta} + \dots.$$

The differential equation is

$$t\frac{dv}{dt} = 3v_1 + 3v_2 + 3v_3 + 3v_4 + \dots + bt$$

$$+ (a_2,\ b_2,\ c_2\!\!\!\;)\!\!\;(v_1 + v_2 + \dots,\ t)^2$$

$$+ (a_3,\ b_3,\ c_3,\ d_3\!\!\!\;)\!\!\;(v_1 + v_2 + \dots,\ t)^3 + \dots.$$

Hence

$$-2v_1 = bt,$$

so that

$$a_1 = -\tfrac{1}{2}b\ ;$$

so that

$$-v_2 = (a_2,\ b_2,\ c_2\!\!\!\;)\!\!\;(v_1,\ t)^2,$$

$$-a_2 = (a_2,\ b_2,\ c_2\!\!\!\;)\!\!\;(a_1,\ 1)^2\ ;$$

so that

$$t^3B = 2v_2\,(a_2,\ b_2\!\!\!\;)\!\!\;(v_1,\ t) + (a_3,\ b_3,\ c_3,\ d_3\!\!\!\;)\!\!\;(v_1,\ t)^3,$$

$$B = 2a_2\,(a_2,\ b_2\!\!\!\;)\!\!\;(a_1,\ 1) + (a_3,\ b_3,\ c_3,\ d_3\!\!\!\;)\!\!\;(a_1,\ 1)^3\ ;$$

and so on.

The constant K remains undetermined and arbitrary; the coefficients which occur in G are all affected by the coefficients in v_3, that is, they involve K. Thus

$$v_4 + t\frac{\partial v_4}{\partial \theta} = 2v_3\,(a_2,\ b_2\!\!\!\;)\!\!\;(v_1,\ t) + a_2v_2v_1{}^2$$

$$+ 3v_2\,(a_3,\ b_3,\ c_3\!\!\!\;)\!\!\;(v_1,\ t)^2$$

$$+ (a_4,\ b_4,\ c_4,\ d_4,\ e_4\!\!\!\;)\!\!\;(v_1,\ t)^4,$$

leading to equations for the coefficients of the various combinations of t and θ. And so for the rest. In particular, $a_2' = 0$, and all the coefficients a' are zero.

COROLLARY. Clearly the condition, in the general case for an integral value of a other than unity, or in the particular case when $a = 3$, that the equation should possess a regular integral vanishing with t is that B should vanish: this corresponding to the condition (§ 69) that $b = 0$ when $a = 1$ for the specified property, and to the condition that

$$(a_2,\ b_2,\ c_2\!\!\!\;)\!\!\;(b,\ -1)^2 = 0$$

when $a = 2$, as may easily be verified.

74. In all the investigations relating to the equation

$$t\frac{dv}{dt} = \phi_1(v,\ t),$$

one assumption has been made tacitly, viz. that some term or terms occur on the right-hand side which do not contain v, in other words, it has been assumed that $\phi_1(0, t)$ does not vanish for all values of t. The alternative must now be considered.

Accordingly, if $\phi_1(0, t)$ is such as to vanish for all values of t, it follows that $\phi_1(v, t)$ contains some factor which is a power of v only; because a is not zero in the present case, it follows that this power is the first, so that

$$\phi_1(v, t) = v(a + \text{powers of } t \text{ and } v)$$
$$= vG(v, t),$$

where G is a regular function that does not vanish when $v = 0$, $t = 0$.

The following are the results which apply to this equation.

When the real part of a is positive but a is not itself a positive integer, the only regular integral of the equation which vanishes with t is $v = 0$. The number of non-regular integrals vanishing with t is unlimited; each of them can be expressed as a regular function of t and t^a; and the form of the integral, when $|t|$ is sufficiently small, is

$$Ct^a,$$

where C, in the limit when $|t| = 0$, is an arbitrary constant.

When a is a positive integer, the number of regular integrals vanishing with t is unlimited, because a regular integral can be obtained involving an arbitrary constant; and all the integrals that vanish with t are regular.

When the real part of a is negative, whether a be complex or be purely real, the only regular integral which vanishes with t is $v = 0$; and there is no non-regular integral which vanishes with t.

75. *Ex.* 1. Consider the equation

$$t\frac{dv}{dt} = p_0 + p_1 v + p_2 v^2,$$

where

$$p_0 = \qquad b_0 t + c_0 t^2 + \dots,$$
$$p_1 = a_1 + b_1 t + c_1 t^2 + \dots,$$
$$p_2 = a_2 + b_2 t + c_2 t^2 + \dots,$$

and assume that a_2 is not zero : then the equation takes the form

$$t\frac{dv}{dt} = a_1 v + b_0 t + f(v, t).$$

Let

$$v = -\frac{t}{p_2}\frac{1}{V}\frac{dV}{dt};$$

then the equation for V is

$$\frac{d^2V}{dt^2} - \frac{dV}{dt}\left(\frac{p_1}{t} + \frac{1}{p_2}\frac{dp_2}{dt} - \frac{1}{t}\right) + \frac{p_0 p_2}{t^2}V = 0,$$

that is,

$$\frac{d^2V}{dt^2} - \frac{dV}{dt}\left(\frac{a_1-1}{t} + b_1 + \frac{b_2}{a_2} + \text{powers}\right) + V\left(\frac{a_2 b_0}{t} + a_2 c_0 + b_0 b_2 + \ldots\right) = 0.$$

The results of the general theory of linear differential equations will be assumed for our immediate purpose.

The determining (indicial) equation for this linear equation of the second order is

$$m(m-1) - m(a_1-1) = 0,$$

having 0, a_1 for its roots.

In the first place, if a_1 be not an integer, two independent solutions are given by

$$V = V_1, \quad V = t^{a_1}V_2,$$

where V_1 and V_2 are regular functions of t in the vicinity of $t = 0$ which do not vanish there ; and the most general solution is

$$V = AV_1 + Bt^{a_1}V_2,$$

where A and B are arbitrary constants.

Substituting this value of V in the expression for v, we have

$$v = -\frac{1}{p_2}\frac{At\dfrac{dV_1}{dt} + t^{a_1}B\left(t\dfrac{dV_2}{dt} + a_1 V_2\right)}{AV_1 + t^{a_1}BV_2}.$$

When the real part of a_1 is positive, then v tends to the limit zero when $t = 0$: it contains an arbitrary constant $\dfrac{A}{B}$: it is a regular function of t and t^{a_1}, for V_1 is not zero when $t = 0$: it is not a regular function of t in general. Hence when a_1 is not a positive integer, but has its real part positive, there is an infinitude of integrals of the equation vanishing with t.

Next, let B be zero ; then

$$v = -\frac{t}{p_2}\frac{1}{V_1}\frac{dV_1}{dt},$$

which, since neither V_1 nor p_2 vanishes with t, is a regular function of t in the vicinity of $t = 0$, and it vanishes at $t = 0$. It is the unique regular integral vanishing with t, known to be possessed by the equation ; that it is unique is manifest from the complete integral which is not regular unless $B = 0$.

Lastly, let A be zero ; then

$$v = -\frac{1}{p_2}\left(1 + \frac{t}{V_2}\frac{dV_2}{dt}\right),$$

undoubtedly a regular integral : but it does not vanish with t, and therefore it does not satisfy the assigned conditions.

In the second place, let a_1 be an integer : so that the roots of the indicial equation 0, a_1 are both integers.

When a_1 is the greater of the two, that is, when a_1 is a positive integer, then two independent integrals of the equation in V are

$$V = t^{a_1} V_1, \quad V = V_2 + K V_1 t^{a_1} \log t,$$

where V_1, V_2 are regular functions of t that do not vanish with t, and where K is a determinate constant that may, in particular cases, be zero; the most general integral then is

$$A t^{a_1} V_1 + B (V_2 + K V_1 t^{a_1} \log t).$$

Substituting this value in v, we have

$$v = -\frac{1}{p_2} \frac{A t^{a_1} \left(t \dfrac{dV_1}{dt} + a_1 V_1 \right) + B \left[t \dfrac{dV_2}{dt} + K \left\{ V_1 t^{a_1} + V_1 a_1 t^{a_1} \log t + t^{a_1+1} \dfrac{dV_1}{dt} \log t \right\} \right]}{A t^{a_1} V_1 + B (V_2 + K V_1 t^{a_1} \log t)}.$$

When t is 0, then p_2, V_1, V_2 are all finite and different from zero ; hence v tends to the value 0 as t tends to zero ; it contains an arbitrary constant $\dfrac{A}{B}$; it is a regular function of t and $t \log t$: it is not a regular function of t in general. Hence when a_1 is a positive integer, there is an infinitude of integrals of the equation vanishing with t ; they can be expressed as regular functions of t and $t \log t$.

Next, let B be zero, which is the only way (unless K should happen to be zero) of securing a regular integral ; then

$$v = -\frac{1}{p_2} \left(a_1 + \frac{t}{V_1} \frac{dV_1}{dt} \right),$$

a regular function of t ; but it does not vanish when $t = 0$. Hence, in general, when a_1 is a positive integer, the equation possesses no regular integral vanishing with $t = 0$.

If however $K = 0$, which effectively is a relation among the coefficients of the equation, then

$$v = -\frac{1}{p_2} \frac{A t^{a_1} \left(t \dfrac{dV_1}{dt} + a_1 V_1 \right) + B t \dfrac{dV_2}{dt}}{A t^{a_1} V_1 + B V_2},$$

and V_2 is not zero when $t = 0$. In this case, v vanishes when $t = 0$; it is a regular function of t ; and it contains an arbitrary constant $\dfrac{A}{B}$. Hence if the condition be satisfied, the equation possesses an infinitude of regular integrals vanishing with t.

The case when $A = 0$ and $B \neq 0$ is of no special interest.

When a_1 is the smaller of the two integers 0 and a_1, that is, when a_1 is a negative integer, $= -s$ say, the two independent integrals of the equation in V are

$$V = V_1, \quad V = t^{-s} V_2 + J V_1 \log t,$$

where J is a determinate constant that may be zero; the most general integral is

$$V = A V_1 + B (t^{-s} V_2 + J V_1 \log t).$$

The value of v then is

$$v = -\frac{t}{p_2} \frac{A \dfrac{d V_1}{dt} + B \dfrac{d}{dt} \{t^{-s} V_2 + J V_1 \log t\}}{A V_1 + B (t^{-s} V_2 + J V_1 \log t)}.$$

In general, the only way of securing a regular integral from this value of v is to make $B = 0$; we then obtain a regular integral vanishing when $t = 0$. It may happen that $J = 0$; in that case, v becomes a regular function containing an arbitrary constant $\dfrac{A}{B}$; but when $t = 0$, the value of v is $\dfrac{s}{a_2}$, not zero. Hence when a_1 is a negative integer, the equation has only one regular integral vanishing with t.

Whether J is, or is not, zero, the value of the general expression of v for $t = 0$ is $\dfrac{s}{a_2}$, that is, not zero: thus when a_1 is a negative integer, there is no integral, except the regular integral, which vanishes when $t = 0$.

In the more special instance when a_1 is unity, the roots of the indicial equation are 0, 1. One independent integral of the equation for V is

$$V = V_1 = t P(t),$$

where $P(t)$ is a regular function of t such that

$$P(t) = 1 + Bt + \ldots,$$

and

$$2B - \left(b_1 + \frac{b_2}{a_2}\right) + a_2 b_0 = 0.$$

To determine another, without assuming the results of the general theory of linear differential equations, let

$$V = u V_1,$$

so that

$$\frac{d^2 u}{dt^2} V_1 + 2 \frac{dv}{dt} \frac{d V_1}{dt} - V_1 \frac{du}{dt} \left(b_1 + \frac{b_2}{a_2} + \text{positive powers}\right) = 0,$$

and therefore

$$\frac{du}{dt} V_1^2 = A e^{\left(b_1 + \frac{b_2}{a_2}\right) t + \text{higher powers}},$$

where A is an arbitrary constant. Thus

$$\frac{du}{dt} = \frac{A}{t^2} \frac{e^{\left(b_1 + \frac{b_2}{a_2}\right) t + \ldots}}{(1 + Bt + \ldots)^2}$$

$$= \frac{A}{t^2} \left\{1 + t\left(-2B + b_1 + \frac{b_2}{a_2}\right) + \ldots\right\}$$

$$= \frac{A}{t^2} \{1 + a_2 b_0 t + \ldots\},$$

$$u = K - \frac{A}{t} + A a_2 b_0 \log t + \text{positive powers},$$

so that the new solution can be taken

$$V_2 = \text{coefficient of } A \text{ in } u V_1$$
$$= a_2 b_0 P(t) t \log t + R(t),$$

where $R(t)$ is a regular function of t such that

$$R(t) = -1 + \text{positive powers.}$$

The general integral of the equation in V is therefore

$$Bt P(t) + C\{R(t) + a_2 b_0 P(t) t \log t\},$$

and the corresponding value of v can be deduced at once. Reverting to the discussion of the case when a_1 is a positive integer, we see that (unless the logarithmic term disappears from this expression) the equation in v possesses no regular integral vanishing with t; and that, if the logarithmic term does disappear, the equation in v possesses an infinitude of regular integrals vanishing with t. The condition for the necessary disappearance is that $a_2 b_0 = 0$ or, as a_2 is not zero, the condition is

$$b_0 = 0.$$

The general proposition of § 69 is accordingly verified for the special case.

Finally, we must take into account the case when there is no term on the right-hand side which is free from v; this occurs when all the coefficients in p_0 vanish, so that $p_0 = 0$, and then the equation is

$$t \frac{dv}{dt} = p_1 v + p_2 v^2,$$

which can be integrated by quadratures. We have

$$\frac{d}{dt}\left(\frac{1}{v}\right) + \frac{p_1}{t} \frac{1}{v} = -\frac{p_2}{t},$$

and therefore

$$\frac{1}{v} e^{\int \frac{p_1}{t} dt} = A - \int \frac{p_2}{t} e^{\int \frac{p_1}{t} dt} dt.$$

Now

$$\int \frac{p_1}{t} dt = \int \frac{a_1}{t} dt + b_1 t + \tfrac{1}{2} c_1 t^2 + \dots$$
$$= a_1 \log t + S(t),$$

where $S(t)$ is a regular function vanishing with t. Also

$$\int \frac{p_2}{t} e^{\int \frac{p_1}{t} dt} dt = \int \left(\frac{a_2}{t} + b_2 + \dots\right) t^{a_1} e^{S(t)} dt$$
$$= \frac{a_2}{a_1} t^{a_1} G(t),$$

where $G(t)$ is a regular function of t such that $G(0) = 1$. Thus

$$\frac{1}{v} t^{a_1} e^{S(t)} = A - \frac{a_2}{a_1} t^{a_1} G(t),$$

and therefore, writing $AC = 1$, we have

$$v = \frac{C t^{a_1} e^{S(t)}}{1 - C \dfrac{a_2}{a_1} t^{a_1} G(t)}.$$

The various inferences in § 74 can be verified.

Ex. 2.   The equation

$$\frac{du}{dz} = \frac{\sin(z-u)}{z},$$

with the condition that $u=0$ when $z=0$, is discussed by Briot and Bouquet, (in their memoir, cited p. 40).

We have

$$z\frac{du}{dz} = z - u - \frac{1}{6}(z-u)^3 + \ldots,$$

so that $b=1$, $a=-1$. Since the coefficient of u is a negative integer, the equation possesses a regular integral that vanishes with z; and this is the only integral vanishing with z that is possessed by the equation.

Moreover, for every finite value of z other than zero, the function $\frac{\sin(z-u)}{z}$ is definite; and it cannot become infinite. Hence the regular integral exists over the whole finite part of the plane; and therefore the power-series which represents the integral in the immediate vicinity of the origin converges absolutely for all finite values of z.

By the substitution $\tan\frac{1}{2}u=v$, the equation can be made an instance of the preceding example. It can also be discussed differently, for it can be definitely integrated in terms of known functions as follows.

Take $v=z-u$; then

$$z\frac{dv}{dz} + \sin v = z.$$

Writing $z=e^{\theta}$, we have

$$\frac{dv}{d\theta} + \sin v = e^{\theta}.$$

Take

$$-\frac{2}{w}\frac{dw}{d\theta} = \eta = e^{\theta}\tan\frac{v}{2};$$

then

$$\frac{d\eta}{d\theta} = e^{\theta}\left(\tan\frac{v}{2} + \frac{1}{2}\sec^2\frac{v}{2}\frac{dv}{d\theta}\right)$$

$$= \tfrac{1}{2}e^{\theta}\sec^2\frac{v}{2}\left(\sin v + \frac{dv}{d\theta}\right)$$

$$= \tfrac{1}{2}e^{2\theta}\sec^2\frac{v}{2}$$

$$= \tfrac{1}{2}(e^{2\theta}+\eta^2),$$

which is a form of Riccati's equation. Also

$$-\frac{2}{w}\frac{d^2w}{d\theta^2} + \frac{2}{w^2}\left(\frac{dw}{d\theta}\right)^2 = \frac{d\eta}{d\theta} = \tfrac{1}{2}e^{2\theta} + \frac{2}{w^2}\left(\frac{dw}{d\theta}\right)^2,$$

and therefore

$$\frac{d^2w}{d\theta^2} + \tfrac{1}{4}e^{2\theta}w = 0.$$

Now Bessel's function $J_0(x)$ satisfies

$$x \frac{d}{dx}\left(x \frac{dy}{dx}\right) + x^2 y = 0 \; ;$$

taking $x = \frac{1}{2} e^\theta$, the equation becomes

$$\frac{d^2 y}{d\theta^2} + \frac{1}{4} e^{2\theta} y = 0.$$

Hence

$$w = A J_0\left(\tfrac{1}{2} e^\theta\right) + B Y_0\left(\tfrac{1}{2} e^\theta\right)$$
$$= A J_0\left(\tfrac{1}{2} z\right) + B Y_0\left(\tfrac{1}{2} z\right).$$

Consequently,

$$\tan \tfrac{1}{2} v = -\frac{2}{w} \frac{dw}{e^\theta d\theta}$$

$$= -\frac{2}{w} \frac{dw}{dz}$$

$$= -\frac{J_0' + c Y_0'}{J_0 + c Y_0} \, ,$$

where c is an arbitrary constant. Now*

$$J_0(x) = 1 - \tfrac{1}{4} x^2 + \tfrac{1}{32} x^4 - \ldots,$$
$$Y_0(x) = J_0 \log x + 2 J_2 - J_4 + \ldots$$
$$= J_0 \log x + \tfrac{1}{4} x^2 - \tfrac{3}{128} x^4 + \ldots \; ;$$

and therefore, for small values of $|x|$, we have

$$J_0 = 1 - \ldots, \qquad Y_0 = \log x + \text{small quantities},$$

$$J_0' = -\tfrac{1}{2} x + \ldots, \qquad Y_0' = \frac{1}{x} + \text{small quantities} \; ;$$

so that, for small values of $|z|$, we have

$$\tan \tfrac{1}{2} v = -\frac{-\tfrac{1}{2} z + \ldots + \dfrac{c}{z} + \ldots}{1 - \ldots + c \log z + \ldots}.$$

Now u is to vanish with z, and therefore $\frac{1}{2} v$ vanishes with z; consequently, the fraction on the right-hand side also must vanish with z. If c is not zero, this fraction

$$= -\frac{c + \text{positive powers of } z}{cz \log z + \text{positive powers of } z}$$

$$= -\frac{1 + \ldots}{z \log z + \ldots} \, ,$$

which becomes infinite in the limit when $z = 0$, instead of zero as it should be. Hence $c = 0$; and we have

$$\tan \tfrac{1}{2} v = -\frac{J_0'}{J_0} = \frac{J_1\left(\tfrac{1}{2} z\right)}{J_0\left(\tfrac{1}{2} z\right)} \, ,$$

which gives

$$u = z - 2 \tan^{-1} \left\{ \frac{J_1\left(\tfrac{1}{2} z\right)}{J_0\left(\tfrac{1}{2} z\right)} \right\} \, ,$$

where that branch of the inverse function is to be chosen which vanishes with its argument.

* See my *Treatise on Differential Equations*, p. 164.

Note. If, instead of the assigned condition that u is to vanish with z, there be a condition that u is to be different from zero when $z = 0$, then $\tan \frac{1}{2} v$ does not become zero with z ; the constant c does not then vanish, and we have

$$\tan \left(-\tfrac{1}{2} u_0 \right) = \text{infinity},$$

that is, u_0 tends to an odd multiple of π. But then we revert to the equation

$$\frac{du}{dz} = \frac{\sin (z - u)}{z},$$

where the value $u =$ odd multiple of π for $z = 0$ makes the right-hand side $= -1$; this combination of values is an ordinary combination for the function.

Ex. 3. Discuss the integral of the equation

$$t \frac{dv}{dt} = av^2 + bt,$$

where a and b are constants.

THE SECOND TYPICAL REDUCED FORM*.

76.   The second of the typical forms to which an equation can be reduced is

$$t \frac{dv}{dt} = gv^m + bt + cvt + \dots$$

$$= \phi_m (v, t),$$

where m is a positive integer greater than unity, and $\phi_m (v, t)$ is a regular function of v and t in the immediate vicinity of 0, 0. Moreover, we shall assume in the first place that $\phi_m (0, t)$ is not zero for all values of t: that is, $\phi_m (v, t)$ contains terms in t which are independent of v.

The argument of § 64, which established the existence of a regular integral for the equation

$$t \frac{dv}{dt} = av + bt + \dots = \phi_1 (v, t),$$

can be applied here, with the sole limitation that $a = 0$. The change that must be made is as regards the quantity θ, which was the least value of $| m - a |$ for $m = 1, 2, 3, \dots$; manifestly, θ must be replaced by unity for the present case.

The dominant function V is now given by

$$V = M \left\{ \frac{1}{\left(1 - \dfrac{t}{r} \right) \left(1 - \dfrac{V}{\rho} \right)} - 1 - \frac{V}{\rho} \right\},$$

---

* The only discussion of this form which is known to me is the discussion in the memoir by Briot and Bouquet, cited p. 40 (*l.c.*, pp. 178—181).

that root of the quadratic being chosen which $= 0$ when $t = 0$: so that

$$2\frac{M+\rho}{\rho^2} V = 1 - \left\{1 - 4M\frac{t}{r-t}\frac{\rho+M}{\rho^2}\right\}^{\frac{1}{2}},$$

which can be represented as a regular function for values of t within the circle defined by

$$\left|\frac{t}{r-t}\right| = \frac{\rho^2}{4M(M+\rho)}.$$

The substitution of unity for θ does not affect the essence of the remaining part of the discussion: and therefore it is inferred that *an integral of the equation*

$$t\frac{dv}{dt} = \phi_m(v, t)$$

exists, which vanishes when $t = 0$ and is a regular function of t in the immediate vicinity of the origin.

Suppose that in $\phi_m(v, t)$ the term, which is independent of v and contains the lowest power of t, is $a_{0,n}t^n$, so that

$$\phi_m(0, t) = a_{0,n}t^n (1 + \text{powers of } t);$$

then it is easy to see that the first term in the expansion of this regular integral, the existence of which has been established, is

$$\frac{1}{n}a_{0,n}t^n.$$

Let u denote this regular integral.

77. In order to investigate whether the regular integral thus obtained is unique as a regular integral and whether other integrals exist, which vanish with t but are not regular, let

$$v = u + U,$$

so that U (whether a regular function of t or not) is to vanish with t. We have

$$t\frac{dU}{dt} = \phi_m(u + U, t) - \phi_m(u, t)$$
$$= U\{gU^{m-1} + \dots + mu^{m-1}g + ct + \dots\},$$

where the unexpressed terms after ct are of higher order in U, u, t combined than those which are retained.

Now U is to vanish with t. If, for sufficiently small values of $|t|$, U can be expressed as of order θ in powers of t, then

$t \dfrac{dU}{dt}$ is also of order θ in powers of t for that range. Now m is at least equal to 2, and u is a regular function of t which vanishes with t; so that the successive terms on the right-hand side are of order

$$m\theta, \ (m-1)\theta + \delta, \ (m-2)\theta + 2\delta, \ldots, \ \theta + (m-1)\delta, \ \theta + 1, \ldots,$$

where δ is the order of the lowest term in u, so that $\delta \geqslant 1$. Consequently,

$$\theta = m\theta, \ \text{or} \ \theta = \theta + 1,$$

according as $m\theta$ or $\theta + 1$ is the smaller. The second alternative makes θ infinite; so that, if

$$U = t\eta,$$

then η is also of infinite order in powers of t and it vanishes with t. The equation for η is

$$t \frac{d\eta}{dt} = \eta\left(-1 + g\eta^{m-1}t^{m-1} + \ldots + mu^{m-1}g + ct + \ldots\right).$$

The theorems of § 74 shew that the only solution of this equation, which vanishes when $t = 0$, is $\eta = 0$; and then $U = 0$. Hence no integral of the original equation, other than the regular integral, can be deduced by assuming an infinite value for θ.

The other alternative, viz. $\theta = m\theta$ where $m\theta < \theta + 1$, must be considered. Since the integer $m \geqslant 2$, it follows that $\theta = 0$, which satisfies the condition $m\theta < \theta + 1$; and then the order of U in powers of t would be zero. To obtain the significance of this result, let

$$t = \psi U^m,$$

so that, as U is of zero order in powers of t, the new variable ψ vanishes with t; it is of order unity in powers of t and therefore is of infinite order in powers of U. Moreover, the regular function u vanishes with t; hence, when this substitution is made for t in the expression

$$U\left(gU^{m-1} + \ldots + mu^{m-1}g + ct + \ldots\right),$$

the new form is divisible by U^m; and so

$$\frac{U^m}{\psi} \frac{d\psi}{dU} = -mU^{m-1} + \frac{1}{g + c\psi U + \ldots}$$
$$= g_0 + \text{powers of } U + \psi P(\psi, U)$$
$$= g_0 + UQ(U) + \psi P(\psi, U),$$

where Q and P are regular functions of their arguments.

Let
$$g_0 + UQ(U) = g_0 + g_1 U + g_2 U^2 + \ldots + g_{m-1} U^{m-1} + g_m U^m + \ldots,$$
where some of the coefficients after g_0 may be zero; and let
$$\phi(U) = \frac{g_0}{(m-1) U^{m-1}} + \frac{g_1}{(m-2) U^{m-2}} + \ldots + \frac{g_{m-2}}{U}.$$
Then
$$\frac{1}{\psi}\frac{d\psi}{dU} + \frac{d\phi(U)}{dU} = \frac{g_{m-1}}{U} + g_m + \text{powers of } U + \frac{\psi}{U^m} P(\psi, U),$$
the last term of which is infinitesimal for small values of U on account of the relative orders of ψ and U. Taking
$$\psi = \mu e^{-\phi(U)} U^{g_m-1},$$
the equation for μ is
$$\frac{d\mu}{dU} = \mu \left[g_m + g_{m+1} U + \ldots + \mu e^{-\phi(U)} U^{g_m-1-m} P(\psi, U) \right].$$

As ψ tends towards zero, μ can tend to an arbitrary value different from zero, say μ_0, only if $e^{-\phi(U)} U^{g_m-1}$ tends to zero with U; and because
$$t = \mu e^{-\phi(U)} U^{m+g_m-1},$$
t also tends towards zero in that case; but, as is not difficult to see, its argument becomes infinite. The preceding equation defines μ as a function of U, which acquires a value μ_0 as U tends to zero.

The relation between the integral and the equation is therefore
$$v = u + U,$$
$$t = \mu e^{-\phi(U)} U^{m+g_m-1}.$$

In order that v may vanish with t, it is necessary that the argument of t should increase indefinitely as its modulus decreases to zero; and then μ tends to an arbitrary value with decrease of U. *It is only for such variations of t that any integral of the equation, vanishing with t and distinct from the regular integral, can exist; but for general variations of t from zero, there is no integral other than the regular integral which vanishes with t.*

In order that the integral v may vanish with t, in accordance with the prescribed condition, U also must vanish with t. Now clearly $U = 0$ is an essential singularity of the function which is equal to t; and it is a known property[*] that, in the immediate vicinity of an essential singularity, a function acquires every possible value and therefore also a zero value. The function however is not regular at the point; in fact, for all integrals other than the regular integral, the value $t = 0$ is a point of indeterminateness.

To examine the relation between t and U more closely, let $U = re^{\theta i}$, $g_n = \Gamma_n e^{i\gamma_n}$, for $n = 0, 1, \ldots, \mu = Me^{\lambda i}$. Then the modulus of t is

$$Mr^m e^{-\frac{\Gamma_0 \cos\{(m-1)\theta - \gamma_0\}}{(m-1)r^{m-1}}(1+\epsilon) + \Gamma_{m-1}(\cos\gamma_{m-1}\log r - \theta\sin\gamma_{m-1})},$$

where ϵ is an infinitesimal quantity vanishing with r. The argument of t is

$$\lambda + (m + \Gamma_{m-1}\cos\gamma_{m-1})\,\theta + \Gamma_{m-1}\sin\gamma_{m-1}\log r$$
$$+ \sum_{n=1}^{m-1} \frac{\Gamma_{n-1}}{(m-n)} \frac{\sin\{(m-n)\theta - \gamma_{n-1}\}}{r^{m-n}},$$

so that the condition, necessary to secure that $|t|$ vanishes, is that $\cos\{(m-1)\theta - \gamma_0\}$ is positive.

If g_{m-1} is real so that γ_{m-1} is a multiple of π, and if at the same time $(m-n)\theta - \gamma_{n-1}$ is a multiple of π for all the $m-1$ values of n, then (and only then) does the argument of t cease to be infinite and become finite. Also $|t|$ will still tend to zero with U, and the argument of t will tend to a finite value, if the value of θ be such as to give

$$\theta = \frac{\gamma_0}{m-1} + r^{m-1}P(r),$$

where $P(r)$ is a uniform finite function of r. Hence, if U approaches a zero value along any curve determined by the last equation, an integral of the equation can exist which vanishes with t and is distinct from the regular integral.

78.   The preceding discussion, which accords with that given by Briot and Bouquet (*l.c.*), is rather unsatisfactory: for no account is effectively taken of the part

$$\frac{\psi}{U^m} P(\psi, U),$$

or its equivalent, in the expression for $\dfrac{d\mu}{dU}$: certainly it cannot be considered that the discussion of the non-regular integrals of

$$t\frac{dv}{dt} = \phi_m(v, t)$$

has been so completely carried out as the discussion of the non-regular integrals of

$$t\frac{dv}{dt} = \phi_1(v, t).$$

It may prove possible, in spite of obvious initial difficulties, to proceed from the case of

$$t\frac{dv}{dt} = \phi_1(v, t),$$

where a is a small positive quantity—so that the non-regular integrals are regular functions of t and t^a—to the limit when a becomes zero; but I have not carried out the investigation. External considerations suggest that there is substantial difference between the case $m = 2$, and the cases $m > 2$.

79. As an example, consider the equation

$$t\frac{dv}{dt} = p_0 + p_1 v + p_2 v^2,$$

where

$$p_0 = \qquad b_0 t + c_0 t^2 + \dots,$$
$$p_1 = \qquad b_1 t + c_1 t^2 + \dots,$$
$$p_2 = a_2 + b_2 t + c_2 t^2 + \dots,$$

and assume that a_2 is not zero ; then the integer m is 2, so that the equation takes the form

$$t\frac{dv}{dt} = a_2 v^2 + b_0 t + \dots.$$

Let

$$v = -\frac{t}{p_2}\frac{1}{V}\frac{dV}{dt},$$

so that the equation for V is

$$\frac{d^2 V}{dt^2} + \frac{dV}{dt}\left(\frac{1}{t} - \frac{p_1}{t} - \frac{1}{p_2}\frac{dp_2}{dt}\right) + \frac{p_0 p_2}{t^2}V = 0,$$

that is,

$$\frac{d^2 V}{dt^2} + \frac{dV}{dt}\left(\frac{1}{t} - b_1 - \frac{b_2}{a_2} + \dots\right) + \left(\frac{a_2 b_0}{t} + a_2 c_0 + b_0 b_2 + \dots\right)V = 0.$$

The indicial equation for the singularity $t = 0$ of this linear equation of the second order is

$$m(m-1) + m = 0,$$

that is, $$m^2 = 0.$$

There is one regular integral, viz.

$$V_1 = 1 - a_2 b_0 t + \text{higher powers of } t.$$

To obtain the other independent integral, write

$$V = V_1 W,$$

so that

$$V_2 \frac{d^2 W}{dt^2} + 2 \frac{dV_1}{dt} \frac{dW}{dt} + V_1 \frac{dW}{dt} \left(\frac{1}{t} - \frac{p_1}{t} - \frac{1}{p_2} \frac{dp_2}{dt} \right) = 0.$$

Now

$$\frac{p_1}{t} = b_1 + c_1 t + \dots = \frac{dP}{dt},$$

say, where P is a regular function that vanishes with t; then

$$V_1{}^2 \frac{dW}{dt} \frac{t}{p_2} e^{-P} = A,$$

that is,

$$\frac{dW}{dt} = A \frac{p_2}{t} e^P V_1{}^{-2}$$

$$= A \left\{ \frac{a_2}{t} + \text{powers of } t \right\},$$

and therefore

$$W = B + A \{ a_2 \log t + R(t) \},$$

where $R(t)$ is a regular function of t, vanishing when $t = 0$. Thus

$$V = V_1 W$$

$$= B V_1 + A V_1 \{ a_2 \log t + R(t) \}$$

$$= B V_1 + A (a_2 V_1 \log t + V_2),$$

where V_2 is a regular function of t, vanishing when $t = 0$. The value of v therefore is

$$v = - \frac{t}{p_2} \frac{1}{V} \frac{dV}{dt}$$

$$= - \frac{1}{p_2} \frac{Bt \dfrac{dV_1}{dt} + A \left\{ a_2 V_1 + a_2 t \dfrac{dV_1}{dt} \log t + t \dfrac{dV_2}{dt} \right\}}{B V_1 + A a_2 V_1 \log t + A V_2}.$$

When $A = 0$, this value of v is a regular function of t which vanishes with t; it is the regular integral which vanishes with t; and manifestly it is the unique regular integral.

When A is not zero, the value of v is not a regular function of t. If a branch of the logarithmic function be taken which has only a finite argument, then the value of v tends to the limit zero as t tends to 0; for the numerator of v is finite, and the denominator tends to an infinite limit, as t becomes infinitesimal. It therefore follows that there are branches of the non-regular integral which tend to the value 0 as t tends to the origin. The origin is, in fact, a point of indeterminateness for the integrals that are not regular functions of t.

We may, by another method of considering this example, follow more closely the argument for the more general case. We have

$$t \frac{dv}{dt} = a_2 v^2 + b_0 t + c_0 t^2 + b_1 vt + \dots;$$

the regular integral which vanishes with t is

$$u = b_0 t + \tfrac{1}{2}(a_2 b_0^2 + b_1 b_0 + c_0) t^2 + \dots$$

To find the other integrals, if any, which vanish with t, we take

$$v = u + U,$$

where U is to vanish with t; so that

$$t \frac{dU}{dt} = U(p_1 + 2p_2 u + p_2 U)$$

$$= U(a_2 U + \lambda_1 t + b_2 t U + \dots),$$

where the unexpressed terms are higher powers of t, or are combinations of U and higher powers of t; and where

$$\lambda_1 = b_1 + 2a_2 b_0.$$

Let $t = \psi U^2$; then
$$\frac{1}{\psi}\frac{d\psi}{dU} + \frac{2}{U} = \frac{1}{(a_2 + \lambda_1 \psi U + \dots)\, U^2}$$

$$= \frac{1}{a_2 U^2} - \frac{\lambda_1}{a_2^2}\frac{\psi}{U} + \dots,$$

and therefore, taking

$$\psi U^2 e^{\frac{1}{a_2 U}} = \mu,$$

we have

$$\frac{1}{\mu}\frac{d\mu}{dU} = -\frac{\lambda_1}{a_2^2}\frac{\psi}{U} P(\psi,\ U),$$

where P is a regular function of ψ and U. Also

$$t = \mu e^{-\frac{1}{a_2 U}},$$

and the equation for μ is

$$\frac{d\mu}{dU} = -\frac{\lambda_1}{a_2^2}\frac{\mu^2}{U^3} e^{-\frac{1}{a_2 U}} P\left(\frac{\mu}{U^2} e^{-\frac{1}{a_2 U}},\ U\right).$$

Evidently $U = 0$ is an essential singularity of t, while μ does not necessarily tend to an infinitesimal value when $t = 0$; integrals U, not zero for values of t other than 0 and distinct from the regular integral but vanishing when $t = 0$, exist; but they acquire the zero value only as one among an infinite number of others connected with the essential singularity. The point $t = 0$ is a point of indeterminateness for all integrals of the equation other than the regular integral.

80.　It has been assumed that, in the equation

$$t \frac{dv}{dt} = \phi_m(v,\ t),$$

terms occur on the right-hand side independent of v, so that $\phi_m(0,\ t)$ does not vanish for all values of t. In the cases where this assumption is not justified, v (or some positive power of v) is a factor of $\phi_m(v, t)$, so that we can have

$$t \frac{dv}{dt} = v^n \psi(v, t),$$

where $\psi(0, t)$ does not vanish for all values of t. The simplest case is naturally that for which $n = 1$; it is the form which has arisen in the discussion of the function U.

The only regular integral, which vanishes when $t = 0$, is given by $v = 0$. The origin, in general, is a point of indeterminateness; and there can be branches of non-regular integrals which tend to the value zero as t tends to its origin.

As an example, consider

$$t \frac{dv}{dt} = P_1 v + P_2 v^m,$$

where

$$P_1 = \qquad b_1 t + c_1 t^2 + \ldots,$$
$$P_2 = a_2 + b_2 t + c_2 t^2 + \ldots.$$

The equation can be integrated by quadratures; we have

$$t \frac{d}{dt} \left(\frac{1}{v^{m-1}} \right) + (m-1) P_1 \frac{1}{v^{m-1}} = -(m-1) P_2.$$

Let

$$Q_1 = \int (m-1) \frac{P_1}{t} dt$$

$$= (m-1) b_1 t + \tfrac{1}{2} (m-1) c_1 t^2 + \ldots,$$

a regular function of t; then

$$\frac{1}{v^{m-1}} e^{Q_1} = A - (m-1) \int \left(\frac{a_2}{t} + b_2 + \ldots \right) e^{Q_1} dt$$

$$= A - (m-1) a_2 \log t + R(t),$$

where $R(t)$ is a regular function of t. Hence writing $AC = 1$, we have

$$v^{m-1} = \frac{C e^{Q_1}}{1 - C(m-1) a_2 \log t + CR(t)}.$$

Manifestly, the only regular integral vanishing with t is obtained by putting $C = 0$; it is an integral which is steadily zero in the vicinity of t.

It is clear that, as t tends to the value zero, branches of the non-regular integral tend to the value zero. The point $t = 0$ is, for the non-regular integrals, a point of indeterminateness with definite branching if $m > 2$; if $m = 2$, the point $t = 0$ is a point of indeterminateness, but without cyclical branching.

THE THIRD TYPICAL REDUCED FORM*.

81. The next of the reduced typical forms that arises for consideration is

$$t^s \frac{dv}{dt} = av + bt + \dots$$

$$= \phi_1 (v, t),$$

where the integer s is greater than unity.

It is not difficult to infer that, in general, the point $t = 0$ is a point of indeterminateness for the integral. We associate a new variable η with the equation, determined by

$$\eta = e^{\frac{1}{1-s} \frac{1}{t^{s-1}}},$$

and we have the system of equations

$$\frac{d\eta}{\eta} = \frac{dt}{t^s} = \frac{dv}{\phi_1 (v, t)};$$

with this system, we associate the partial differential equation

$$\eta \frac{\partial U}{\partial \eta} + t^s \frac{\partial U}{\partial t} + \phi_1 (v, t) \frac{\partial U}{\partial v} = 0.$$

If $U(v, t, \eta)$ be any integral of this equation, then

$$U(v, t, \eta) = \text{constant},$$

---

* In their memoir, which has frequently been quoted, Briot and Bouquet advert only very briefly to this form. The only papers known to me, which deal with it, are :—

> Horn, *Ueber das Verhalten der Integrale von Differentialgleichungen bei der Annäherung der Veränderlichen an eine Unbestimmtheitsstelle*, Crelle, t. cxviii (1897), pp. 257—274 : *ib.*, t. cxix (1898), pp. 196—209, 267—291 : (in the first part, an infinite value of the independent variable is considered chiefly in connection with the Riccati equation
> $$x^{-k} \frac{dy}{dx} = Ay^2 + By + C,$$
> where A, B, C are regular functions of $1/x$; in the later parts, the variables are restricted to be real); and

> Bendixson, *Sur les points singuliers des équations différentielles*, Stockh. Öfver., t. lv (1898), pp. 69—85; this paper restricts the discussion to real variables, and is a development of the investigations quoted at the end of the discussion of Ex. 1 in § 60.

is an integral of the two equations in differential elements; when we combine the relation between η and t with the relation $U = $ constant, it gives a solution of the original equation

$$t^s \frac{dv}{dt} = \phi_1(v, t).$$

Now the general existence-theorem of partial differential equations shews that integrals of the form $U(v, t, \eta)$ exist. When any one such has been obtained involving v, then the equation

$$U(v, t, \eta) = \text{constant},$$

will in general—that is, save for special values of the constant or for special forms of U—determine v as a function of t and η, with one or more values. For each value, $t = 0$ is a point of indeterminateness of the solution obtained for v, because it is an essential singularity of η; that is, $t = 0$ is, in general, a point of indeterminateness for the solution of the original equation.

It may happen that, for particular equations, there are special values of the constant in

$$U(v, t, \eta) = \text{constant},$$

or there may be special forms of U, such that the relation determines v as a function of t only and not of η; for each such determination, the point $t = 0$ ceases to be a point of indeterminateness for the equation.

Examples will be given immediately.

82. Briot and Bouquet assert that the equation cannot, in general, have a regular integral by shewing that, even when taken in its simplest form, it does not unconditionally possess a regular integral. The form adopted for this purpose is

$$t^2 \frac{dv}{dt} = av + P(t),$$

where $P(t)$ is a regular function of t. It can be integrated by quadratures in the form

$$v = A e^{-\frac{a}{t}} + e^{-\frac{a}{t}} \int e^{\frac{a}{t}} \frac{1}{t^2} P(t) \, dt$$

$$= A e^{-\frac{a}{t}} - \frac{1}{a} \left\{ P(t) + \left(\frac{t^2}{a} \frac{d}{dt} \right) P(t) + \left(\frac{t^2}{a} \frac{d}{dt} \right)^2 P(t) + \ldots \right\}$$

Let A be different from zero; then, even without regarding the infinite series, the point $t = 0$ is an essential singularity of the integral, so that the point is one of indeterminateness for the general integral.

Let A be zero; the integral then obtained is a regular integral provided the series converges; otherwise there is no regular integral vanishing with t.

To discuss the convergence of the series, let

$$- \frac{1}{a} \left\{ 1 + \frac{t^2}{a} \frac{d}{dt} + \left(\frac{t^2}{a} \frac{d}{dt} \right)^2 + \ldots \right\} P(t) = \sum_{n=1}^{\infty} C_n t^n,$$

and suppose that

$$P(t) = \sum_{m=1}^{\infty} c_m t^m.$$

Then

$$\frac{t^2}{a} \frac{d}{dt} P(t) = \frac{1}{a} \sum_{m=1}^{\infty} m c_m t^{m+1},$$

$$\left(\frac{t^2}{a} \frac{d}{dt} \right)^2 P(t) = \frac{1}{a^2} \sum_{m=1}^{\infty} m(m+1) c_m t^{m+2},$$

and so on; so that

$$- a C_n = c_n + (n-1) \frac{c_{n-1}}{a} + (n-1)(n-2) \frac{c_{n-2}}{a^2} + \ldots,$$

and therefore

$$\frac{a^n}{(n-1)!} C_n = \frac{c_n a^{n-1}}{(n-1)!} + \frac{c_{n-1} a^{n-2}}{(n-2)!} + \ldots + c_2 a + c_1$$

$$= S_n,$$

say. The series $\sum_{n=1}^{\infty} C_n t^n$ must converge absolutely in order that the regular integral may exist; in order that this may be the case, the limit* of the quantity $|C_n|^{\frac{1}{n}}$, as n becomes infinitely great, must be finite; that is,

$$\text{Lim}_{n=\infty} |n! \, S_{n+1}|^{\frac{1}{n}}$$

* Chrystal's *Algebra*, vol. II, p. 107.

must be finite. But, for very large values of n, Stirling's theorem[*] gives

$$n! = (2\pi n)^{\frac{1}{2}} n^n e^{-n} \left\{ 1 + \frac{1}{12n} + \dots \right\},$$

so that

$$|n!|^{\frac{1}{n}} = \frac{1}{e} (2\pi n)^{\frac{1}{2n}} n \left\{ 1 + \frac{1}{12n} + \dots \right\}^{\frac{1}{n}}.$$

$$> \frac{n}{e},$$

whence

$$\underset{n=\infty}{\text{Lim}} |n! \, S_{n+1}|^{\frac{1}{n}} > \underset{n=\infty}{\text{Lim}} \frac{n}{e} |S_{n+1}|^{\frac{1}{n}}.$$

Now the quantity on the left-hand side is to remain finite in the limit when n is infinite, say C; thus

$$\underset{n=\infty}{\text{Lim}} |S_{n+1}|^{\frac{1}{n}} = \underset{n=\infty}{\text{Lim}} \frac{eC}{n} = 0,$$

and therefore

$$\underset{n=\infty}{\text{Lim}} S_{n+1} = 0.$$

The condition is necessary: it is also sufficient, when $|t|$ is small enough. For because

$$P_1(a) = c_1 + c_2 a + \dots + \frac{c_n}{(n-1)!} a^{n-1} + \frac{c_{n+1}}{n!} a^n + \dots = 0,$$

it follows that

$$C_n = \frac{c_{n+1}}{n} + \frac{a c_{n+2}}{n(n+1)} + \dots.$$

Let $|a| = \alpha$; and let β denote the greatest value of $|c_m^{\frac{1}{m}}|$ for all values of m, so that β is a finite quantity. Then

$$|C_n| < \frac{\beta^n}{n} + \frac{\alpha \beta^{n+1}}{n(n+1)} + \frac{\alpha^2 \beta^{n+2}}{n(n+1)(n+2)} + \dots$$

$$< \frac{\beta^n}{n} + \frac{\alpha \beta^{n+1}}{n^2} + \frac{\alpha^2 \beta^{n+2}}{n^3} + \dots$$

$$< \frac{\beta^n}{n} \frac{1}{1 - \frac{\alpha \beta}{n}}.$$

For some finite value of n and for all subsequent values of n, $1 - \frac{\alpha\beta}{n}$ is positive and it increases with increasing values of n.

[*] Boole's *Finite Differences*, p. 94.

Let p be some finite value such that $1 - \dfrac{\alpha\beta}{p} = \theta$, where θ is positive; then for all values of n equal to p or greater than p,

$$|C_n| < \frac{1}{\theta} \frac{\beta^n}{n}.$$

Hence

$$\left| \sum_{n=p}^{\infty} C_n t^n \right| < \frac{1}{\theta} \sum_{n=p}^{\infty} \frac{1}{n} \beta^n |t|^n,$$

a series which converges if $\beta|t| < 1$. Also S_n is a finite quantity for all values of $n \leqslant p$, and therefore also C_n is finite for those values; consequently, for all values of t such that $|t| < \dfrac{1}{\beta}$, the series

$$\sum_{n=1}^{\infty} C_n t^n$$

converges, provided that a satisfies the equation

$$P_1(a) = 0.$$

A regular integral then exists which vanishes with t. But if a does not satisfy this critical equation, then no regular integral exists, for the series $\sum\limits_{n=1}^{\infty} C_n t^n$ does not converge.

It will be seen that, if $c_1 = 0$, then $a = 0$ is a solution of

$$P_1(a) = 0.$$

The particular equation then becomes

$$t^2 \frac{dv}{dt} = P(t) = t^2(c_2 + c_3 t + \ldots),$$

which is trivial; but there is thus suggested the form

$$t^2 \frac{dv}{dt} = bv^2 + t^2(c_0 + c_1 t + \ldots)$$

$$= bv^2 + t^2 P(t),$$

which will be discussed later (§ 83).

Ex. 1. Discuss, in the same way, the equation

$$t^\kappa \frac{dv}{dt} = av + P(t),$$

when κ is an integer greater than 2.

Ex. 2. As a similar example, consider $t^s \dfrac{dv}{dt} = av + bt + cvt$, where $s \geqslant 2$.

If a regular integral exists, vanishing with t, we must have

$$av + bt = A_2 t^2 + A_3 t^3 + \ldots,$$

and therefore

$$t^s \{ -b + 2A_2 t + 3A_3 t^2 + \ldots \} = a (A_2 t^2 + A_3 t^3 + A_4 t^4 + \ldots)$$
$$+ ct (-bt + A_2 t^2 + A_3 t^3 + \ldots).$$

If $s = 2$, we have

$$nA_n = aA_{n+1} + cA_n, \quad -b = aA_2 - bc,$$

so that

$$a^n A_{n+1} = - (n - c)(n - 1 - c) \ldots (2 - c)(1 - c) b \, ;$$

the series $A_2 t^2 + A_3 t^3 + \ldots$ becomes

$$-ba \left[(1 - c) \frac{t^2}{a^2} + (1 - c)(2 - c) \frac{t^3}{a^3} + (1 - c)(2 - c)(3 - c) \frac{t^4}{a^4} + \ldots \right]$$

which is certainly a diverging series unless c is a positive integer : but if c is a positive integer, the series terminates with a term in $\dfrac{t^c}{a^c}$.

Consequently, the equation has a regular integral only if c be a positive integer, when $s = 2$. This result, with this condition, appears also as follows. We have

$$\frac{dv}{dt} - v \left(\frac{a}{t^2} + \frac{c}{t} \right) = \frac{b}{t},$$

so that

$$v e^{\frac{a}{t}} t^{-c} - A = \int \frac{b}{t^{1+c}} e^{\frac{a}{t}} \, dt.$$

Now, integrating by parts, we have

$$\int \frac{1}{t^{1+c}} e^{\frac{a}{t}} \, dt = - \frac{1}{a} \frac{e^{\frac{a}{t}}}{t^{c-1}} - \frac{(1 - c)}{a} \int \frac{e^{\frac{a}{t}}}{t^c} \, dt,$$

$$\int \frac{e^{\frac{a}{t}}}{t^c} \, dt = - \frac{1}{a} \frac{e^{\frac{a}{t}}}{t^{c-2}} - \frac{2 - c}{a} \int \frac{e^{\frac{a}{t}}}{t^{c-1}} \, dt,$$

and so on in succession, the series of operations being finite in number when c is a positive integer. The regular integral is given by taking $A = 0$; for the general integral, the point $t = 0$ is an essential singularity.

If $s > 2$, there is no regular integral vanishing with t, unless b and c vanish ; the point $t = 0$ is a point of indeterminateness for the integral.

Ex. 3. Shew that, for the equation

$$t^s \frac{dv}{dt} = av + bt + kvt^{s-1},$$

there is a regular integral vanishing with t, if k is a positive integer multiple of $s - 1$.

Ex. 4.   Consider, more generally, the equation

$$t^s \frac{dv}{dt} = p_0 + p_1 v + p_2 v^2,$$

where $s > 1$, and p_0 does not vanish.   Take

$$v = -\frac{t^s}{p_2} \frac{1}{V} \frac{dV}{dt};$$

then the equation for V is

$$\frac{d^2 V}{dt^2} + \frac{dV}{dt}\left(-\frac{p_1}{t^s} + \frac{s}{t} - \frac{1}{p_2}\frac{dp_2}{dt}\right) + \frac{p_0 p_2}{t^{2s}} V = 0.$$

For the present purpose,

$$p_0 = \quad b_0 t + c_0 t^2 + \ldots,$$
$$p_1 = a_1 + b_1 t + c_1 t^2 + \ldots,$$
$$p_2 = a_2 + b_2 t + c_2 t^2 + \ldots;$$

so that, as $s \geqslant 2$, the numbers for the characteristic index are given by

$$0+2, \quad 1+s, \quad 2s-1.$$

When $s > 2$, (and a_2 and b_0 do not vanish), the last of these numbers is the greatest: the determining function is a constant, and the linear equation then has no regular* integrals, that is, integrals such that

$$Vt^{-a} = \text{uniform function of } t.$$

But the equation in V might have sub-regular* integrals.

When $s = 2$ (and a_2 and b_0 do not vanish), the last two numbers are equal; there may be one regular integral of the equation in V.

In particular, take $s = 2$; then the equation for V becomes

$$\frac{d^2 V}{dt^2} - Q_1 \frac{dV}{dt} + Q_2 V = 0,$$

where

$$Q_1 = \frac{a_1}{t^2} + \frac{b_1 - 2}{t} + c_1 + \frac{b_2}{a_2} + \text{positive powers of } t,$$

$$Q_2 = \frac{a_2 b_0}{t^3} + \frac{a_2 c_0 + b_0 b_2}{t^2} + \frac{b_0 c_2 + c_0 b_2 + d_0 a_2}{t} + \text{positive powers.}$$

The equation in V may possibly possess one regular integral, since the lowest index of Q_1 is greater only by 1 than the index of Q_2.   To find it, write

$$V = t^m \Theta;$$

the equation for Θ is

$$\frac{d^2 \Theta}{dt^2} - P_1 \frac{d\Theta}{dt} + P_2 \Theta = 0,$$

where

$$P_1 = Q_1 - \frac{2m}{t} = \frac{a_1}{t^2} + \frac{b_1 - 2 - 2m}{t} + \left(c_1 + \frac{b_2}{a_2}\right) + \ldots,$$

$$P_2 = Q_2 - \frac{m}{t} Q_1 + \frac{m(m-1)}{t^2}$$

$$= \frac{a_2 b_0 - m a_1}{t^3} + \frac{a_2 c_0 + b_0 b_2 - m(b_1 - 2) + m(m-1)}{t^2} + \ldots.$$

* In the same sense as on p. 73.

Let $a_2 b_0 - m a_1 = 0$; then the equation for Θ is

$$t^2 \frac{d^2\Theta}{dt^2} - \frac{d\Theta}{dt}(a_1 + \text{positive powers})$$

$$+ \Theta \{a_2 c_0 + b_0 b_2 - b_1 m + m(m+1) + \text{positive powers}\} = 0.$$

A solution of this, in the form of a regular function of t not vanishing when $t = 0$, is given by

$$\Theta_1 = 1 - \frac{a_2 c_0 + b_0 b_2 - b_1 m + m(m+1)}{a_1} t + \dots,$$

provided this series converges. To settle this proviso, let

$$t^2 P_1 = a_1 + B_1 t + B_2 t^2 + \dots,$$

$$t^2 P_2 = C_0 + C_1 t + C_2 t^2 + \dots,$$

so that

$$t^2 \frac{d^2\Theta}{dt^2} - \frac{d\Theta}{dt}(a_1 + B_1 t + B_2 t^2 + \dots) + \Theta(C_0 + C_1 t + C_2 t^2 + \dots) = 0.$$

Now Θ, if a regular function of t, is expressible in the form

$$\Theta = J_0 + J_1 t + J_2 t^2 + \dots,$$

where the coefficients J are constants. Substituting and making the coefficient of t^n vanish, we find

$$a_1(n+1) J_{n+1} = \{n(n-1) - n B_1 + C_0\} J_n + \sum_{r=1}^{n} \{C_r - B_{r+1}(n-r)\} J_{n-r},$$

for $n = 2, 3, \dots$; while for $n = 0, 1$, we have

$$J_1 a_1 = J_0 C_0,$$

$$2 J_2 a_1 = J_0 C_1 + J_1(C_0 - B_1).$$

When the values of the coefficients J_n are obtained in succession, it is clear that, if the consequent series for Θ proceeds to infinity and if no relations subsist among the coefficients, then the expression for Θ is a diverging series and is consequently of no significance: that is, the regular integral of the equation in V would not exist.

But for special relations the series may converge; particularly if, for all values of n from and after one particular value, the coefficients J vanish, the series converges. There then exists a function Θ, say $\Theta = \Theta_1$; and a regular integral of the equation in V is

$$V = t^{\frac{a_2 b_0}{a_1}} \Theta_1.$$

The equation may possess one sub-regular integral. To find it, write

$$V = e^{\Omega} U,$$

where Ω is a function of t at our disposal; the equation for U is

$$\frac{d^2 U}{dt^2} + (2\Omega' - Q_1)\frac{dU}{dt} + U(\Omega'' + \Omega'^2 - Q_1 \Omega' + Q_2) = 0.$$

We take Ω' an algebraical function of negative powers, with index greater than 1, so as to annihilate the terms of greatest negative index in the coefficient of U; that is, we take

$$\Omega' = \frac{a_1}{t^2},$$

and so

$$\Omega = -\frac{a_1}{t}.$$

With this value of Ω', the equation in U is

$$\frac{d^2U}{dt^2} + R_1\frac{dU}{dt} + R_2 U = 0,$$

where

$$R_1 = \frac{a_1}{t^2} - \frac{b_1 - 2}{t} - \left(c_1 + \frac{b_2}{a_2}\right) - \dots,$$

$$R_2 = \frac{a_2 b_0 - a_1 b_1}{t^3} + \left(a_2 c_0 + b_0 b_2 - a_1 c_1 - \frac{a_1 b_2}{a_2}\right)\frac{1}{t^2} + \dots.$$

The equation in U may have a regular integral in the form

$$U = t^{\frac{a_2 b_0}{a_1} - b_1}\,\Theta_2,$$

where Θ_2 is a regular function of t not vanishing with t; and then a sub-regular integral of the equation in V is

$$V = t^{\frac{a_2 b_0}{a_1} - b_1}\, e^{-\frac{a_1}{t}}\,\Theta_2.$$

Consequently, if the conditions for the convergence of the respective series are satisfied, the general solution of the equation in V is

$$V = t^{\frac{a_2 b_0}{a_1}}\,(A\Theta_1 + Bt^{-b_1}\,e^{-\frac{a_1}{t}}\,\Theta_2);$$

and therefore

$$v = -\frac{t^2}{p_2}\left[\frac{a_2 b_0}{a_1 t} + \frac{A\dfrac{d\Theta_1}{dt} + Bt^{-b_1}e^{-\frac{a_1}{t}}\left\{\left(\dfrac{a_1}{t^2} - \dfrac{b_1}{t}\right)\Theta_2 + \dfrac{d\Theta_2}{dt}\right\}}{A\Theta_1 + Bt^{-b_1}e^{-\frac{a_1}{t}}\Theta_2}\right].$$

When $B = 0$, the value of v is a regular function of t which vanishes when $t = 0$; there is, accordingly, a regular solution of the original equation which vanishes with t; its expansion begins with

$$v = -\frac{b_0}{a_1}t + \dots;$$

and manifestly it is the only regular solution of this kind.

When B and A are distinct from zero, the function, which is equal to v, has $t = 0$ for an essential singularity. Accordingly, among its values will be found zero values, and the integral is not regular; the point $t = 0$ is a point of indeterminateness for the integral.

When B is distinct from zero and A vanishes, then $t = 0$ ceases to be an essential singularity of the function. We then have

$$v = -\frac{a_2 b_0}{a_1 p_2}t - \frac{1}{p_2 \Theta_2}\left\{(a_1 - b_1 t)\,\Theta_2 + t^2\frac{d\Theta_2}{dt}\right\},$$

which is a regular function of t; but it is not zero, in general, for its value when $t=0$ is

$$-\frac{a_1}{a_2}.$$

Hence unless a_1 vanishes, there is no regular integral vanishing with t and distinct from that given by $B=0$. If a_1 vanishes, the form of v is ineffective because it contains infinite coefficients: separate investigation is required.

The point $t=0$ is a point of indeterminateness for the most general integral of the equation.

Next, let $s>2$; it will suffice to take $s=3$ as an illustration. The equation then is

$$\frac{d^2 V}{dt^2} - Q_1 \frac{dV}{dt} + Q_2 V = 0,$$

where

$$Q_1 = \frac{a_1}{t^3} + \frac{b_1}{t^2} + \frac{c_1-3}{t} + d_1 + \frac{b_2}{a_2} + \text{positive powers},$$

$$Q_2 = \frac{a_2 b_0}{t^5} + \frac{a_2 c_0 + b_0 b_2}{t^4} + \dots.$$

The equation has no regular integrals; but it may have sub-regular integrals. To obtain the latter, if they exist, we take

$$V = e^\Omega U,$$

where Ω is a sum of negative powers of t at our disposal; the equation for U is

$$\frac{d^2 U}{dt^2} + (2\Omega' - Q_1)\frac{dU}{dt} + U(\Omega'' + \Omega'^2 - Q_1\Omega' + Q_2) = 0,$$

and Ω is chosen so as to annihilate as many terms as possible in the coefficient of U which have negative indices. Simple calculations shew that

$$\Omega' = \frac{\phi}{t^3} + \frac{\theta}{t^2}$$

is a suitable form, and further that

$$\phi = a_1, \quad a_1\theta = a_1 b_1 - a_2 b_0;$$

the coefficient of $\frac{dU}{dt}$ then is

$$\frac{a_1}{t^3} + \frac{2\theta - b_1}{t^2} + \frac{3 - c_1}{t} + \dots = R_1(t) \text{ say},$$

and that of U is

$$\frac{1}{t^4}\{-3a_1 + \theta^2 - a_1(c_1 - 3) - b_1\theta + (a_2 c_0 + b_0 b_2)\} + \dots$$

$$= \frac{a_1}{t^4} + \dots = R_2(t) \text{ say},$$

so that the equation in U now is

$$\frac{d^2 U}{dt^2} + R_1(t)\frac{dU}{dt} + R_2(t) U = 0.$$

Of this, there may be one regular integral; if so, it is

$$U_1 = t^{-\frac{a_1}{a_1}} \Theta_1,$$

where Θ_1 is a regular function of t, not vanishing when $t=0$. To find another integral, write

$$U = U_1 W;$$

then

$$\frac{1}{\dfrac{dW}{dt}} \frac{d^2 W}{dt^2} + \frac{2}{U_1} \frac{dU_1}{dt} = - R_1(t)$$

$$= -\frac{a_1}{t^3} + \frac{b_1 - 2\theta}{t^2} + \frac{c_1 - 3}{t} + \dots$$

Hence

$$U_1{}^2 \frac{dW}{dt} = A t^{c_1 - 3} e^{\frac{a_1}{2t^2} - \frac{b_1 - 2\theta}{t} + \text{positive powers}},$$

and therefore

$$\frac{dW}{dt} = A' t^{c_1 - 3 + 2\frac{a_1}{a_1}} e^{\frac{a_1}{2t^2} + \frac{b_1'}{t} + \text{positive powers}},$$

where $b' = b_1 - 2\theta$; hence the value of W is

$$W = B + A' \int t^{c_1 - 3 + 2\frac{a_1}{a_1}} e^{\frac{a_1}{2t^2} + \frac{b_1'}{t} + \text{positive powers}} \, dt$$

$$= B + A'\Phi,$$

say. Consequently, we have

$$V = e^{\Omega} U$$

$$= e^{-\frac{a_1}{2t^2} - \frac{\theta}{t}} (B U_1 + A' U_1 \Phi),$$

as the primitive of the equation.

Hence the integral of the original equation in v is

$$v = -\frac{t^3}{p_2} \frac{1}{V} \frac{dV}{dt}$$

$$= -\frac{t^3}{p_2} \left\{ \frac{a_1}{t^3} + \frac{\theta}{t^2} + \frac{B \dfrac{dU_1}{dt} + A' \dfrac{d}{dt}(U_1 \Phi)}{B U_1 + A' U_1 \Phi} \right\}.$$

It is clear that, in general, no regular integral exists which vanishes with t.

It appears as if, taking $a_1 = 0$, we should have a regular integral on making $A' = 0$; but then the index of the leading term in U_1 is $-\frac{a_1}{a_1}$, that is, it would become infinite, and the resulting integral could not then be declared regular.

Consequently, when $s > 2$, there is no regular integral of the equation which vanishes when $t = 0$. The point $t = 0$ is a point of indeterminateness for all the (non-regular) integrals of the equation.

83.   In order to discuss the case omitted from the preceding investigation, and also to discuss the equation suggested at the end of Example 4 in § 82 where a_1 is to be zero, consider

$$t^2 \frac{dv}{dt} = bv^2 + t^2 (c_0 + c_1 t + c_2 t^2 + \dots)$$
$$= bv^2 + t^2 R (t).$$

Write

$$v = - \frac{t^2}{b} \frac{1}{V} \frac{dV}{dt};$$

then the equation for V is

$$\frac{d^2 V}{dt^2} + \frac{dV}{dt} \frac{2}{t} + \frac{b}{t^2} VR (t) = 0.$$

The determining indicial equation is

$$m^2 + m + bc_0 = 0,$$

the roots of which are unequal unless $bc_0 = \frac{1}{4}$.   Let

$$1 - 4bc_0 = 4\theta^2,$$

so that

$$m = - \tfrac{1}{2} \pm \theta.$$

If 2θ be not an integer, there are two regular integrals of the equation in V; and the primitive is

$$V = At^{-\frac{1}{2}-\theta} T_1 + Bt^{-\frac{1}{2}+\theta} T_2,$$

where T_1 and T_2 are regular functions of t that do not vanish when $t = 0$.   The value of v is

$$v = - \frac{t}{b} \frac{At^{-\theta} \left\{ t \dfrac{dT_1}{dt} - (\tfrac{1}{2} + \theta) T_1 \right\} + Bt^\theta \left\{ t \dfrac{dT_2}{dt} - (\tfrac{1}{2} - \theta) T_2 \right\}}{At^{-\theta} T_1 + Bt^\theta T_2}.$$

When $A = 0$, we obtain a regular integral which vanishes when $t = 0$.   When $B = 0$, we obtain another regular integral which vanishes when $t = 0$.   These two integrals are the only regular integrals vanishing with t.

When neither A nor B vanishes, we have an integral which vanishes when $t = 0$ and contains an arbitrary constant in the ratio $\dfrac{A}{B}$.   According as the real part of θ is positive or is negative, the integral is a regular function of t and $t^{2\theta}$ or of t and $t^{-2\theta}$; if the real part of θ be zero, then the integral is a regular function of either combination.   Hence there is an infinitude of non-regular integrals, in addition to the two regular integrals, which vanish when $t = 0$.

If 2θ be an integer, a similar investigation shews that there is one regular integral which vanishes when $t = 0$; and that there is, in addition, an infinitude of non-regular integrals vanishing when $t = 0$, these integrals being regular functions of t and $t \log t$. The simplest case is that in which $c_0 = 0$ and $2\theta = 1$; the equation is

$$t^2 \frac{dv}{dt} = bv^2 + t^3 R(t).$$

Ex. Discuss these equations, by considering the effect of the substitution

$$v = ut.$$

84. There still remains one case for consideration, viz. that in which all the terms of $\phi_1(v, t)$ independent of v disappear; since a is not zero, the equation then is

$$t^s \frac{dv}{dt} = v\psi_1(v, t),$$

where ψ_1 is a regular function of v and t such that $\psi_1(0, 0) = a$, and a is a constant distinct from zero.

The integral $v = 0$ is, in general, the only regular integral which vanishes with t. For all the other (non-regular) integrals it may be proved, as in § 81, that the point $t = 0$ is, in general, a point of indeterminateness.

Note. It may happen for special forms of $\psi_1(v, t)$, such that a is zero though not all the terms in $\psi_1(v, t)$ independent of v vanish, some regular integrals, other than $v = 0$, exist which vanish with t: it may even happen that all the integrals vanishing with t are regular, and are infinite in number.

Ex. 1. Consider the equation

$$t^s \frac{dv}{dt|} = p_1 v + p_2 v^2,$$

where p_1 and p_2 are regular functions of t, say

$$p_1 = a_1 + b_1 t + c_1 t^2 + \ldots$$
$$p_2 = a_2 + b_2 t + c_2 t^2 + \ldots.$$

We have

$$t^s \frac{d}{dt}\left(\frac{1}{v}\right) + p_1 \frac{1}{v} = -p_2,$$

provided v be not steadily equal to zero. (It is evident that $v=0$ is a regular integral of the equation; we shall assume this regular integral as already considered.) Then

$$\frac{1}{v} e^{\int \frac{p_1}{t^s} dt} = A - \int \frac{p_2}{t^s} e^{\int \frac{p_1}{t^s} dt} dt,$$

so that

$$v = \frac{C e^{\int \frac{p_1}{t^s} dt}}{1 - C \int \frac{p_2}{t^s} e^{\int \frac{p_1}{t^s} dt} dt},$$

where $AC=1$. It is obvious that, with general values of p_1 and p_2, the point $=0$ is a point of indeterminateness for the integral: and that the only integral (other than isolated values of the non-regular integrals), which vanishes with t, is the regular integral $v=0$.

But for particular forms, this result does not necessarily hold. Suppose that

$$p_1 = \kappa t^{s-1} + \text{higher powers of } t;$$

then

$$e^{\int \frac{p_1}{t^s} dt} = t^\kappa e^{G(t)},$$

where $G(t)$ is a regular function of t vanishing with t. If κ be such that its real part is positive and greater than $s-1$, say

$$\kappa = s - 1 + \epsilon,$$

where the real part of ϵ is positive, then

$$\frac{p_2}{t^s} e^{\int \frac{p_1}{t^s} dt} = \frac{p_2 e^{G(t)}}{t^{1-\epsilon}}$$

$$= p_2 e^{G(t)} t^{-1+\epsilon};$$

and therefore

$$\int \frac{p_2}{t^s} e^{\int \frac{p_1}{t^s} dt} dt = \frac{a_2}{\epsilon} t^\epsilon H(t),$$

where $H(t)$ is an integral function of t that does not vanish with t. In that case

$$v = \frac{C t^{s-1+\epsilon} e^{G(t)}}{1 - C \frac{a_2}{\epsilon} t^\epsilon H(t)};$$

so that, as C is an arbitrary constant, there is an infinitude of integrals which vanish with t; and except $v=0$, all of them are non-regular when ϵ is distinct from a positive integer. But if ϵ be a positive integer, that is, if κ be a positive integer $\geqslant s$, then there is an infinitude of regular integrals which vanish with t.

In general, the point $t=0$ is a point of indeterminateness for the complete primitive of the equation; and in general, $v=0$ is the only integral which vanishes when $t=0$. These results, however, do not hold for particular forms.

The equation can also be discussed in connection with the method of Ex. 4, § 82. We have $p_0 = 0$; and taking

$$v = \frac{-t^s}{p_2} \frac{1}{V} \frac{dV}{dt},$$

the equation for V is

$$\frac{d^2V}{dt^2} + \frac{dV}{dt}\left(\frac{-p_1}{t^s} + \frac{s}{t} - \frac{1}{p_2}\frac{dp_2}{dt}\right) = 0.$$

The solution of this equation, given by $V = $ constant, leads to the regular integral $v = 0$. The complete solution is

$$V = B + A \int \frac{p_2}{t^s} e^{\int \frac{p_1}{t^s}dt} \, dt \, ;$$

from this, the other solution can be deduced immediately.

Ex. 2. In the same way, discuss the equation

$$t^s \frac{dv}{dt} = p_1 v + p_2 v^n,$$

where n is an integer > 2.

THE REMAINING TYPICAL FORMS.

85. The typical forms that as yet have not been discussed are of less frequent occurrence as initial forms; some of them can be brought into relation with the preceding forms.

One such form is

$$t^2 \frac{dv}{dt} = \phi_m(v, t)$$
$$= gv^m + hv^{m+1} + v^{m+2}Q(v) + t\{b + tR(t)\}$$
$$+ vt\{c + S(v, t)\},$$

where $Q(v)$, $R(t)$ are regular functions such that $Q(0)$, $R(0)$ are not necessarily zero, and $S(v, t)$ is a function such that $S(0, 0) = 0$. Suppose that b does not vanish. The application of a Puiseux diagram to this equation suggests a transformation

$$t = \theta^m, \quad v = u\theta,$$

where u is to be finite (not zero) when $\theta = 0$. Substituting, we find

$$\frac{1}{m} \theta^{m+1}\left(\theta \frac{du}{d\theta} + u\right) = \theta^m gu^m + \theta^{m+1}hu^{m+1} + b\theta^m + cu\theta^{m+1}$$
$$+ \frac{1}{m} \theta^{m+2} P(u, \theta),$$

where $P(u, \theta)$ is a regular function of u and θ; and therefore

$$\theta^2 \frac{du}{d\theta} = mgu^m + bm + (mc - 1) u\theta + mhu^{m+1}\theta + \theta^2 P(u, \theta).$$

Let $u = \rho$ when $\theta = 0$; manifestly

$$g\rho^m + b = 0.$$

In general, therefore, we take

$$u = \rho + U,$$

so that U is to be zero when $\theta = 0$; then

$$\theta^2 \frac{dU}{d\theta} = m^2 g\rho^{m-1} U + \{(mc - 1)\rho + mh\rho^{m+1}\}\theta + \cdots$$

$$= -m^2 \frac{b}{\rho} U + \{(mc - 1)\rho + mh\rho^{m+1}\}\theta + \cdots,$$

which is an instance of the third typical form.

In general, $\theta = 0$, that is, $t = 0$, is a point of indeterminateness for U and therefore also for the original dependent variable v. But for particular forms of the equation, it has been seen to be possible that integrals exist, which vanish with θ and are regular functions of θ. For all such equations, it follows that there are integrals of the original equation vanishing with t; they are regular functions of $t^{\frac{1}{m}}$, that is, the point $t = 0$ is an algebraic critical point for the integrals, the branches forming one cycle round the point.

Ex. 1.   As an example, consider

$$t^2 \frac{dv}{dt} = v^2 - a^2 t + cvt.$$

We take $t = \theta^2, \quad v = u\theta,$

so that the equation for u is

$$\tfrac{1}{2}\theta^3 \left(\theta \frac{du}{d\theta} + u\right) = u^2\theta^2 - a^2\theta^2 + cu\theta^3,$$

and therefore

$$\theta^2 \frac{du}{d\theta} = -2a^2 + 2u^2 + (2c - 1) u\theta.$$

As u is not to vanish with θ, its value is manifestly a for $\theta = 0$; take

$$u = a + U,$$

where the new variable U must vanish for $\theta = 0$. The equation for U is

$$\theta^2 \frac{dU}{d\theta} = (2c - 1) a\theta + U\{4a + (2c - 1)\theta\} + 2U^2,$$

an equation of the type considered in § 82, Ex. 4. We have $s = 2$, $p_2 = 2$, $p_0 = (2c - 1) a\theta$, $p_1 = 4a + (2c - 1)\theta$; and we know that an integral of the equation may exist, which vanishes with θ and is a regular function, its expression being

$$U = -\frac{\theta^2}{a_2}\frac{a_2 b_0}{a_1 \theta} - \frac{\theta^2}{a_2}\frac{1}{\Phi}\frac{d\Phi}{d\theta}$$

$$= -\frac{1}{4}(2c-1)\theta - \frac{1}{2}\theta^2\frac{1}{\Phi}\frac{d\Phi}{d\theta},$$

where Φ is a regular function of θ which is equal to 1 when $\theta = 0$.

When the value of U is substituted, it appears that Φ satisfies the equation

$$\frac{1}{2}\theta^2\frac{d^2\Phi}{d\theta^2} - (2a-\theta)\frac{d\Phi}{d\theta} - \frac{1}{8}(3-8c+4c^2)\Phi = 0.$$

First, suppose that the coefficient of Φ does not vanish, that is, that c is neither $\frac{1}{2}$ nor $\frac{3}{2}$; then if

$$\Phi = 1 + \sum_{n=1}^{\infty} A_n \theta^n,$$

the equation determining the coefficients is

$$(m+1)\frac{A_{m+1}}{A_m} = \frac{1}{16a}[4m^2 + 4m - (3 - 8c + 4c^2)],$$

which gives the convergence-ratio for the series. It is clear that the corresponding series diverges : and therefore the postulated condition of the example quoted is not satisfied, and then the function $\Phi(\theta)$ does not exist if determined by the condition that $\Phi(0) = 1$.

Next, let $3 - 8c + 4c^2 = 0$; then

$$\Phi = B + A\int e^{-\frac{4}{\theta}}\frac{1}{\theta^2}d\theta = B + \frac{A}{4a}e^{-\frac{4a}{\theta}}$$

Taking $B = 0$, the value of U gives a regular function of θ; but it does not vanish with θ (unless $a = 0$), and therefore it is not suitable.

Taking $A = 0$, the value of U is $-\frac{1}{4}(2c-1)\theta$. If we take $2c = 3$, this gives $U = -\frac{1}{2}\theta$, an integral vanishing with θ. Accordingly, we infer that the equation

$$t^2\frac{dv}{dt} = v^2 - a^2 t + \frac{3}{2}vt$$

has an integral

$$v = a\theta - \frac{1}{2}\theta^2 = at^{\frac{1}{2}} - \frac{1}{2}t,$$

which vanishes with t. The verification is immediate.

If we take $2c - 1 = 0$, this gives $U = 0$, an integral vanishing with t. Accordingly, we infer that the equation

$$t^2\frac{dv}{dt} = v^2 - a^2 t + \frac{1}{2}vt$$

has an integral

$$v = a\theta = at^{\frac{1}{2}},$$

which vanishes with t. The verification again is immediate.

Ex. 2.   Apply the method to the equation

$$t^\kappa \frac{dv}{dt} = \phi_m(v, t),$$

where $\kappa > 2$.

86.   The form that remains for consideration is

$$t^\kappa \frac{dv}{dt} = \frac{\phi_n(v, t)}{\phi_m(v, t)}.$$

When $\kappa = 0$, it has been discussed at length in §§ 38 sqq.   When $\kappa = 1$, so that the equation is

$$t \frac{dv}{dt} = \frac{av^n + bt + cvt + dv^2 t + \dots}{\alpha v^m + \beta t + \gamma vt + \dots},$$

one method of proceeding is that which is indicated in Ex. 5 of § 60.   If $m \geqslant n$, we take

$$t = \theta^n, \quad v = u\theta,$$

so that

$$\frac{1}{n}\theta\left(\theta\frac{du}{d\theta} + u\right) = \frac{b + au^n + cu\theta + du^2\theta^2 + \dots}{\beta + \alpha u^m \theta^{m-n} + \gamma u\theta + \dots}.$$

Take

$$u = \rho + K\theta + V\theta,$$

where V vanishes with θ, and K is a constant: let

$$b + a\rho^n = 0.$$

First, let $m = n$; then choose K so that

$$\frac{1}{n}\rho(\beta + \alpha\rho^n) = na\rho^{n-1}K + c\rho.$$

The equation becomes, after straightforward reductions,

$$\theta^2 \frac{dV}{d\theta} = \frac{n^2 a\rho^{n-1}}{\beta + \alpha\rho^n} V + C\theta + \dots,$$

where

$$C = -2K + \frac{\frac{1}{2}n^2(n-1)a\rho^{n-2}K^2 + n^2 bK\frac{\alpha}{a} + ncK + \rho^2(nd - \gamma)}{\beta + \alpha\rho^n};$$

this is a typical form already considered.

Next, let $m \geqslant n + 1$; then choose K so that

$$\frac{1}{n}\rho\beta = na\rho^{n-1}K + c\rho.$$

After corresponding reductions, the equation becomes

$$\theta^2 \frac{dV}{d\theta} = \frac{n^2 a \rho^{n-1}}{\beta} V + C'\theta - \alpha \rho^{m+1} \theta^{m-n} + \ldots,$$

where

$$C' = -2K + \frac{1}{\beta} \left[\tfrac{1}{2} n^2 (n-1) a \rho^{n-2} K^2 + ncK + \rho^2 (nd - \gamma) \right];$$

the term in θ^{m-n} coalescing with the term $C'\theta$ if $m - n = 1$, and with other terms if $m - n > 1$. Again, the typical form has already been considered.

Next, let $m < n$. First, take $m = n - 1$: and again, let

$$t = \theta^n, \quad v = u\theta;$$

so that

$$\frac{1}{n} \theta \left(\theta \frac{du}{d\theta} + u \right) = \frac{(b + au^n) \theta^n + cu\theta^{n+1} + du^2 \theta^{n+2} + \ldots}{au^{n-1} \theta^{n-1} + \beta \theta^n + \gamma u \theta^{n+1} + \ldots},$$

and therefore

$$\theta \frac{du}{d\theta} + u = n \frac{b + au^n + cu\theta + du^2 \theta^2 + \ldots}{au^{n-1} + \beta \theta + \gamma u \theta^2 + \ldots}.$$

Let

$$\rho = n \frac{b + a\rho^n}{a\rho^{n-1}},$$

so that

$$\rho^n = \frac{nb}{\alpha - na};$$

then if $u = \rho + V$, we have

$$\theta \frac{dV}{d\theta} = \left(n^2 \frac{a}{\alpha} \frac{n-1}{\rho} - 1 \right) V + \frac{nc - \beta}{\alpha \rho^{n-2}} \theta + \ldots,$$

a special instance of the first typical form.

Next, take

$$m < n - 1, \quad = n - 1 - \sigma,$$

where σ is an integer $\geqslant 1$. We assume

$$t = \theta^{n-\sigma}, \quad v = u\theta = (\rho + V)\theta,$$

choosing ρ so that

$$\alpha \rho^{n-\sigma} = (n - \sigma) b;$$

then, after reductions, the equation becomes

$$\theta \frac{dV}{d\theta} = -(n - \sigma) V + \frac{c(n - \sigma) - \beta}{\alpha \rho^{n-2-\sigma}} \theta + \frac{a\rho^{n+1}}{b} \theta^\sigma + \ldots,$$

another special instance of the typical form. Since the coefficient of V in the expression for $\theta \dfrac{dV}{d\theta}$ is a negative integer, there is only one integral of this equation which vanishes with θ; it is the regular integral, the first term of which is

$$\frac{c\,(n-\sigma)-\beta}{(n-\sigma)(n-\sigma+1)}\,\frac{\rho^2}{b}\,\theta$$

if $\sigma > 1$, and is

$$\left\{\frac{c\,(n-\sigma)-\beta}{n-\sigma}\,\frac{\rho^2}{b}+\frac{a}{b}\,\rho^{n+1}\right\}\,\frac{\theta}{n-\sigma+1}$$

if $\sigma = 1$.

87.   When $\kappa > 1$, the Puiseux diagram shews that the appropriate substitution is

$$t = \theta^n, \quad v = u\theta,$$

so that

$$\theta^{n(\kappa-1)+1}\left(\theta\,\frac{du}{d\theta}+u\right) = n\,\frac{(au^n+b)\,\theta^n + cu\theta^{n+1} + du^2\theta^{n+2}+\dots}{\alpha u^m\,\theta^m + \beta\theta^n + \gamma u\theta^{n+1}+\dots}$$

If $m > n$, we take

$$a\rho^n + b = 0, \quad K = -\frac{c}{na\rho^{n-2}};$$

and we substitute

$$u = \rho + K\theta + V\theta;$$

then the quantity V, which must vanish with θ, is given by

$$\theta^{n(\kappa-1)+2}\,\frac{dV}{d\theta} = \frac{n^2 a\rho^{n-1}}{\beta}\,V + \frac{n\,(cK+d\rho^2)}{\beta}\,\theta + \dots,$$

one of the typical forms already considered.

If $m = n$, we still take

$$a\rho^n + b = 0, \quad K = -\frac{c}{na\rho^{n-2}},$$

and

$$u = \rho + K\theta + V\theta;$$

the equation determining V is

$$\theta^{n(\kappa-1)+2}\,\frac{dV}{d\theta} = \frac{n^2 a\rho^{n-1}}{\beta - \alpha\dfrac{b}{a}}\,V + \frac{n\,(cK+d\rho^2)}{\beta - \alpha\dfrac{b}{a}}\,\theta + \dots,$$

a satisfactory reduction to a typical form already considered unless $\alpha\beta - ab = 0$.

If $m < n, = n - \sigma$, we still take

$$a\rho^n + b = 0, \quad K = -\frac{c}{na\rho^{n-2}},$$

$$u = \rho + K\theta + V\theta;$$

the equation determining V is

$$\theta^{n(\kappa-1)+2-\sigma}\frac{dV}{d\theta} = \frac{n^2a\rho^{\sigma-1}}{\alpha}V + \frac{n(cK+d\rho^2)}{a\rho^{n-\sigma}}\theta + \dots,$$

again one of the typical forms already considered.

Ex. 1.   Discuss the equation

$$t^\kappa\frac{dv}{dt} = \lambda\frac{u^n + bt + cut + \dots}{u^n + bt + aut + \dots},$$

where c and a are unequal.

Ex. 2.   Discuss the equation

$$t^\kappa\frac{dv}{dt} = v^r\frac{\phi_n(v, t)}{\phi_m(v, t)},$$

where r is an integer that may be positive or negative, so as to obtain the integrals v (if any) which vanish when $t = 0$; the integer κ being positive.

Summary.

88.   We may now, very briefly, summarise the results of the investigation of the behaviour of the integral of the equation

$$\frac{dw}{dz} = f(w, z),$$

in the immediate vicinity of a combination of values $w = \alpha$, $z = c$, which constitute an accidental singularity of the second kind for the function $f(w, z)$: the integral being further defined by the condition that it must assume the value α when $z = c$.

In the first place, it was shewn that, by an appropriate algebraical transformation

$$z - c = x = t^q, \quad w - \alpha = y = (\rho + v)\,t^p,$$

(which could, in general, be definitely determined so that p and q are positive integers, ρ is a constant different from zero, and v is a new variable that vanishes when $t = 0$), the original equation could be replaced by a differential equation determining v and belonging to one or other of a limited number of ascertained types. In many cases, it happens that a number of distinct reductions of the same broad character are possible in connection with the original equation, each such reduction leading to a differential equation for v.

In some cases, it happens that a steady zero value of y while x varies, is the only integral possible: for them, a constant value $w = \alpha$ in the immediate vicinity of the singularity is the only solution which satisfies the conditions. In other cases, it happens that a steady zero value of x, while y varies, provides the only solution of the equation; for them, there is no integral of the equation which satisfies the assigned conditions. But in all the remaining cases, the character of the integral of the original equation is subject to the character of the quantity v.

Various possibilities arise in connection with the ultimate forms to which the equation has been, or can be, reduced. In some instances, there is only one integral v which is a regular function of t; in some, there is an infinitude of regular integrals; in some, there is an infinitude of non-regular integrals, the deviations from regularity being of specified types; in some, there are no integrals of a non-regular class; in some, the point is an essential singularity of an otherwise regular integral; in some, it is a point of more general indeterminateness, with or without definite branching. For the various instances, the respective tests have been given in connection with the typical reduced forms as they were discussed.

CHAPTER VII.

Essential Singularities of a Single Equation of the First Order*.

89. VERY little can at present be said in general discussion of the integral of a differential equation

$$\frac{dw}{dz} = f(w, z),$$

for values of the variables in the immediate vicinity of an essential singularity of the function $f(w, z)$; because there is no generally adopted type (or set of types) of expression of the function in the vicinity. Such an expression as

$$\sum_{-\infty}^{\infty} \sum_{\infty}^{\infty} A_{m, n} (w - \alpha)^m (z - c)^n$$

would have significance (if at all), only for values of w in a ring-space in the w-plane round α as centre, and for values of z in a ring-space in the z-plane round c as centre; but it is not possible to make the expression effective when the variables are made to approach any immediate vicinity of α and c.

In the case of the accidental singularity of the second kind, it proved possible to obtain definite expression for $\frac{dw}{dz}$ in the immediate vicinity, though the value was indefinite actually at the singularity: and even so, it appeared that for many cases—indeed, for the most general cases, because the detailed results

---

* For reasons indicated in § 89, little is said in discussion of the general equation. With regard to particular illustrations, reference may be made to Painlevé's Stockholm Lectures, *Sur la théorie analytique des équations différentielles*, (Paris, Hermann, 1897), pp. 1—16.

14

obtained frequently arose through the assignment of special numerical values or limitations upon coefficients—the singularity of the function $f(w, z)$ was a point of indeterminateness for the complete integral; though, for special branches of the complete integral, the singularity might be an algebraical branch-point or even an ordinary point. It is therefore not unreasonable to expect that, when the combination $w = \alpha$, $z = c$ is an essential singularity of the function $f(w, z)$, the point $z = c$ will certainly be a point of indeterminateness for the integral of the equation.

When individual instances are propounded, it may be possible to discuss them; but the natural objection, quite apart from the difficulty of the discussion, is that the instances may not be typical of any important class of cases. Thus, it might be that the function $f(w, z)$ is a regular function of w, z, and $z \log z$, for some finite simply-connected region not enclosing the origin; it is manifest that continuations of that regular function, when completely carried out by variations in the z-plane alone, would lead to an unlimited number of values of $\dfrac{dw}{dz}$, even supposing that at any point in the w-plane the value of w is definite. Though such an expression would doubtless cover one class of essential singularity, there appears no indication of its relative importance.

All that seems possible is therefore to deal with such instances, as they arise and when they arise, by any method that may be found appropriate to their case: it does not, in the present stage of knowledge of essential singularities of functions of two variables, appear possible to initiate any useful general discussion of the integral (if any) of the differential equation, which is to be determined by a condition that occasions an essential singularity of the expression for the derivative.

Simple examples suffice to shew, even in the class of cases indicated, that the mode of functional occurrence of the singularity in the differential equation may be replaced by one of entirely distinct functional occurrence in the integral equation.

Thus for the equation

$$\frac{dw}{dz} = \frac{az + az \log z - w}{z \log z},$$

the combination $w = 0$, $z = 0$ is an essential singularity of the expression for $\dfrac{dw}{dz}$. There is a complete integral of the equation, which can be obtained by quadratures in the form

$$e^{\frac{A}{w-az}} = z,$$

an equation which is uniform so far as regards the occurrence of the variables.

Similarly, $w = 0$, $z = 0$ is an essential singularity of the equation

$$\frac{dw}{dz} = \frac{w}{z} \cdot \frac{az\,(\log w + \log z) - 1}{w\,(\log w + \log z) + w - az};$$

there is a complete integral in the form

$$e^{\frac{A}{w-az}} = wz,$$

again an equation which is uniform so far as regards the occurrence of the variables.

PAINLEVÉ'S THEOREM.

90. In the case of an equation

$$\frac{dw}{dz} = f(w, z),$$

where f is rational in w and is uniform in z, there are no combinations of values of w and z which give rise to branch-points of the function $f(w, z)$; it is, in fact, a uniform function of its arguments.

The exceptional points of $f(w, z)$ have been taken into consideration; and the integral of the equation has been discussed for values of the variable in the vicinity of the different classes of exceptional points in turn. At and near some points, the integral has been regular; some points have been algebraical critical points of the integrals; some points have been points of indeterminateness of the integrals; and all the points have been suggested by the equation.

It may, however, happen that the integral of an equation possesses singularities at points not suggested by the equation itself; thus an accidental singularity of the first kind (which was seen to be, not an isolated combination, but a continuous aggregate) leads to a parametric critical point of an algebraical nature. All

14—2

the other singularities, that have arisen, have been fixed combinations and have led to no parametric critical points. Moreover, every combination of values, which is not a singularity of the function $f(w, z)$, however arbitrary in character subject to this negative condition, is proved by Cauchy's theorem to be an ordinary combination for the integral determined by those values as initial values. It might therefore be inferred from these results that, so far as concerns a single equation, all the points of indeterminateness of the integral have been obtained: the inference is confirmed by the theorem that *the points of indeterminateness, in particular, the essential singularities of the integral of the equation, are fixed points determined by the equation itself.* To the establishment of this theorem, which is due to Painlevé*, we now proceed.

It has been seen that essential singularities of the integral can be provided by essential singularities of $f(w, z)$ and by accidental singularities of $f(w, z)$ of the second kind; and that (§ 22) these singularities are isolated points. Let $w = \alpha$, $z = c$ denote an ordinary combination of values for the equation and therefore for the function $f(w, z)$; and through c let a curve be drawn in the z-plane, so that no point of it lies within an infinitesimal distance of any one of the two classes of exceptional points of $f(w, z)$, which can lead to the singularities of the integral indicated in the theorem. Part at least of this curve lies in the domain of the point c, connected with the region of existence of the first element of the integral determined by the initial value α; and therefore continuations of the integral, leading to successive elements, can be constructed by taking successive points along the curve, leading in each instance to new domains. Unless the domain of at least one point on the curve becomes infinitesimal, there is no limit to the curve other than that which already has been imposed; all the points of the plane, not belonging to some one or other of the selected classes of exceptional points, are then ordinary points of the integral.

Accordingly, let it be supposed that there is a point Z on the curve such that, as the variable z approaches Z, the successive

---

* Painlevé, *Sur les lignes singulières des fonctions analytiques*, (Thèse pour le doctorat, 1887), pp. 38—40; see also his Stockholm Lectures, *Sur la théorie analytique des équations différentielles*, (1897), pp. 23—26. An exposition of the theorem is given by Picard, *Cours d'Analyse*, t. II, pp. 324—329; it is upon Picard's exposition that the proof in the text is based.

domains become so small that the continuation of the integral beyond Z is impossible along the curve. When z tends to coincide with Z, the limiting value of w must be either indeterminate or determinate; if determinate, it must be either finite or infinite.

If the limiting value of w for $z = Z$ be determinate and finite, it follows that, as the integral cannot be continued beyond Z, the function $f(w, z)$ becomes indeterminate or becomes a determinate infinity. The former alternative is impossible, because the combination of values would be either an accidental singularity of the second kind or an essential singularity of $f(w, z)$, and the curve in question is drawn so as to be at a finite distance from any such point in the z-plane; hence $f(w, z)$ is there determinately infinite, and the point arises through an accidental singularity of the first kind of $f(w, z)$. Such a point is known (§§ 24, 25) to be an algebraical critical point of the integral.

If the limiting value of w for $z = Z$ be determinate and infinite, we take $ww_1 = 1$. A similar argument to that adopted in the last case shews that, as the integral w_1 cannot be continued beyond Z and as its value at Z is a determinate zero, the function $w_1{}^2 f\left(\dfrac{1}{w_1}, z\right)$ becomes determinately infinite there, and the point in the z-plane arises through an accidental singularity of the first kind. Such a point is known (§ 26) to be an algebraical critical point of the integral.

If w does not tend to a determinate value as z tends to coincide with Z, then as $f(w, z)$ is a uniform function of z and a rational function of w, $f(w, z)$ cannot have one determinate finite value at $z = Z$. It therefore either must have one of a number of determinate finite values, or it must be indeterminate, or it must be infinite. (If it have one of a number of determinate finite values, then $f(w, z)$ is really a branch of a multiform function and this branch is uniform in the immediate vicinity of $z = Z$; hence the continuation of the integral could be effected beyond Z for each such branch, contrary to the initial hypothesis. In reality, however, this case does not arise at present, for $f(w, z)$ has been supposed uniform.) If it were indeterminate for $z = Z$, then the point—or the possible combination of values, if w has one of a limited number of definite values for $z = Z$—must be either an essential singularity or an accidental singularity of the second kind: each of these results is excluded by the course of the curve as drawn, and therefore $f(w, z)$ cannot be indeterminate. The only possibility therefore is that $f(w, z)$ should be infinite when $z = Z$, though w does not tend to a determinate value as z tends to Z, and therefore that

$$\frac{1}{f(w, Z)} = 0.$$

Now the function f is rational in w, and therefore the equation

$$\frac{1}{f(w, Z)} = 0$$

has a limited number of roots; let them be $v_1, \ldots, v_n$, any one or more than one of which may be a repeated root. All of these will be assumed finite; the equation, being algebraical in w, may be regarded as also possibly having an infinite root.

Let a small circle of radius r be drawn round Z; then, as z moves within or on this circle, the roots of the equation

$$\frac{1}{f(w, z)} = 0,$$

which are finite when $z = Z$, remain in the immediate vicinity of $v_1, \ldots, v_n$. Consequently small closed curves, say circles of radii $\rho_1, \ldots, \rho_n$, can be drawn in the w-plane round $v_1, \ldots, v_n$ respectively, such that these roots lie within the respective curves when z lies within its circle. As for the root of the equation which is infinite when $z = Z$, it has a modulus which tends to become infinite as z approaches Z; hence a circle of very large radius R can be drawn in the w-plane such that this root of the equation lies outside the circle.

The adopted hypothesis being that $f(w, z)$ tends to become infinite, though w does not tend to a determinate value as z tends to Z, consider points such that $|z - Z| \leqslant r$. On this hypothesis, a corresponding value of w could not lie within one of the circles round one of the v-points; if it could, then as z tends to Z, the variable w would tend to the corresponding value v, that is, to a determinate value. Nor could the corresponding value of w lie entirely without the circle of radius R; if it could, then as z tends to Z, the variable w would tend to an infinite value, and $\frac{1}{w}$ would acquire a definite zero value. Hence as z varies within or on the circle $|z - Z| = r$, the point w must lie within the circle of radius R and without the circles round the points v. But for all such points w, and for values of z, which (i) do not lie within an infinitesimal distance of either an essential singularity or an accidental singularity of the second kind (a condition rendered possible in this case, for r can be made small and Z is not within an infinitesimal distance of a singularity of either class), and (ii) are not an accidental singularity of the first kind (a condition also which can be regarded as satisfied, for the influence of such a point has already been taken into consideration), there exists a regular integral of the differential equation. This regular integral can be expressed as a power-series, converging within a circle (§ 10) of radius

$$\sigma = d \left(1 - e^{-\frac{D}{2Md}} \right),$$

where d is the distance of the z-point from the nearest exceptional point, D the distance of the corresponding value of w from the associated exceptional point in the w-plane, and M is the maximum value of $|f(w, z)|$ for the regions of variation indicated. Now D is not zero, because on the one hand z is moving along a curve that does not approach infinitesimally near an exceptional point of $f(w, z)$, and on the other hand w lies within the circle of radius R and without the circles round the points v. For all such points, M is finite, and it has been pointed out that ρ does not vanish; hence σ is different from zero, being a finite quantity.

Accordingly, when a point z within or on the circle $|z - Z| = r$ is taken as the centre of a new domain, this domain will certainly extend as far as the circle of radius σ. The domain will certainly include Z and points of the variation-curve of z beyond Z, because σ is finite. All points included in the circle are ordinary points of the differential equation, and continuations of the integral beyond points lying within the circle can be made; therefore the integral can be continued beyond the point Z, that is, in the circumstances, Z is not a point beyond which continuation of the integral is impossible.

It therefore follows that, on the preceding hypothesis, Z is not a point of indeterminateness of the integral of the equation.

It has been seen that the integral may become infinite at a point in the plane, but the point is then an algebraical critical point of the integral, or it may even be an ordinary point of the integral; it is not a point of indeterminateness of the integral.

Consequently, all the points of indeterminateness (if any) of the integral of the equation

$$\frac{dw}{dz} = f(w, z),$$

where f is rational in w and is uniform in z, are to be found among the essential singularities of f and among its accidental singularities of the second kind; that is, they are fixed points, determined by the equation itself.

91.   It is an immediate inference that, so far as concerns an equation

$$\frac{dw}{dz} = f(w, z),$$

parametric singularities of the integral function may be poles and may be branch-points; they cannot be points of indeterminateness, in particular, they cannot be essential singularities of the integral. It will be seen hereafter (§ 110) that this property belongs also to any equation of the first order

$$F\left(\frac{dw}{dz}, w, z\right) = 0,$$

which is algebraical in w and $\frac{dw}{dz}$, and is uniform in z.

In this respect, there is a fundamental distinction in character between an equation of the first order and an equation of higher order or, what is effectively the same thing, between an equation of the first order and a system of simultaneous equations of the first order. As an example, consider the integral

$$w^2 = a + \log (z - b)$$

of the system

$$\left.\begin{array}{l} \dfrac{dw}{dz} = w_1 \\[2ex] \dfrac{dw_1}{dz} = -\dfrac{w_1{}^2 + 2w^2w_1{}^2}{w} \end{array}\right\} ;$$

there is a parametric singularity at b. This particular system can be changed so as to depend upon the equation

$$\frac{dw_1}{dw} = -\frac{w_1{}^2 + 2w^2w_1{}^2}{ww_1},$$

expressing w_1 as a function of w; the argument of the preceding sections could be applied to obtain w_1 as a function of the (temporarily independent) variable w, only on the supposition that the independent variable is definite at each point in its plane. This supposition, however, in the present case is the whole matter at issue; and the result shews that the supposition would not be justified, for the temporarily independent variable is not definite at, or in the immediate vicinity of, its parametric essential singularity.

This substantial difference of property, between a single equation and a system of equations, must not be supposed necessarily to apply to all systems. Thus in the case of a system which is the equivalent of an ordinary linear equation of order n, every exceptional point of the integral (whether it be an accidental singularity, or a branch-point, or an essential singularity as of a uniform function, or a more complicated point of indeterminateness) is found to be a fixed point, that is, a point determined by the functions that occur in the differential equation: all values, parametric in regard to the differential equation, are ordinary points for its integral. But it is only to special systems of equations, and not to a system of any unlimitedly general type, that the property belongs.

The mere fact that all the critical points of a system of equations, that lead to a single linear equation, are fixed points, coupled with the property that an equation of the first order has no parametric points of indeterminateness, suggests an investigation of those equations of the first order the integrals of which have none of their exceptional points parametric. The consideration of this question will be undertaken later (Chap. IX), after the branch-points of an equation of the first order but not of the first degree have been discussed.

92.   One or two results of particular equations may be useful illustrations of this general theorem*. In the first place, consider equations of the first order. The equation

$$2zw \frac{dw}{dz} = 1$$

* See Painlevé, *Stockholm Lectures*, pp. 5—12.

has its integral in the form

$$w = (a + \log z)^{\frac{1}{2}},$$

where a is an arbitrary parameter. The points of indeterminate-ness of the integral (they are points with indefinite branching) are $z = 0$ and $z = \infty$; both of these are fixed points. The other critical point is $z = e^{-a}$; in its immediate vicinity, let

$$z = e^{-a} + Z,$$

so that

$$w = e^{\frac{1}{2}a} Z^{\frac{1}{2}} + \ldots,$$

that is, the point is an algebraical branch-point. It is the only parametric non-ordinary point for the integral.

The equation

$$\frac{dw}{dz} + w^2 = 0$$

has

$$w = \frac{1}{z - a}$$

for its integral; the only singularity of the integral is the para-metric point a, and it is a pole.

The equation

$$n \frac{dw}{dz} + w^{n+1} = 0$$

has

$$w^n = \frac{1}{z - a}$$

for its integral; the only non-ordinary points for the integral are $z = \infty$ (a branch-point, and a fixed point) and $z = a$ (a branch-point; it is a parametric point, and it is not a point of indeterminateness).

The equation

$$z \frac{dw}{dz} + w^2 = 0$$

has

$$w = \frac{1}{a + \log z}$$

for its integral; the points of indeterminateness are $z = 0$, $z = \infty$, both fixed points; the only parametric non-ordinary point is $z = e^{-a}$, and it is a pole.

The equation

$$z^2 \frac{dw}{dz} + w = 0$$

has

$$w = ae^{\frac{1}{z}}$$

for its integral; the only non-ordinary point of the integral is $z = 0$, which is a fixed point.

The equation

$$z^2 \left(\frac{dw}{dz}\right)^2 + w^2 - w^4 = 0$$

has

$$w = \operatorname{cosec}\left(a + \log z\right)$$

for its integral; the parametric point $z = e^{-a}$ is a pole; the only other non-regular points are the points of indeterminateness $z = 0$, $z = \infty$, both of them fixed points.

As regards equations of the second order—or a system of two equations of the first order—with parametric points of indeterminateness, they can be deduced from equations of the first order, by differentiating the latter and eliminating z between the equation and the derivative. Because z occurs only in the element dz in the differential coefficients of the eliminant equation, the integral of the new equation will involve z in the form $z - c$, where c is a parameter; hence all the exceptional points of the integral may be parametric. Thus from the equation

$$z^2 w_1^2 = w^4 - w^2,$$

where $w_1 = \dfrac{dw}{dz}$, we have

$$z = \frac{w\,(w^2 - 1)^{\frac{1}{2}}}{w_1},$$

so that

$$1 = (w^2 - 1)^{\frac{1}{2}} + \frac{w^2}{(w^2 - 1)^{\frac{1}{2}}} - \frac{w\,(w^2 - 1)^{\frac{1}{2}}}{w_1^2} \frac{dw_1}{dz};$$

and therefore the equations

$$\left.\begin{array}{c} \dfrac{dw_1}{dz} = \dfrac{w_1^2}{w\,(w^2 - 1)^{\frac{1}{2}}} \left\{2w^2 - 1 - (w^2 - 1)^{\frac{1}{2}}\right\} \\[2mm] \dfrac{dw}{dz} = w_1 \end{array}\right\}$$

have

$$w = \operatorname{cosec}\left\{a + \log\left(z - c\right)\right\}$$

for their integral; the parametric point $z = c + e^{-a}$ is a pole; the parametric point $z = c$ is an essential singularity; the fixed point $z = \infty$ is an essential singularity.

93.   One more illustration, in connection with an equation of the third order, may be taken as shewing that more complicated singularities can occur than are even suggested by a single equation of the first order.

Consider the two quantities

$$2K = \int_0^1 \{z(1-z)(1-cz)\}^{-\frac{1}{2}} dz,$$

$$2K' = \int_0^1 \{z(1-z)(1-c'z)\}^{-\frac{1}{2}} dz,$$

where $c + c' = 1$; then K and K' are independent solutions of the equation

$$cc'\frac{d^2y}{dc^2} + (c'-c)\frac{dy}{dc} - \tfrac{1}{4}y = 0.$$

From the known theory of elliptic functions (or as a deduced property of the independent solutions of the differential equation), it is known that

$$K\frac{dK'}{dc} - K'\frac{dK}{dc} = -\frac{\pi}{4cc'}.$$

Now, when $|c|$ is a small quantity, we have

$$K = \frac{\pi}{2}\{1 + cP(c)\},$$

where $P(c)$ is a regular function of c in the vicinity of $c = 0$; consequently

$$K' = -\frac{K}{\pi}\log c + R(c),$$

where $R(c)$ is a regular function of c in the vicinity of $c = 0$. Hence $c = 0$ is a logarithmic singularity for the function K'/K.

Similarly, $c = 1$ is a logarithmic singularity for the function $\dfrac{K}{K'}$, and therefore also for $\dfrac{K'}{K}$.

For large values of $|c|$, let $c\gamma = 1$, $cz = \zeta$; then

$$c^{\frac{1}{2}}\int \{z(1-z)(1-cz)\}^{-\frac{1}{2}} dz = \int \{\zeta(1-\zeta)(1-\gamma\zeta)\}^{-\frac{1}{2}} d\zeta.$$

Let Γ and Γ' be periods of the latter integral ; then*

$$c^{\frac{1}{4}}(K + iK') = \Gamma,$$

$$c^{\frac{1}{4}}K = \Gamma + c\Gamma',$$

and therefore $c^{\frac{1}{4}}K' = -\Gamma'.$

Now $\gamma = 0$ is a logarithmic singularity for Γ'/Γ; and therefore $c = \infty$ is a logarithmic singularity for Γ'/Γ, and therefore also for K'/K.

Now consider the function of c defined by

$$v = \frac{K'}{K}.$$

The points $c = 0, 1, \infty$ are logarithmic singularities of v; all other points in the c-plane are ordinary points of K' and K; and therefore in the domain of any such point (that is, a simply-connected region not enclosing $0, 1, \infty$), the function v is uniform.

But the function v ceases to be uniform in the domain of any of the points $c = 0, 1, \infty$. When the variable is made to describe paths round each of them any number of times in any sequence and in either trigonometrical direction, there arises an infinitude of values of the function represented by

$$\frac{\alpha K' + \beta K}{\gamma K' + \delta K},$$

say $$V = \frac{\alpha v + \beta}{\gamma v + \delta},$$

where β, γ are even integers, and γ, δ are odd integers such that

$$\alpha\delta - \beta\gamma = 1 ;$$

and each of these arises for a definite value of c, the particular form that occurs depending upon the path by which the argument attains its value c.

Consequently, c is a function of v—it is called a modular function—such that, when v is given, there is a single value of c, that is, c is a uniform function of v; it is such as to remain unchanged for the transformations

$$V = \frac{\alpha v + \beta}{\gamma v + \delta};$$

* *Th. Fns.*, § 303.

and therefore, if we write $c = \phi(v)$, we have

$$c = \phi(v) = \phi\left(\frac{\alpha v + \beta}{\gamma v + \delta}\right).$$

Now the line that is transformed into itself by the infinite group of transformations

$$\left(v, \ \frac{\alpha v + \beta}{\gamma v + \delta}\right),$$

is the axis of real quantities in the v-plane, the coefficients $\alpha, \beta, \gamma, \delta$ all being real; hence, by the known theory of (automorphic) modular functions, the line $v = 0$ is a line of essential singularities for c, regarded as a function of v.

The invariant of the equation

$$cc'\frac{d^2y}{dc^2} + (c' - c)\frac{dy}{dc} - \tfrac{1}{4}y = 0$$

is given by

$$I = \frac{1}{4}\left\{\frac{1}{c^2} + \frac{1}{c'^2} + \frac{1}{cc'}\right\}$$

$$= \frac{1}{4}\left\{\frac{1}{c^2} + \frac{1}{(1-c)^2} + \frac{1}{c(1-c)}\right\};$$

and therefore*, if $\{v, c\}$ denote the Schwarzian derivative of v with regard to c, where v denotes the quotient of two independent solutions of the equation, we have

$$\{v, \ c\} = 2I.$$

But

$$\{v, \ c\} = -\left(\frac{dv}{dc}\right)^2 \{c, \ v\};$$

consequently one integral of the equation

$$\{c, \ v\} + \frac{1}{2}\left\{\frac{1}{c^2} + \frac{1}{(1-c)^2} + \frac{1}{c(1-c)}\right\}\left(\frac{dc}{dv}\right)^2 = 0$$

has the axis of real quantities in the v-plane for a line of essential singularity. If an integral be

$$c = \phi(v),$$

the general integral of the differential equation is

$$c = \phi\left(\frac{Av + B}{Cv + D}\right);$$

the differential equation of course does not restrict the arbitrary constants A, B, C, D to be real integers. It is easy to see that

* *Treatise on Differential Equations*, p. 92.

this general integral has a circle in the v-plane for a line of essential singularity.

These examples may suffice to shew the substantial difference, between a single equation of the first order and a system of equations each of the first order. For the former, the essential singularities (or points of indeterminateness of a more general character than essential singularities) are fixed points which are isolated; for the latter this is not necessarily the case, and the points of indeterminateness may be parametric isolated points: they may even form a continuous aggregate in the plane of the variable.

CHAPTER VIII.

BRANCH-POINTS OF AN EQUATION OF THE FIRST ORDER
AND ANY DEGREE, AS DETERMINED BY THE EQUATION:
SINGULAR AND PARTICULAR SOLUTIONS*.

94. ONE class of exceptional points of the equation

$$\frac{dw}{dz} = f(w, z)$$

still remains unconsidered, viz. those at or near which the function
$f(w, z)$ does not admit of expression as a uniform function, yet in
such a way that, for any combination of values of w and z, the
number of distinct values which it can assume is limited. It was
pointed out (§ 23) that, taking account of all possible values

---

* This chapter deals with equations of the first order and of degree higher than
the first. In so far as the form of the equation gives rise to singularities in the
vicinity of which the various values of w' are uniform, the appropriate discussion
has been effected in preceding chapters and is not repeated. The chapter is
specially devoted to the discussion of the integral function in the vicinity of
those combinations of values which give branchings of w' as determined by the
equation itself.

In such a discussion, the subject of singular solutions is bound to arise. It
will be found that the subject is discussed almost entirely from the point of view
of functional relation: and that, with relatively slight exceptions, the geometrical
theory (in which both variables, are restricted to real values) is omitted. Such
developments are undoubtedly interesting in themselves and form the material of
a large mass of literature; the reason for the omission is, that my purpose is
chiefly the discussion of functional relations with unrestricted values of the
variables, and deviation, only infrequent, from this range into (purely real)
geometrical theory is made almost entirely for purposes of illustration.

As regards the matter included in the chapter, references are given in various
connections. More general reference may be made to the frequently cited memoir
by Briot and Bouquet (p. 40, note), pp. 191 et seq. (l.c.); to a memoir by Hamburger,
quoted in the note at the end of § 102; to a memoir by Fine, *Amer. Journ. Math.*,
t. XI (1889), pp. 317—320; and to Painlevé's *Stockholm Lectures*, quoted p. 209.

obtained by making the variables w and z describe all conceivable paths, f is then the root of an equation of finite order n, which is irreducible when f is regarded as the variable to be determined. Let this equation be

$$F(z, w, f) = 0 ;$$

then writing W for $\dfrac{dw}{dz}$, the differential equation is one branch of the irreducible equation

$$F(z, w, W) = 0,$$

where the left-hand side is a polynomial in W of degree n, the coefficients of which are uniform functions of z and w. All the exceptional points of the equation $W = f$, which are of the indicated type, will arise in connection with the full investigation of the integral of this differential equation of the first order and nth degree, represented by

$$A_0 W^n + A_1 W^{n-1} + \dots + A_{n-1} W + A_n = 0,$$

where $A_0, A_1, \dots, A_{n-1}, A_n$ are uniform functions of z and w, which have no common factor. It will, for many purposes, be assumed that the coefficients A are algebraical in w, so that all of them (on account of the presence of A_0) can be regarded as polynomials in w, the coefficients of the various powers of which are uniform functions of z, regular in the vicinity of the point considered; but this assumption will not initially be made.

The first question to be discussed is the existence of an integral of the equation and of the initial conditions which determine that integral: and therefore, taking account of the existence-theorem for the equation of the first degree, we have to consider the existence of an integral or integrals (if any) of the equation

$$F\left(z, w, \frac{dw}{dz}\right) = 0,$$

which become equal to α when $z = c$.

For this purpose, consider the equation

$$F(c, \alpha, p) = 0,$$

which is of degree n in p, unless c, α be such as to make $A_0 = 0$; let $p = \gamma$ be a root of the equation, and suppose that γ is finite.

Then if we take

$$z = c + x, \quad w = \alpha + y, \quad \frac{dw}{dz} = \gamma + U,$$

in connection with the original equation, where U vanishes with x and y, we have, on substitution,

$$\frac{\partial F}{\partial c} x + \frac{\partial F}{\partial \alpha} y + \frac{\partial F}{\partial \gamma} U + \ldots = 0,$$

which is algebraical in U. Now for sufficiently small values of $|x|$ and $|y|$, we can make $|U|$ infinitesimal; hence

$$w - \alpha - \gamma (z - c)$$

can be made infinitesimal for those values, that is, we can take

$$y = \gamma x + v,$$

where v (not necessarily a uniform function of x) is of order higher than the first in powers of x, when $|x|$ is sufficiently small. We then have

$$\gamma + U = \frac{dw}{dz} = \frac{dy}{dx},$$

so that $$U = \frac{dv}{dx};$$

and therefore the equation for v is

$$\left(\frac{\partial F}{\partial c} + \gamma \frac{\partial F}{\partial \alpha}\right) x + \frac{\partial F}{\partial \alpha} v + \frac{\partial F}{\partial \gamma} \frac{dv}{dx} + \ldots = 0,$$

where the unexpressed terms are of the second and higher orders in x, v, $\frac{dv}{dx}$; and both v and $\frac{dv}{dx}$ must vanish when $x = 0$.

95. First, let γ be a simple root of

$$F(c, \alpha, \gamma) = 0,$$

so that $\frac{\partial F}{\partial \gamma}$ is not zero. Then if $\frac{\partial F}{\partial c} + \gamma \frac{\partial F}{\partial \alpha}$ is not zero, the preceding equation gives

$$\frac{dv}{dx} = \phi_1(v, x)$$

$$= Av + B_1 x + \text{higher powers of } v \text{ and } x,$$

where ϕ_1 is a regular function of v and x; the coefficients are

$$B_1 = \frac{\dfrac{\partial F}{\partial c} + \gamma \dfrac{\partial F}{\partial \alpha}}{-\dfrac{\partial F}{\partial \gamma}}, \quad A = \frac{\dfrac{\partial F}{\partial \alpha}}{-\dfrac{\partial F}{\partial \gamma}},$$

and B_1, according to the hypotheses made, does not vanish. To this equation, we can apply Cauchy's existence-theorem. The initial condition is that $v = 0$ when $x = 0$; and therefore there exists a unique regular function of x, which vanishes with x and satisfies the equation. There may be other integrals of the equation which are non-regular functions of x vanishing with x, if $x = 0$ is a point of indeterminateness for the equation

$$\frac{dZ}{dx} = \phi_1 (v + Z, x) - \phi_1 (v, x);$$

but unless this is the case, the unique regular integral is the only integral that vanishes with x, and its first term clearly is

$$\tfrac{1}{2} B_1 x^2.$$

It therefore follows that, in the vicinity of $z = c$, there exists a branch of the integral which is a regular function in that vicinity, associated with a simple root γ of the equation

$$F (c, \alpha, p) = 0,$$

provided $\dfrac{\partial F}{\partial c} + \gamma \dfrac{\partial F}{\partial \alpha} \neq 0$. This is true for each simple root of the equation; and therefore if all the roots of the equation

$$F (c, \alpha, p) = 0$$

are simple, the integral has n branches, each of them a regular function in the vicinity of $z = c$.

If however $\dfrac{\partial F}{\partial c} + \gamma \dfrac{\partial F}{\partial \alpha}$ is zero, so that the term in the first power of x alone is absent from ϕ_1, then the coefficient of x^2 in ϕ_1 is

$$\frac{1}{2} \frac{1}{-\dfrac{\partial F}{\partial \gamma}} \left(\frac{\partial}{\partial c} + \gamma \frac{\partial}{\partial \alpha} \right)^2 F = B_2,$$

say. When this is different from zero, the conclusion is the same as before, except that the leading term in v is

$$\tfrac{1}{3} B_2 x^3.$$

More generally, if

$$-\frac{1}{\dfrac{\partial F}{\partial \gamma}} \left(\frac{\partial}{\partial c} + \gamma \frac{\partial}{\partial \alpha} \right)^{s} F$$

vanishes for $s = 1, 2, \ldots, m - 1$, but not for $s = m$, then the conclusion is the same as before, except that the leading term in v is

$$\frac{x^{m+1}}{(m+1)!} \frac{1}{\dfrac{\partial F}{\partial \gamma}} \left(\frac{\partial}{\partial c} + \gamma \frac{\partial}{\partial \alpha} \right)^{m} F.$$

Hence when γ is a simple root of F, so that $\dfrac{\partial F}{\partial \gamma}$ is not zero, there exists, associated with it, a branch of the integral which, in the immediate vicinity of $z = c$, is a regular function of $z - c$, unless the values of c, α, γ are such that

$$\left(\frac{\partial}{\partial c} + \gamma \frac{\partial}{\partial \alpha} \right)^{s} F = 0,$$

for all positive integral values of s. For general values, the latter possibility can happen only if F is a function of $\alpha - c\gamma$ alone, or is a function of γ alone, or is a function of $\alpha - c\gamma$ and γ alone; so that, remembering that F is algebraical of order n in γ, the original equation* would be

$$G \left(w - z \frac{dw}{dz}, \frac{dw}{dz} \right) = 0,$$

where G is an algebraical function of both its arguments such that $\dfrac{dG(\alpha - c\gamma, \gamma)}{d\gamma}$ is not zero. As this equation is merely a generalised form of Clairaut's equation, the treatment of which belongs to the elements of the subject, we may suppose that it has been considered and therefore it can be placed on one side; accordingly, we are justified in assuming that

$$\left(\frac{\partial}{\partial c} + \gamma \frac{\partial}{\partial \alpha} \right)^{s} F$$

does not vanish for all integer values of s.

---

* That is, if c and α (and consequently γ) are parametric quantities; but if they are pure constants, the form in the text is not necessarily that of the equation. The latter alternative is discussed later (§ 98).

If all the roots of $F(c, \alpha, \gamma) = 0$, an algebraical equation in γ, are simple, the preceding result applies to each of them; the necessary and sufficient condition is that the root γ shall keep $\dfrac{\partial F}{\partial \gamma}$ distinct from zero.

96. Next, let γ be a multiple root of F of order m, say, so that $\dfrac{\partial^r F}{\partial \gamma^r} = 0$ for $r = 1, \ldots, m-1$. With the same notation and substitutions as before, we have

$$\left(\frac{\partial F}{\partial c} + \gamma \frac{\partial F}{\partial \alpha} \right) x + \frac{\partial F}{\partial \alpha} v + \frac{1}{m!} \frac{\partial^m F}{\partial \gamma^m} \left(\frac{dv}{dx} \right)^m + \ldots = 0,$$

where the unexpressed terms are of higher order than unity in x and v alone or combined, and of order higher than m in $\dfrac{dv}{dx}$ alone.

When $\dfrac{\partial F}{\partial c} + \gamma \dfrac{\partial F}{\partial \alpha}$ for this multiple root of F is not zero, let $x = t^m$; then we find

$$v = t^{m+1} P(t),$$

where P is a regular function of t in the vicinity of $t = 0$, such that

$$P(0) = \frac{m}{m+1} \left\{ m! \, \frac{\dfrac{\partial F}{\partial c} + \gamma \dfrac{\partial F}{\partial \alpha}}{-\dfrac{\partial^m F}{\partial \gamma^m}} \right\}^{\frac{1}{m}}, \quad = \frac{m}{m+1} \vartheta,$$

say; and therefore

$$w - \alpha = (z - c)\gamma + v$$
$$= (z - c)\gamma + (z - c) P_1 \{(z - c)^{\frac{1}{m}}\},$$

where P_1 is a regular function of $(z - c)^{\frac{1}{m}}$ in the vicinity of $z = c$, such that

$$P_1 \{(z - c)^{\frac{1}{m}}\} = (z - c)^{\frac{1}{m}} P(0) + \ldots.$$

Consequently, there are m branches of the integral of the original equation which become equal to α when $z = c$; the point is an algebraical branch-point, and the m branches form a single cycle. The condition for this result is that, γ being a multiple root of order m, the quantity $\dfrac{\partial F}{\partial c} + \gamma \dfrac{\partial F}{\partial \alpha}$ is not zero.

To settle whether this cycle of integrals is the complete set of integrals vanishing with x (and therefore with t), let

$$v = t^{m+1} P(t) + Z,$$

where Z should vanish with t. Then we have

$$\frac{\partial F}{\partial \alpha} Z + \frac{1}{m!} \frac{\partial^m F}{\partial \gamma^m} m \left(\frac{dv}{dx}\right)^{m-1} \frac{dZ}{dx} + \dots = 0,$$

where the unexpressed terms are, in part, higher powers of Z, higher powers of $\frac{dZ}{dx}$, combinations of Z and $\frac{dZ}{dx}$, and terms in the first power of $\frac{dZ}{dx}$ involving higher powers of $\frac{dv}{dx}$ than the $(m-1)$th. Now

$$m \left(\frac{dv}{dx}\right)^{m-1} \frac{dZ}{dx} = \frac{1}{\left(\frac{dx}{dt}\right)^m} m \left(\frac{dv}{dt}\right)^{m-1} \frac{dZ}{dt}$$

$$= \frac{dZ}{dt} (\vartheta^{m-1} + \text{positive powers of } t),$$

and so for the other terms; thus the equation for Z is

$$\frac{\partial F}{\partial \alpha} Z + \frac{1}{m!} \frac{\partial^m F}{\partial \gamma^m} \vartheta^{m-1} \frac{dZ}{dt} + \dots = 0.$$

Unless $t = 0$ is a point of indeterminateness for this equation, the only integral which vanishes when $t = 0$ is a steady zero; and the test, as to whether the point is one of indeterminateness or not, depends largely upon the terms

$$\frac{\partial F}{\partial \alpha} Z + \frac{1}{2!} \frac{\partial^2 F}{\partial \alpha^2} Z^2 + \frac{1}{3!} \frac{\partial^3 F}{\partial \alpha^3} Z^3 + \dots,$$

which involve powers of Z alone (§§ 29, 32).

If the point $t = 0$ is not a point of indeterminateness for the equation in Z, then the group of m branches which are regular functions of t are the only branches of the integral which vanish when $z = c$.

97. Next, suppose that

$$\frac{\partial F}{\partial c} + \gamma \frac{\partial F}{\partial z} = 0$$

for the multiple root γ under consideration; but for the same
reasons as before, assume that

$$\left(\frac{\partial}{\partial c} + \gamma \frac{\partial}{\partial \alpha}\right)^s F$$

does not vanish for all values of $s = 1, 2, 3, \ldots$, ad inf.: let $s = n$ be
the first value for which it does not vanish, so that the term
involving the lowest power of x alone has index n. There is a
term in $\left(\frac{dv}{dx}\right)^m$; so that the equation takes the form

$$\left(\frac{dv}{dx}\right)^m + \Sigma C_{ijk} x^i v^j \left(\frac{dv}{dx}\right)^k + A x^n + \ldots = 0,$$

where the unexpressed terms are powers of x alone higher than
the nth, powers of $\frac{dv}{dx}$ alone higher than the mth, and terms of order
that obviously is higher than that of every term retained in the
triple summation.

Now v is to vanish with x, as must also $\frac{dv}{dx}$; so that, if v be of
order μ in powers of x for small values of x, $\frac{dv}{dx}$ will be of order
$\mu - 1$ in powers of x, and therefore μ must be greater than 1.
Accordingly, let

$$v = x^{\rho+1} \left(\frac{\alpha}{\rho + 1} + \xi\right),$$

where α is a constant different from zero, ρ is a real positive
quantity, and ξ vanishes with x. Then the dimension in x of
the lowest term in $C_{ijk} x^i v^j \left(\frac{dv}{dx}\right)^k$ is

$$i + j(\rho + 1) + k\rho$$
$$= i + j + (k + j)\rho.$$

When the value of v is substituted in the equation, the latter
must be satisfied identically; the terms of lowest order must therefore disappear. If two such equal orders arise from the terms in
C_{ijk} and $C_{i'j'k'}$, we have

$$i' + j' + (k' + j')\rho = i + j + (k + j)\rho,$$

and therefore

$$\rho = \frac{i + j - (i' + j')}{k' + j' - (k + j)}.$$

To determine the appropriate values of ρ, we use a Puiseux diagram in the same way as in Chapter V; we mark the points $(k+j, i+j)$, corresponding to all the terms in the equation; if a line joining the two points $(k+j, i+j)$ and $(k'+j', i'+j')$ make an angle $\tan^{-1}\rho$ with the negative sense of the axis $O\xi$, then ρ is the magnitude given by the above fraction. Clearly there will be one (and only one) point on the axis $O\xi$, viz. $(m, 0)$; and there will be one (and only one) point on the axis $O\eta$, viz. $(0, n)$. We take a line through this point on the axis $O\eta$, make it turn in a counterclockwise sense until it meets some of the points in the tableau; then make it turn about the last of these until it meets others, and so on: the last direction of the line passes through the single point on the axis $O\xi$. Every portion of this broken line gives a positive value to ρ; the points lying on any portion correspond with the terms in the equation which, for the substitution adopted, give rise to the terms in x that then are of lowest order.

The various quantities ρ are real positive commensurable magnitudes; for any portion of the line, let

$$\rho, = \frac{i+j-(i'+j')}{(k'+j')-(k+j)},$$

be equal to $\dfrac{p}{q}$ when expressed in its lowest terms, where the two points may be supposed the extreme points on that portion. Take

$$x = t^q, \quad v = t^{p+q}\left(\frac{\alpha q}{p+q} + \xi\right),$$

where ξ vanishes with t, so that

$$\frac{dv}{dx} = t^p\left(\alpha + \frac{t}{q}\frac{d\xi}{dt} + \frac{p+q}{q}\xi\right)$$
$$= t^p(\alpha + \eta),$$

where η vanishes with t. Then substituting, we have

$$0 = \Sigma C_{ijk}\, t^{q(i+j)+p(j+k)}\left(\frac{\alpha q}{p+q}\right)^{j-1} \alpha^{k-1}\left(\frac{\alpha q}{p+q} + j\xi\right)(\alpha + k\eta)$$

+ terms involving higher powers of t in combination with constants and powers and products of ξ and η.

Choose α so that

$$\Sigma C_{ijk}\left(\frac{\alpha q}{p+q}\right)^j \alpha^k = 0;$$

on dividing out by the lowest power of t, the equation takes the form

$$\Sigma C_{ijk} \left(\frac{\alpha q}{p+q}\right)^{j-1} \alpha^k \left\{ j\xi + k\,\frac{q}{p+q}\,\eta \right\} + At + tP_1(\xi,\,\eta,\,t) = 0,$$

where $P_1(\xi,\,\eta,\,t)$ is a regular function vanishing with ξ, η, t. Now

$$j\xi + k\,\frac{q}{p+q}\,\eta = (j+k)\,\xi + \frac{k}{p+q}\,t\,\frac{d\xi}{dt};$$

and therefore the equation takes the form

$$\lambda t\,\frac{d\xi}{dt} = \mu\xi + At + tP\,(\xi,\,\eta,\,t),$$

where

$$\lambda = \ \Sigma C_{ijk} \left(\frac{\alpha q}{p+q}\right)^{j-1} \alpha^k\,\frac{k}{p+q},$$

and

$$\mu = -\,\Sigma C_{ijk} \left(\frac{\alpha q}{p+q}\right)^{j-1} \alpha^k\,(j+k).$$

The general character of the solution of this equation in the vicinity of $t=0$ is known (Chap. VI). If $\frac{\mu}{\lambda}$ is not a positive integer, there is a regular function of t, vanishing with t, which satisfies the equation; one such regular function arises for each value of α connected with the determinate quantity $\frac{p}{q}$; and each such function leads to a branch of the integral of the original equation, for which therefore the point is a branch-point of algebraical type; there are as many groups of cycles of integrals at the branch-point as there are distinct values of $\frac{p}{q}$; in each group of cycles, there are as many cycles as there are simple roots α of the equation

$$\Sigma C_{ijk} \left(\frac{\alpha q}{p+q}\right)^{j} \alpha^k = 0\,;$$

and each cycle contains q members which circulate round the point $x=0$, these being regular functions of $x^{\frac{1}{q}}$.

If $\frac{\mu}{\lambda}$, though not a positive integer, has its real part positive, then, in addition to these regular functions of $x^{\frac{1}{q}}$ which are integrals of the equation, there is an infinitude of other integrals which are regular functions of $x^{\frac{1}{q}}$ and $x^{\frac{\mu}{\lambda q}}$. All these quantities ξ vanish with x, as do the regular functions of $x^{\frac{1}{q}}$.

If $\dfrac{\mu}{\lambda}$, not being a positive integer, has its real part negative, then the regular functions of $x^{\frac{1}{q}}$ are the only integrals of the equation of the required type.

If $\dfrac{\mu}{\lambda}$ is a positive integer, there is no regular integral of the equation in ξ which vanishes with t unless $A = 0$; but if $A = 0$, there is an infinitude of such integrals, which thus lead to integrals of the original equation and are regular functions of $x^{\frac{1}{q}}$. When $A \neq 0$, there is an infinitude of integrals of the equation in ξ, which vanish with x and are regular functions of $x^{\frac{1}{q}}$ and $x^{\frac{1}{q}} \log x$.

It is unnecessary to recite the corresponding results for alternative cases not quoted in the above list. The form of the equation obtained is one of the typical forms which have been discussed at length in a preceding chapter; the results apply, *mutatis mutandis*, to the present case.

Similarly, when λ is zero, the equation has the form

$$\lambda_1 t^\theta \frac{d\xi}{dt} = \mu\xi + At + tP\,(\xi,\,\eta,\,t),$$

where the integer θ is > 1, and all the terms in $tP\,(\xi,\,\eta,\,t)$ are of higher order than $t^\theta \dfrac{d\xi}{dt}$: again, the properties of the integral of this equation have been obtained.

If α is not a simple root of the algebraical equation which determines it, corresponding investigations lead to others of the typical forms previously discussed. In every case where the method can be applied, the first necessity is the reduction to one of the typical forms: the properties of the branches of the integral are then known.

98. What has thus far been considered has been the possibility that, for the original equation of degree n, the values $w = \alpha$

and $z = c$ give for $\dfrac{dw}{dz}$ a value γ, which is finite and may be either a simple root or a multiple root of

$$F(c, \ \alpha, \ \gamma) = 0.$$

But in the first place, it should be noticed that the equation may become evanescent for these initial values; thus the equation

$$hz^3 + gz^2 \frac{dw}{dz} + fw^2 \left(\frac{dw}{dz}\right)^2 + z^2 w \left(\frac{dw}{dz}\right)^3 = 0$$

ceases to determine any quantity γ if $w = 0$, $z = 0$ are initial values. In such a case, there may exist an integral which vanishes with z; if, when $|z|$ is very small, it can have the form $w = \sigma z^\mu + \ldots$, so that μ is a positive quantity, it still does not follow that μ is greater than unity or even is at least as great as unity. Such cases must be considered separately.

Further, one general class of equations has been put on one side, viz. those for which

$$\left(\frac{\partial}{\partial c} + \gamma \frac{\partial}{\partial \alpha}\right)^s F = 0$$

is satisfied for all values of $s = 1, \ 2, \ \ldots$; the corresponding functional form has been indicated when the quantities c, α, γ are of a parametric character, and the equation then falls under a recognised type of forms already discussed. But it may easily happen that the foregoing relation is satisfied for all values of s, the constants c, α and γ not being parametric. For instance, if F is explicitly independent of z, and if the value of γ is zero, then the condition would be satisfied: such a case arises for an equation

$$\left(\frac{dw}{dz}\right)^m \{1 + wG(w)\} + w^{m'+m} H(w) = 0,$$

where G and H are regular functions of w, and m and m' are positive integers. It may happen for some such equations that the Puiseux-diagram will determine suitable values of ρ; this does not occur for the equation written down, because no value of ρ that is positive can be obtained from the diagram. Equations which do not explicitly involve the independent variable will be considered separately (§§ 127—130).

Lastly, it may be the case that the combination $w = \alpha$, $z = c$ gives only $n - r$ finite values of $\dfrac{dw}{dz}$, while the original equation is of degree n; the implication is that $\dfrac{dw}{dz}$ has r infinite values. To discuss the corresponding branches of the integral, we should interchange the variables, so that $\dfrac{dz}{dw}$ then is a zero-root of multiplicity r; and the relation of z to w would be obtained as that of w to z is obtained in § 96.

It therefore remains to discuss the equation, whether in w and z, or in v and x, for these omitted cases; for this purpose it can be taken in the form

$$\Sigma\Sigma\Sigma A_{ijk}\, z^i w^j \left(\frac{dw}{dz}\right)^k = 0,$$

such that some factor free from $\dfrac{dw}{dz}$ occurs on the left-hand side, and the equation is irreducible in $\dfrac{dw}{dz}$; the complete generality of the form justifies the assumption that $w = 0$, $z = 0$ are initial values. It will not be assumed that terms involving z alone are absent: in fact, this presence of terms involving z alone is sometimes needed for those equations which do not determine any value of $\dfrac{dw}{dz}$ for $w = 0$, $z = 0$.

As w is to vanish with z, let its order in powers of z be μ when $|z|$ is sufficiently small; and let

$$w = \theta z^\mu + \ldots = v z^\mu,$$

where θ is a constant; then $\mu > 0$, though it is not necessarily > 1. The dimension of the term $A_{ijk}\, z^i w^j \left(\dfrac{dw}{dz}\right)^k$ in z is

$$i + \mu j + (\mu - 1)k = i - k + \mu(j + k).$$

If the dimension of this term be the same as that of

$$A_{i'j'k'}\, z^{i'} w^{j'} \left(\frac{dw}{dz}\right)^{k'},$$

then

$$i' - k' + \mu(j' + k') = i - k + \mu(j + k),$$

and therefore

$$\mu = \frac{i - k - (i' - k')}{j' + k' - (j + k)}.$$

Accordingly, we mark in a plane, referred to two perpendicular axes, all the points $(j + k, \ i - k)$; all of these have positive abscissæ, unless there exist terms such that $j = 0$, $k = 0$, and then the corresponding points lie (for the various values of i) on the axis $O\eta$. In the latter case, we choose the point on $O\eta$ nearest to O (its ordinate is positive) as the initial point; in the former case, we take the points which have the smallest abscissa and among them, if there be more than one, that which has the smallest ordinate, and choose this last as the initial point.

Through this initial point, we draw a line parallel to $O\eta$, and make it turn in the counterclockwise sense until it meets points in the tableau; and in the customary manner we construct the broken line, continuing it so long as it is inclined at an acute angle to the negative direction of $O\xi$.

If in one extreme limit, there be a value $\mu = \infty$, the implication is that $w = 0$ while z varies; that is, w is a constant zero for a small region round the origin.

If at the other extreme limit, there be a value $\mu = 0$, the implication is that z is a constant zero while w varies; that is, w is not a function of z.

If there be no value of μ which is positive, the implication is that no variable function w exists satisfying the equation and vanishing with z; there then is no term in the equation involving z alone. An instance of such an equation is

$$\left(\frac{dw}{dz}\right)^2 + 2\frac{dw}{dz}w^2 + w^3 = 0 \ ;$$

that $w = 0$ satisfies the equation is obvious; that no other function, which vanishes with z, satisfies the equation, can easily be verified.

If there be one (or more than one) value of μ which is positive, it is a real commensurable quantity, because it is the quotient of two integers; let it be $\frac{p}{q}$, when expressed in its lowest terms. (If $p > q$, then $\frac{dw}{dz}$ vanishes with z; if $p < q$, $\frac{dw}{dz}$ is infinite when $z = 0$; if $p = q$, $\frac{dw}{dz}$ is a finite quantity different from zero when $z = 0$.) We take

$$z = t^q, \quad w = vt^p = (\rho + V)t^p,$$

where ρ is a finite constant, and $V = 0$ when $t = 0$; and then

$$\frac{dw}{dz} = \frac{1}{q} t^{p-q} \left(p\rho + pV + t \frac{dV}{dz} \right).$$

When these values are substituted, one set of terms is of common order lower than all others, namely the set corresponding to the points on the portion of the broken line which gives the value $\frac{p}{q}$ of μ adopted; when these are made to vanish, there arises an algebraical equation for the determination of ρ.

The subsequent analysis leads to a form, of the same character as the typical reduced forms for the earlier investigations; the nature and number of the integrals, regular and non-regular, which vanish with t, are determined by using the known results relating to the integrals of those typical reduced forms. In some cases, the non-regular integrals can be made regular by an algebraical substitution of the type $x = t^q$; in some cases, the non-regular integrals are regular functions of t and t^a, where a is a constant having its real part positive but not an integer, or are regular functions of t and $t \log t$; in other cases, the point can be a point of indeterminateness of character less simple than these, including the possibility of belonging to one of various classes of essential singularities. In every case, the points considered are isolated points: and they are fixed points, determined by the equation.

Ex. 1. Consider the equation

$$hz^3 + gz^2 \frac{dw}{dz} + fw^2 \left(\frac{dw}{dz} \right)^2 + z^2 w \left(\frac{dw}{dz} \right)^3 = 0,$$

so as to obtain the integrals (if any) which vanish when $z = 0$.

The points $(j+k, i-k)$ are $(0, 3)$; $(1, 1)$; $(4, -2)$; $(4, -1)$; there are two values of μ, viz. $\mu = 2$ and $\mu = 1$.

First, let $\mu = 2$; and write

$$w = \rho z^2 + \ldots = (\rho + V) z^2,$$

no change of the independent variable being necessary; here ρ is a finite constant, and V vanishes with z. We choose

$$\rho = -\frac{h}{2g};$$

and then we find

$$z \frac{dV}{dz} = -2V - 4f \frac{\rho^4}{g} z^3 - \ldots.$$

Of this equation, there is one and only one integral vanishing with z; it is the regular function

$$-\frac{4}{5}\frac{f}{g}\rho^4 z^3 + \text{higher powers of } z.$$

Thus corresponding to the value $\mu = 2$, there is one (and only one) integral w which vanishes with z; its value is

$$w = -\frac{hz^2}{2g} - \frac{1}{20}\frac{h^4 f}{g^5} z^5 + \dots.$$

Secondly, let $\mu = 1$; and write

$$w = \rho z + \dots = (\rho + V) z,$$

again no change in the independent variable being necessary: as before, ρ is a finite constant and V vanishes with z. Substituting, we choose

$$f\rho^3 + g = 0,$$

so as to annihilate the lowest terms in z; and then we have, after reduction,

$$z\frac{dV}{dz} = -3V + \left(\frac{h}{g} - \frac{\rho}{f}\right) z + \dots.$$

There are three distinct values of ρ. For each value of ρ, there is one and only one integral of this equation vanishing with z; it is the regular function

$$V = \frac{1}{4}\left(\frac{h}{g} - \frac{\rho}{f}\right) z + \dots.$$

Thus corresponding to the value $\mu = 1$, there are three (and only three) integrals of the equation which vanish with z; each of them is a regular function of z; and their analytical expression is

$$w = \rho z + \frac{1}{4}\left(\frac{h}{g} - \frac{\rho}{f}\right) z^2 + \dots,$$

where ρ is any one of the three roots of the equation

$$f\rho^3 + g = 0.$$

Ex. 2. Discuss the foregoing equation, (i) when $g = 0$, (ii) when $f = 0$, (iii) when $h = 0$.

99. The preceding results indicate the character of the integral as determined by an arbitrarily assigned pair of initial values, $w = a$, $z = c$; and they shew that, for the simplest cases, there is a fundamental distinction according as all the roots γ of

$$F(c, a, \gamma) = 0$$

are simple or not. Each simple root γ determines a regular function of $z - c$, which is equal to a when $z = c$ and is an integral of the equation; and there may be other non-regular functions of $z - c$, also equal to a when $z = c$, each of them an integral of the equation. Each multiple root γ, say of multiplicity m, determines a group of m integrals of the equation, each of which is equal to α when $z = c$; they are regular functions of $(z - c)^{\frac{1}{m}}$ in the immediate vicinity of the point, and they form a single cycle for circulation of z round c.

If, then, $F = 0$ has all its roots simple, the point is an ordinary point of the equation; each of the n integrals, corresponding to the n determined values of γ, is a regular function of $z - c$. If, however, only some of its roots are simple, and the rest are multiple, then the integrals corresponding to the simple roots are regular functions of $z - c$; and the remaining integrals can be arranged in cycles, each cycle corresponding to one value of α (multiple) root and the members of the cycle circulating round the point. The point is a branch-point of the equation.

If a value of γ is infinite, so that $\dfrac{dw}{dz}$ is infinite for $w = \alpha$, $z = c$, then $\dfrac{dz}{dw}$ is zero for those values. Should the infinity be a simple infinity, then $z - c$ is a regular function of $w - \alpha$ in the vicinity, and as $\dfrac{dz}{dw}$ is zero for $w = \alpha$, $z = c$, the expansion of $z - c$ begins with a term in $(w - \alpha)^m$, where $m \geqslant 2$; consequently, reversing the series, $w - \alpha$ is a regular function of $(z - c)^{\frac{1}{m}}$, that is, the point is a branch-point of the integral. Should the infinity be a multiple infinity of order m, then $z - c$ is a regular function of $(w - \alpha)^{\frac{1}{m}}$ in the vicinity; as $\dfrac{dz}{dw}$ is zero for $w = \alpha$, $z = c$, the expansion of $z - c$ begins with a term in $(w - \alpha)^{1 + \frac{n}{m}}$, where $n \geqslant 1$; consequently, reversing the series, $w - \alpha$ is a regular function of $(z - c)^{\frac{1}{m+n}}$, the first term in its expansion being the term in $(z - c)^{\frac{m}{m+n}}$: the point is a branch-point of the integral.

If for $w = \alpha$, $z = c$, values of $\dfrac{dw}{dz}$ are not determinate, then we proceed, as in § 98, to obtain the expression in the vicinity of $w = \alpha$, $z = c$; the character of the integral depends upon the typical reduced forms.

100. Reverting then to the equation

$$F(c, \alpha, \gamma) = 0,$$

the fundamental distinction for finite values of γ lies between simplicity and multiplicity of occurrence of a root γ. Construct the discriminant of $F = 0$, and denote it by Δ (c, α); then as F is of degree n in γ, Δ is of degree $2n - 2$ in the coefficients of F. If these coefficients are algebraical in α, say of degree $\leqslant m$, then Δ is of degree $\leqslant m$ $(2n - 2)$, that is, algebraical, in α; and as F is merely analytical in c, then Δ also is analytical in c.

Now when each root of F is simple, Δ does not vanish; and when Δ does vanish, then at least one root of F is multiple. In the former case, the point is an ordinary point for the differential equation, and the n integrals are regular functions; in the latter case, only some of the integrals are regular functions, the remainder being non-regular functions that range themselves in cycles.

If N, $\leqslant m$ $(2n - 2)$, be the degree of Δ in α, then for general values of c there are values of α, $\leqslant N$ in number, for which Δ vanishes; they are, of course, the roots of Δ. Each such combination gives at least one multiple root γ of F; every other value of α, combined with c, gives simple roots of F.

These results can be illustrated geometrically, by regarding z, w as the coordinates of a point in a plane; the illustration, of course, implies restrictions as regards functionality and variation. When the discriminant of $F\left(z, w, \dfrac{dw}{dz}\right) = 0$ is formed, let it be denoted by Δ (z, w). When $z = c$, let $\beta_1, \beta_2, \ldots$ be the corresponding values of w given by

$$\Delta(c, \beta) = 0;$$

and let α denote a general parametric value of w when it is not a root of $\Delta = 0$.

Let a point move along the line $z = c$ parallel to the w-axis. For all except a limited number of positions, α is a value of w; and, for each of these, the n integrals of the equation are regular functions, determined by the n (different) values of γ, each simple. Through each such position, there pass n distinct curves; and as the point moves along the ordinate, the n curves move with it, changing continuously their directions. When the point comes to a position β, then some root γ is repeated, it may be, more than one root; say γ_1 repeated λ_1 times, γ_2 repeated λ_2 times, and the rest simple. Through this position there pass n curves; λ_1 of these have a common tangent at the point, its direction determined by γ_1: other λ_2 of them have a common tangent at the point, its direction determined by γ_2; the remaining $n - \lambda_1 - \lambda_2$ do not touch one another or either of the two sets. Moreover, the position lies on the curve

$$\Delta\,(z,\,w) = 0,$$

of which it may be an ordinary point or a singular point; but there is nothing to shew that any one of the directions of the integral curves (there being $n - \lambda_1 - \lambda_2 + 2$ of such directions) at the point coincides with the direction of the tangent to the discriminant-curve: though, on the other hand, there is no obvious general necessity which could prevent the coincidence from sometimes occurring. When the moving point has passed through the β-position, then w resumes an α-value; and the n integral curves are again distinct. And so on, in succession.

Ex. 1.   As an example, consider

$$F = \left(\frac{dw}{dz}\right)^3 - z\,\frac{dw}{dz} + w = 0.$$

For $z = a$, $w = a$, the values of $\dfrac{dw}{dz}$ are the roots of

$$\gamma^3 - \gamma a + a = 0.$$

The discriminant is $a^2 - \tfrac{4}{27}a^3$; and therefore we have

$$\beta^2 - \tfrac{4}{27}a^3 = 0.$$

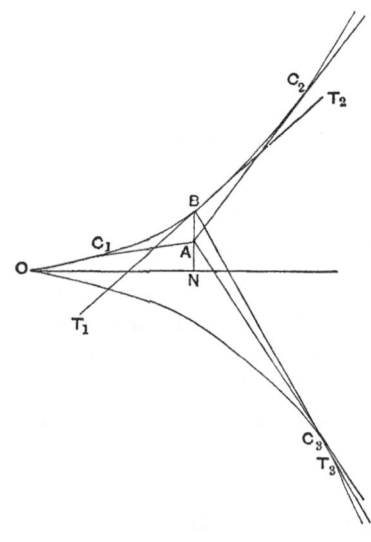

Take any value $a = ON$, and draw this curve; let the ordinate through N meet it in B.

At any point A, where $a = AN$, not on the curve, the three values of γ are distinct; through A there are three directions AC_1, AC_2, AC_3; and the branches of the integral through A are, in fact, three straight lines. They remain distinct as A moves on the line NB.

When the moving point comes to B, where $\beta = BN$, a point on the curve, one value of γ is repeated simply; the other value is distinct from it. The three directions through B are BT_1, BT_2 (which two are in the same straight line), and BT_3. The branches of the integral through A are three straight lines; two of them coincide there, having T_1BT_2 for their common direction; the third of them is BT_3.

When the moving point passes beyond B, the three values of γ again become distinct; two of them are conjugate complex values, and the remaining one is real. The full geometrical representation is now not possible; but for each such position, the three branches of the integral are regular functions determined by the three distinct values of γ.

It is easy to verify for the present instance :—

(i) that each of the three directions at any point, such as A, touches the discriminant-curve;

(ii) that at a point B on the discriminant-curve, the common direction of the two coincident branches of the integral coincides with the tangent of the discriminant-curve.

The general integral of the equation is

$$w = Az - A^3,$$

where the arbitrary constant A, if determined by initial values, is such that

$$a = Aa - A^3;$$

manifestly, there are three branches of the integral. There is a solution (which will be recognised as the "singular solution") given by

$$w^2 = \tfrac{4}{27} z^3;$$

being determined by initial values a and β, they must be such as to satisfy

$$\beta^2 = \tfrac{4}{27} a^3.$$

The statement, that the discriminant-curve in this case provides a solution of the differential equation, is justified for the following reasons.

At every point on the discriminant-curve, the value of $\dfrac{dw}{dz}$ determined by that curve is the same as the value of $\dfrac{dw}{dz}$ determined by the common direction of the two coincident branches of the integral that pass through the point. Hence, at the point of the discriminant-curve, the values of w and z are the same as for all the integral curves; and the value of $\dfrac{dw}{dz}$ is the same as that for two of the integral curves. These values satisfy the equation for the integral curves; accordingly, when regarded as belonging to the discriminant-curve, they satisfy the equation, that is, the discriminant-curve can be regarded as providing a solution of the equation.

It is not difficult to see that the discriminant-curve is not an integral curve, in the preceding sense of the words: it provides a value of $\dfrac{dw}{dz}$, with a value of w and a value of z which, taken together, satisfy the equation.

Ex. 2. As another example, consider

$$3c\left(\frac{dw}{dz}\right)^2 - 2z\frac{dw}{dz} + w - b = 0.$$

For $z = a$, $w = a$, the values of $\frac{dw}{dz}$ are the roots of

$$3c\gamma^2 - 2a\gamma + a - b = 0.$$

The discriminant is $a^2 - 3c\,(a - b)$; and therefore we have

$$a^2 = 3c\,(\beta - b).$$

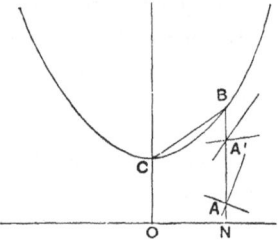

Draw this curve, manifestly a parabola; let $ON = a$, and let the ordinate meet it in B.

At a point A, where a is less than b, one of the values of γ is negative, the other positive: there are two integral curves through A, touching the lines indicated in the figure. At a point A', where a is greater than b, both values of γ are positive, but they are unequal; there are again two integral curves through A', touching the lines indicated in the figure.

When the moving point comes to B, the two values of γ are equal; and the common direction is easily proved to be that of BC. The two integral curves touch at B; and BC is their common tangent.

When the moving point passes beyond B, the two values of γ again become distinct; as they are conjugate complex quantities, the geometrical representation is now not possible. For each such position beyond B, the two branches of the integral are regular functions of z, determined by the two distinct values of γ.

It is easy to verify for the present instance :—

(i) that neither of the two directions at a point A touches the discriminant-curve except for the special point such that $a = b$; when $a = b$, one direction touches the discriminant-curve, C being its point of contact ;

(ii) that at any point B on the discriminant-curve, the common tangent of the two touching branches of the integral curve does not coincide with the tangent of the discriminant-curve: the sole exception being for the vertex C of the discriminant-curve.

The general integral is

$$\{2z^3 - 9cz\,(w - b) + A\}^2 = 4\,\{z^2 - 3c\,(w - b)\}^3,$$

where the arbitrary constant A, if determined by initial values a and a, is such that

$$\{2a^3 - 9ca\,(a - b) + A\} = 4\,\{a^2 - 3c\,(a - b)\}^3.$$

101.   Thus far we have been considering the character of the integral of the equation $F = 0$, as determined by assigned initial values of the variables; and it has been seen, in general, that values of the variables, which do not make the discriminant of F vanish, determine n regular functions, each of them an integral: while values of the variables, which do make that discriminant vanish, determine n functions, some of them   erhaps regular, some of them certainly non-regular, but such that the point is a branch-point for these non-regular integrals. (In geometrical phraseology, the discriminant provides a locus of singular points for some of the branches of the integral.)

If the discriminant of $F\left(z,\ w,\ \dfrac{dw}{dz}\right) = 0$ be $\Delta\,(z,\ w)$, then the value of $\dfrac{dw}{dz}$ provided by

$$\Delta\,(z,\ w) = 0,$$

together with a value of w satisfying $\Delta = 0$, does not in general satisfy the differential equation.   From the first example considered, however, it was seen that this statement does not hold in all particular instances; that, in fact, the combination of $\Delta = 0$ with the deduced value of $\dfrac{dw}{dz}$ does make the equation satisfied, so that then it may be considered as providing a solution of the equation.

From the second example considered, it was seen that the statement is true for a general range of values of the independent variable; but that there was one set of values provided by $\Delta\,(z, w) = 0$ with the deduced value of $\dfrac{dw}{dz}$ which made the original differential equation satisfied.   In this case, the differential equation is satisfied only by a number of isolated values connected with $\Delta = 0$; and $\Delta = 0$ cannot then be considered as providing a solution of the equation.

It therefore is necessary to investigate the nature of the functional relation of the values of w, determined by

$$\Delta\,(z,\ w) = 0,$$

to the integral (or integrals) of the original equation.

102. That the value of w furnished by the discriminant-equation $\Delta (z, w) = 0$, together with the value of $\dfrac{dw}{dz}$ given by

$$\frac{\partial \Delta}{\partial z} + \frac{\partial \Delta}{\partial w} \frac{dw}{dz} = 0,$$

does not, in general, satisfy the equation can be seen as follows.

Writing W for $\dfrac{dw}{dz}$, the original equation is

$$F(z, w, W) = 0.$$

The discriminant is the eliminant of F and $\dfrac{\partial F}{\partial W}$; and the vanishing of the discriminant is the condition that $F = 0$ and $\dfrac{\partial F}{\partial W} = 0$ have a common root W, or that this root W is a multiple root of $F = 0$. Taking the simplest case, when the multiple root is of order 2, so that it is a simple root of $\dfrac{\partial F}{\partial W} = 0$, let it be denoted by W_1; and let the other $n - 2$ roots of $\dfrac{\partial F}{\partial W} = 0$ be

$$V_1, V_2, ..., V_{n-2}.$$

When these values of W are substituted in F, let it become $G_1, ..., G_{n-2}$ respectively; and let F_1 be the (zero) value of F for $W = W_1$. Then the analytical expression of the discriminant, save as to a factor which is a power of the coefficient of W^n in F, is given by

$$\Delta = F_1 . G_1 G_2 ... G_{n-2}.$$

Hence taking complete derivatives with regard to z, we have

$$\frac{d\Delta}{dz} = \frac{dF_1}{dz} . G_1 G_2 ... G_{n-2} + F_1 \frac{d}{dz} (G_1 G_2 ... G_{n-2}).$$

When $W = W_1$, the discriminant is zero; F_1 is zero: and $G_1 G_2 ... G_{n-2}$ are not zero. Consequently

$$\frac{d\Delta}{dz} = \frac{dF_1}{dz} \times \text{a factor which does not vanish.}$$

But the value of $\dfrac{dw}{dz}$ from the discriminant-equation is given by

$$\frac{d\Delta}{dz} = \frac{\partial \Delta}{\partial z} + \frac{\partial \Delta}{\partial w} \frac{dw}{dz} = 0;$$

consequently it also is given by

$$0 = \frac{dF_1}{dz}$$

$$= \frac{\partial F_1}{\partial z} + \frac{\partial F_1}{\partial w}\frac{dw}{dz} + \frac{\partial F_1}{\partial W_1}\frac{dW_1}{dz}$$

$$= \frac{\partial F_1}{\partial z} + \frac{\partial F_1}{\partial w}\frac{dw}{dz},$$

because $\frac{\partial F_1}{\partial W_1} = 0$. Now for the particular value of w, the value of $\frac{dw}{dz}$ must be such as to satisfy $F(z, w, W) = 0$, if the discriminant-equation leads to a solution of the original equation: that is, $\frac{dw}{dz}$ must be either W_1, the repeated root of $F = 0$, or it must be one of the remaining $n - 2$ roots of $F = 0$; and therefore we then should have

$$F\left(z,\ w,\ -\frac{\partial F_1}{\partial z}\Big/\frac{\partial F_1}{\partial w}\right) = 0,$$

satisfied when the value $W = W_1$ is inserted in the derivatives of F_1; and it must be satisfied either identically, or in virtue of

$$\Delta\,(z,\ w) = 0.$$

When the substitution takes place in $F\left(z,\ w,\ -\frac{\partial F_1}{\partial z}\Big/\frac{\partial F_1}{\partial w}\right)$, let the value be $H(z, w)$; then

$$H(z, w) = 0$$

either is an identity or is satisfied in virtue of $\Delta = 0$. It is clear that, in general, the elimination of W_1 between

$$F(z, w,\ W_1) = 0,$$

and $\qquad F\left(z, w, -\frac{\partial F_1}{\partial z}\Big/\frac{\partial F_1}{\partial w}\right) = 0,$

does not lead to an identically nul-equation; it is also clear that, in general, it does not lead to an equation which is the same as, or involves, or is involved by, the equation obtained on the elimination of W_1 between

$$F(z, w, W_1) = 0, \quad \frac{\partial F_1}{\partial W_1} = 0.$$

Consequently, the condition necessary that the value of $\frac{dw}{dz}$ provided by $\Delta = 0$ should satisfy the original equation is not fulfilled in general; and therefore the value of w provided by $\Delta = 0$ is not, in general, a solution of the original equation.

For instance, in the case of the equation

$$W^n - mWz + w = 0,$$

where m is any constant different from unity, we have

$$-\frac{\dfrac{\partial F_1}{\partial z}}{\dfrac{\partial F_1}{\partial w}} = mW,$$

so that $H(z, w) = 0$ is the result of eliminating W between

$$\left.\begin{array}{c} W^n - mWz + w = 0 \\ (mW)^n - m^2 Wz + w = 0 \end{array}\right\}.$$

Because m is not equal to unity, the eliminant is not an identically nul-equation: it is, in fact,

$$H(z, w) = w^{n-1} - m^{2n-1}\frac{(m^{n-1}-1)^{n-1}(m-1)}{(m^n - 1)^n} z^n = 0.$$

The discriminant of $W^n - mWz + w = 0$, equated to zero, is

$$\Delta(z, w) = w^{n-1} - \frac{m^n}{n^n}(n-1)^{n-1} z^n = 0.$$

These equations are only the same if $m = 1$, a value that has been excluded; hence $\Delta(z, w) = 0$ does not provide a solution of the original equation.

But though this is true of the general function given by $\Delta(z, w) = 0$, there may be special values of z (and the associated values of w) for which the value of $\frac{dw}{dz}$ satisfies the equation. The necessary condition is

$$H(z, w) = 0;$$

and therefore, for values of w and z that satisfy the equations $H = 0$, $\Delta = 0$ simultaneously, the value of $\dfrac{dw}{dz}$ given by $\Delta = 0$ satisfies the equation. Manifestly, these values (both of w and of z) are isolated and, in general, do not form a continuous aggregate.

Thus for the above instance, the values $z = 0$, $w = 0$ lead from $\Delta(z, w) = 0$ to $\dfrac{dw}{dz} = 0$; the combination of these values satisfies the original differential equation. It manifestly is an isolated combination of values.

If, however, $H(z, w) = 0$ were satisfied in virtue of $\Delta(z, w) = 0$, the value of w provided by $\Delta = 0$ would then be a solution of the equation. Instead of discussing the analytical conditions for the equivalence, partial or complete, of these two equations, we investigate the relation of values of w, which are provided by $\Delta = 0$, to integrals of the differential equation, as follows*.

103. Writing w' in place of $\dfrac{dw}{dz}$, we have the equation in the form

$$F(z, w, w') = a_0 w'^n + a_1 w'^{n-1} + \ldots + a_{n-1} w' + a_n = 0,$$

where a_0, a_1, ..., a_n are uniform functions of w and z without a common factor, these functions being generally algebraical in w and analytical in z. Moreover, the equation is supposed irreducible in w'.

Let the discriminant of F, in regard to w', be $\Delta(z, w)$; if

$$\Delta(z, w) = 0,$$

then some value w', which satisfies $F = 0$, satisfies also

$$\frac{\partial F}{\partial w'} = 0,$$

and one root (or more than one root) of $F = 0$ is repeated. Because the coefficients a_0, a_1, ..., a_n are algebraical in w, the discriminant Δ also is algebraical in w; let $w = \eta$ be one of its roots. Suppose that the substitution of η for w does not make the

---

* Hamburger, *Ueber die singulären Lösungen der algebraischen Differential-gleichungen erster Ordnung*, Crelle, t. cxii (1893), pp. 205—246. See also Fuchs, *Ueber Differentialgleichungen deren Integrale feste Verzweigungspunkte besitzen*, Berl. Sitzungsber. (1884), pp. 700—705.

coefficient a_0 vanish identically, so that it remains a function of z after the substitution; and denote by c a constant quantity distinct from the roots of the equation $a_0(z, \eta) = 0$. Then

$$F(z, \eta, w') = 0$$

determines n functions w'; some of them may be distinct from one another, but some of them will certainly coincide with one another; none of them will be infinite in the vicinity of $z = c$. Let p roots w' of $F(z, \eta, w') = 0$ be equal to ζ.

Now in the equation $F(z, w, w') = 0$, $w' = \zeta$ is a root of multiplicity p when $w = \eta$; let

$$w' - \zeta = v, \quad w - \eta = u,$$

so that, in the equation,

$$F(z, u, v) = 0,$$

there are p roots v equal to zero when $u = 0$; hence it must have the form

$$v^p h(z, u, v) + u^q k(z, u) = 0,$$

where q is an integer $\geqslant 1$, and neither $h(z, 0, 0)$ nor $k(z, 0)$ vanishes identically. Assuming that the value $z = c$ is chosen so that neither $h(c, 0, 0)$ nor $k(c, 0)$ vanishes, we have

$$v = g_0 u^{\frac{\kappa}{\alpha}} + g_1 u^{\frac{\kappa+1}{\alpha}} + \dots,$$

where α either is p or is a factor of p, κ is an integer which is positive because $v = 0$ when $u = 0$; and the functions $g_0, g_1, \dots$ are regular functions of z in the vicinity of $z = c$, the first of them g_0 not being zero there; consequently

$$w' - \zeta = g_0 (w - \eta)^{\frac{\kappa}{\alpha}} + g_1 (w - \eta)^{\frac{\kappa+1}{\alpha}} + \dots.$$

104   Now η is a function of z; so that $w = \eta$ may, or may not, constitute a solution of the differential equation.

First, suppose that $w = \eta$ does not constitute a solution of the equation; then, because $w' = \zeta$ is a value of w' that satisfies $F(z, \eta, w') = 0$, it follows that ζ and $\dfrac{d\eta}{dz}$ are not equal to one another, and therefore $\zeta - \dfrac{d\eta}{dz}$ is not identically zero. Being a function of z, it can have isolated zeros; it might happen that $z = c$ is one of them; whether this is so or not, a point b in the

vicinity of $z = c$ can be chosen which is not a zero of $\zeta - \dfrac{d\eta}{dz}$.
Now take

$$w - \eta = u = y^{\alpha},$$

so that

$$\alpha y^{\alpha-1} \frac{dy}{dz} = w' - \frac{d\eta}{dz}$$

$$= \zeta - \frac{d\eta}{dz} + g_0 y^{\kappa} + g_1 y^{\kappa+1} + \dots,$$

and therefore

$$\frac{dz}{dy} = \frac{\alpha y^{\alpha-1}}{\zeta - \dfrac{d\eta}{dz} + g_0 y^{\kappa} + g_1 y^{\kappa+1} + \dots}.$$

Now as $z = b$ is not a zero of $\zeta - \dfrac{d\eta}{dz}$, let the value of $\zeta - \dfrac{d\eta}{dz}$ at $z = b$ be B, where B is not zero; thus

$$B \frac{dz}{dy} = \alpha y^{\alpha-1} G(z - b, y),$$

where $G(z - b, y)$ is a regular function of $z - b$ and y, such that $G(0, 0) = 1$. At the point under consideration, η is the value of w, though $\dfrac{d\eta}{dz}$ is not the value of w'; and therefore, when $z = b$, we have $w = \eta$, that is, we have $y = 0$. Consequently $z = b$, $y = 0$ are initial values for the last equation; and therefore we have

$$B(z - b) = y^{\alpha} + \kappa_1 y^{\alpha+\kappa} + \kappa_2 y^{\alpha+\kappa+1} + \dots,$$

where κ_1, κ_2, ... are constants, and the function on the right-hand side is a regular function of y. Hence

$$y = \{B(z - b)\}^{\frac{1}{\alpha}} + \kappa_1'(z - b)^{\frac{2}{\alpha}} + \kappa_2'(z - b)^{\frac{3}{\alpha}} + \dots,$$

on reversing the series; and therefore

$$w - \eta = y^{\alpha}$$

$$= B(z - b) + \gamma_1 (z - b)^{1+\frac{1}{\alpha}} + \gamma_2 (z - b)^{1+\frac{2}{\alpha}} + \dots.$$

We thus have a set of α branches of the integral, having b for a common branch-point and $\eta(b)$ for a common value at the branch-point. Moreover,

$$\left(\frac{dw}{dz} - \frac{d\eta}{dz}\right)_{z=b} = B = \zeta(b) - \left(\frac{d\eta}{dz}\right)_{z=b},$$

so that

$$\left(\frac{dw}{dz}\right)_{z=b} = \zeta(b);$$

thus the α branches have a common value $\zeta(b)$ for their first derivative at the point; this common value is different from the derivative of η at that point, where η is the quantity given by the discriminant.

Now let b, so far restricted only to be a non-zero of $\zeta - \dfrac{d\eta}{dz}$ in the vicinity of a point c, which itself is not a zero of $a_0(z, \eta)$, move in the z-plane in the vicinity of c, merely avoiding points which are zeros of $\zeta - \dfrac{d\eta}{dz}$. Then there is a corresponding set (or group of sets) of branches, which have a common value and the derivatives of which also have a common value for each point b; but while the common value of the branches agrees with the value of a function given by the discriminant, the common value of their derivatives does not agree with the derivative of the function given by the discriminant.

A corresponding investigation leads to a similar result for any other root of

$$F(z, \eta, w') = 0,$$

which is a multiple root.

For any root, which is a simple root of

$$F(z, \eta, w') = 0,$$

the corresponding result is obtainable by making $\alpha = 1$ in the preceding investigation; that is, $w - \eta$ is a regular function of $z - b$. Then the single branch of the integral, which thus arises, has the same value at $z = b$ as the function given by the discriminant; but the derivatives of the integral and the discriminant-function are not the same.

105. Thus far it has been assumed that all the roots w' of

$$F(z, \eta, w') = 0$$

are finite. If however $w = \eta$ makes $a_0(z, w)$ vanish for all values of z, then some of the foregoing roots w' will be infinite; but not all of them can be infinite because, since the coefficients $a_0, a_1, a_2, \dots$ have no common factor, they cannot all vanish when $w = \eta$. Of these roots, let there be p which are infinite; then writing

$$w' = v^{-1}, \quad w - \eta = u,$$

so that, in the equation

$$F(z, u, v^{-1}) = 0,$$

there are p roots v zero when $u = 0$, this equation must have the form

$$v^p h(z, u, v) + u^q k(z, u) = 0,$$

where q is an integer $\geqslant 1$, h and k are regular functions of their arguments, and $h(z, 0, 0)$ and $k(z, 0)$ do not vanish identically. As before, we assume that the constant c is chosen so that neither $h(c, 0, 0)$ nor $k(c, 0)$ vanishes; then we have

$$v = g_0 u^{\frac{\kappa}{a}} + g_1 u^{\frac{\kappa+1}{a}} + \dots,$$

where a is either p or a factor of p, κ is a positive integer because $v = 0$ when $u = 0$; and the functions $g_0, g_1, \dots$ are regular functions of z in the vicinity of $z = c$, g_0 not vanishing there: consequently,

$$w' = v^{-1}$$

$$= u^{-\frac{\kappa}{a}} \{ g_0 + g_1 u^{\frac{1}{a}} + \dots \}^{-1}$$

$$= u^{-\frac{\kappa}{a}} \{ h_0 + h_1 u^{\frac{1}{a}} + h_2 u^{\frac{2}{a}} + \dots \},$$

where the quantity, which $u^{-\frac{\kappa}{a}}$ multiplies, is a regular function of $z - c$ and $u^{\frac{1}{a}}$, that does not vanish when $u = 0$, $z = c$. Now let

$$w - \eta = u = y^a,$$

so that

$$a y^{a-1} \frac{dy}{dz} = -\frac{d\eta}{dz} + y^{-\kappa} (h_0 + h_1 y + h_2 y^2 + \dots),$$

and therefore

$$\frac{dz}{dy} = \cfrac{a y^{a+\kappa-1}}{h_0 - y^\kappa \dfrac{d\eta}{dz} + h_1 y + h_2 y^2 + \dots}.$$

Now h_0 is not zero; moreover, under the present hypothesis, $w = \eta$ is not a solution of the equation, so that $\dfrac{d\eta}{dz}$ is not infinite. Hence taking a value b in the vicinity of $z = c$ such that b is not a zero of h_0, the right-hand side can be expanded as a regular function of y and $z - b$, in the form

$$\frac{dz}{dy} = \frac{a}{h_0(b)} y^{a+\kappa-1} + y^{a+\kappa-1} P(z - b, y),$$

where P is a regular function such that $P(0, 0) = 0$. At the point b under consideration, η is the value of w, though $\dfrac{d\eta}{dz}$ is not infinite there; that is, $w = \eta$ when $z = b$; or $z = b$, $y = 0$, are initial values for this equation. Thus

$$z - b = \frac{\alpha}{(\alpha + \kappa)\,h_0(b)}\, y^{a+\kappa} + \beta_1 y^{a+\kappa+1} + \ldots$$

$$= \frac{\alpha g_0(b)}{\alpha + \kappa}\, y^{a+\kappa} + \beta_1 y^{a+\kappa+1} + \ldots,$$

and therefore

$$y = \left\{\frac{\alpha + \kappa}{a g_0(b)}\right\}^{\frac{1}{a+\kappa}} (z - b)^{\frac{1}{a+\kappa}} + \gamma_1 (z - b)^{\frac{2}{a+\kappa}} + \ldots;$$

hence

$$w = \eta + y^a$$

$$= \eta + \left\{\frac{\alpha + \kappa}{a g_0(b)}\right\}^{\frac{a}{a+\kappa}} (z - b)^{\frac{a}{a+\kappa}} P\{(z - b)^{\frac{1}{a+\kappa}}\},$$

where P is a regular function of its argument such that $P(0) = 1$. We thus have a set of $\alpha + \kappa$ branches of an integral, having $z = b$ for a common branch-point and $w = \eta(b)$ for a common value at the branch-point; for each branch, the value of w' at the branch-point is infinite and is different from the value at b of the derivative of $\eta(z)$.

Now κ is a positive integer $\geqslant 1$; also α is a positive integer, which is either p or a factor of p, unity included; hence

$$1 > \frac{\alpha}{\alpha + \kappa} > 0,$$

or the index of the lowest power of $z - b$ in the expansion of $w - \eta$ as a regular function of $(z - b)^{\frac{1}{a+\kappa}}$ is a positive proper fraction.

If the infinity for w', instead of being of multiplicity p, be a simple infinity, all that is necessary is to make α equal to unity through the preceding investigation.

The analysis has implicitly assumed that η is not infinite. As, however, the discriminant-equation, being algebraical in w, might have infinite roots, account must be taken of this possibility. For all such infinite roots, we should assume

$$w = \frac{1}{u};$$

and then it would be necessary to consider zero-roots of the equation obtained by equating to zero the discriminant, with regard to u', of

$$F\left(z, \frac{1}{u}, -\frac{1}{u^2}u'\right) = 0.$$

Lastly, it might happen that the discriminant-equation is satisfied through the vanishing of some factor which involves z alone. If $z = a$ be a value determined by this vanishing factor, so that $z = a$ is (by hypothesis) not a solution of the equation, then $z - a$ would be expressible as a function of w; the necessary analysis is similar to that for the preceding cases.

106. Next, suppose that $w = \eta$, where η is a finite root of the discriminant-equation, provides a solution of the differential equation, so that the equation

$$F\left(z, \eta, \frac{dw}{dz}\right) = 0$$

is satisfied by taking $\dfrac{dw}{dz} = \dfrac{d\eta}{dz}$.

Since η is a root of the discriminant, several of the roots $\dfrac{dw}{dz}$ are equal to $\dfrac{d\eta}{dz}$ for $w = \eta$: let there be a set of α roots, derivable as before, such that

$$\frac{d(w-\eta)}{dz} = g_0(w-\eta)^{\frac{\kappa}{\alpha}} + g_1(w-\eta)^{\frac{\kappa+1}{\alpha}} + \dots,$$

where g_0 is not identically zero, κ is an integer > 0, and κ and α have no common factor. Take

$$w - \eta = y^\alpha,$$

so that

$$\frac{dz}{dy} = \frac{\alpha y^{\alpha-1-\kappa}}{g_0 + g_1 y + g_2 y^2 + \dots}.$$

There are two cases for consideration.

(A) Let $\alpha - 1 - \kappa \geqslant 0$, so that $0 < \kappa \leqslant \alpha - 1$. Hence $\alpha > 1$. Take b an ordinary point of g_0, such that g_0 is not zero at $z = b$: then, as we wish to have the values w equal to η at $z = b$, we find

$$z - b = \frac{\alpha}{\alpha - \kappa} \frac{1}{g_0(b)} y^{\alpha-\kappa} + \beta_1 y^{\alpha-\kappa+1} + \dots.$$

Hence

$$y = \left\{\frac{\alpha - \kappa}{\alpha} g_0(b)\right\}^{\frac{1}{\alpha - \kappa}} (z-b)^{\frac{1}{\alpha-\kappa}} + \gamma_1 (z-b)^{\frac{2}{\alpha-\kappa}} + \ldots,$$

and therefore

$$w - \eta = \left\{\frac{\alpha - \kappa}{\alpha} g_0(b)\right\}^{\frac{\alpha}{\alpha - \kappa}} (z-b)^{\frac{\alpha}{\alpha-\kappa}} [1 + P\{(z-b)^{\frac{1}{\alpha-\kappa}}\}],$$

the index $\dfrac{\alpha}{\alpha - \kappa}$ of the lowest power of $z - b$ on the right-hand side

being always > 1. We thus have a group of $\alpha - \kappa$ $(\geqslant 1)$ branches of
the integral, which are equal to one another at b and have $\eta(b)$ for

their common value there. Let λ be the greatest integer in $\dfrac{\alpha}{\alpha - \kappa}$,

if this be fractional: if not, so that $\alpha - \kappa = 1$, let λ denote $\alpha - 1$:
then

$$\frac{d^s w}{dz^s} = \frac{d^s \eta}{dz^s} \text{ at } z = b, \qquad\qquad (s = 1, \ldots, \lambda)$$

λ being at least 1. The $\alpha - \kappa$ branches, which have $\eta(b)$ for a
common value at $z = b$, have their derivatives, of order up to λ,
equal to one another and also equal to those of η up to that order.

Now let the point b move in the vicinity of the point c, which
is not a zero of $a_0(z, \eta)$; and let it avoid points which are zeros
of a_0. Then there is a corresponding set (or group of sets) of
branches of an integral function, which have a common value at
each such point b; their derivatives also, up to a definite order, also
have common values at the point b. The common value of the
branches at the point is the value, at the point, of the function η
determined by the discriminant-equation; the common value, for
each order, of their derivatives is the value of the derivative of η
of that order, at the point.

If $\alpha - \kappa = 1$, there is only one branch of the integral function;
the branch then becomes a uniform function of z in the vicinity
of $z = b$.

Reverting for the moment to the geometrical statement of the
case, when w and z are restricted to have only real values, we see
that the integral of the equation is composed of a set of $\alpha - \kappa$
different curves, touching one another at any point on the discriminant-curve, and having contact with one another and with
the discriminant-curve up to some definite order. Then $w = \eta$ is

an envelope of each of those $\alpha - \kappa$ families of curves, each family being constituted by one of the curves for all the values of z; and it is a solution of the equation. If $\alpha - \kappa = 1$, there is only one family of curves enveloped.

The solution $w = \eta$ is then a *singular solution* of the equation.

(B) Let $\alpha - 1 - \kappa < 0$, so that $\kappa > \alpha - 1$. Proceeding as before, we have

$$\frac{dy}{dz} = \frac{1}{\alpha} y^{\kappa - (\alpha - 1)} (g_0 + g_1 y + g_2 y^2 + \dots);$$

and the index $\kappa - (\alpha - 1)$ of y is an integer greater than zero.

Because $z = b$ is an ordinary point of $g_0, g_1, g_2, \dots$, and g_0 does not vanish at $z = b$, the only integral of this equation, which is a regular function either of $z - b$, or of a transformed independent variable giving a finite number of branches, is

$$y = 0.$$

Consequently, $w = \eta$ is the sole integral of the differential equation, which at the point $z = b$ is equal to the function given by the discriminant-equation, and which is a regular function either of $z - b$ or of a transformed variable. Hence, for values of $\kappa > \alpha - 1$, the integral $w = \eta$ is a *particular* solution of the original differential equation.

If for the root η of the discriminant-equation, several distinct sets of roots w' of the equation

$$F(z, \eta, w') = 0$$

be equal to one another, it may happen that $\kappa \leqslant \alpha - 1$ for one set, and $\kappa > \alpha - 1$ for another set. The solution $w = \eta$ is singular, in its association with the former set; it is particular, in its association with the latter set.

Further, as the discriminant-equation is an algebraical equation, some of its roots may be infinite; and an infinite root may satisfy the original differential equation. In that case, we should substitute

$$w = \frac{1}{u},$$

and consider roots of the equation obtained by equating to **zero** the discriminant, with regard to u', of

$$F\left(z, \frac{1}{u}, -\frac{1}{u^2}u'\right) = 0 ;$$

the initial values are $u = 0$, $z = 0$.

The discussion of the respective cases is the same as for the case of finite roots of the discriminant-equation. Such a root may be either a singular solution, or a particular solution, or it may be associated with one set of solutions in the former capacity and at the same time with another set in the latter capacity.

Corresponding results hold when the discriminant-equation is satisfied by the vanishing of a function of z alone which also provides an integral of the original equation.

107. A case, intermediate between the two which have been considered, still remains. In the first of these cases, $w = \eta$ is not a solution of the equation, so that $\zeta - \dfrac{d\eta}{dz}$ is not zero for a continuous aggregate of values of z; in the second, $w = \eta$ is a solution, so that $\zeta - \dfrac{d\eta}{dz}$ is identically zero for such an aggregate. It remains to consider the integrals in the vicinity of a point which is an isolated zero of $\zeta - \dfrac{d\eta}{dz}$, say $z = f$; at that point, $w = \eta$, $w' = \dfrac{d\eta}{dz}$ satisfy the equation; but for $z = f + \epsilon$, where $|\epsilon|$ is to be made infinitesimally small, $\zeta - \dfrac{d\eta}{dz}$ is not zero and then $w = \eta$ is not a solution. Let

$$\zeta - \frac{d\eta}{dz} = \lambda (z - f)^s P(z - f)$$

where $P(z - f)$ is a regular function of $z - f$ such that $P(0) = 1$. Then write

$$w - \eta = y^a,$$

as before; the equation for y is

$$a y^{a-1} \frac{dy}{dz} = \lambda (z - f)^s + g_0 y^k + \dots,$$

where the unexpressed terms are of order higher than k in y or of order higher than s in f; what is desired is an integral y that vanishes when $z = f$.

Clearly the general investigation will be of the same character as that adopted in Chapter V for the reduction of a differential equation to a typical reduced form. Without entering on this general investigation, some indication of the character of the result will be obtained by considering the simplest case, viz. $\alpha = 2$, $s = 1$, $k = 1$, so that the equation is

$$2y \, \frac{dy}{dz} = \lambda \, (z - f) + g_0 y + g_1 y^2 + \ldots$$

Let

$$g_0 = f_0 + f_1 (z - f) + \ldots;$$

and take

$$y = (z - f)(\rho + Y),$$

choosing ρ so that

$$2\rho^2 - f_0 \rho - \lambda = 0,$$

and assigning to Y the condition that it shall vanish when $z = f$. The differential equation for Y is

$$(z - f) \, \frac{dY}{dz} = \frac{f_0 - 4\rho}{2\rho} \, Y + \mu \, (z - f) + \ldots,$$

where the unexpressed terms are of higher order than the first in $z - f$ and Y.

Unless $\dfrac{f_0 - 4\rho}{2\rho}$, that is, unless

$$\frac{-2 \, (f_0^2 + 8\lambda)^{\frac{1}{2}}}{f_0 + (f_0^2 + 8\lambda)^{\frac{1}{2}}},$$

be a positive integer or zero, the equation has a regular integral vanishing when $z = f$; while if that quantity be a positive integer, some relation must hold (§ 73) among the constants μ if the equation has a regular integral.

Assuming that a regular integral exists, we have

$$w - \eta = (z - f)^2 (\rho + Y)^2;$$

hence at $z = f$, we have

$$\frac{dw}{dz} = \frac{d\eta}{dz},$$

but this is not the case for points in the immediate vicinity of f.

Hence it follows that, when $w = \eta$ does not provide a solution of the differential equation, though for a particular value of z, say $z = f$, the value $\dfrac{d\eta}{dz}$ at that point, when substituted for w', satisfies the equation, there are integral curves which touch the discriminant-curve at that point; as z moves away from that value, then $\dfrac{d\eta}{dz}$, when substituted for w', no longer satisfies the equation. In other words, the function $w = \eta$ satisfies the differential equation at a particular point, or at a number of points each of which is isolated; but, except at those points, it does not satisfy the equation.

Note 1. The function g_0 in the preceding investigations does not vanish identically. The point $z = b$, chosen as an initial point, is such that g_0 does not vanish there; the alternative should be considered, so as to obtain the relation between the integral of the equation and the discriminant-function in the vicinity of a zero of g_0, when the integral and the function have the same value at that zero. The investigation is left as an exercise.

Note 2. In these cases, it may occur that, though some portions of the integral are expressible in a form which is regular, or can be made regular by an algebraical transformation of the independent variable, yet for the general integral the values $z = f$, $w = \eta(f)$; or $z = b$, $w = \eta(b)$; can be points of indeterminateness. It should be remarked that these are fixed points, that is, points fixed by the equation itself; they are not parametric.

108. Summarising the results obtained in connection with a root $w = \eta$ of the equation

$$\Delta(z, w) = 0,$$

where Δ is the discriminant of $F(z, w, w') = 0$, we have the following cases.

I. When $w = \eta$ is not in general a solution of the equation, so that the substitutions $w = \eta$, $w' = \dfrac{d\eta}{dz}$ do not satisfy the equation, then all the integrals which are equal to $\eta(b)$ at $z = b$, where b is a point in the z-plane in the vicinity of which η is regular, can be represented in the form

$$w - \eta = P(z - b),$$

where P is a series of positive powers of $z - b$, the indices being either integers or commensurable fractions; the lowest index in P is unity, when all the values of $\dfrac{dw}{dz}$ given by

$$F(z, \eta, w') = 0$$

are finite at $z = b$; the lowest index is less than unity for those of the values of $\dfrac{dw}{dz}$ given by $F(z, \eta, w') = 0$ which are infinite at $z = b$.

This restriction as to the magnitude of the lowest index of powers of $z - b$ in P can no longer be imposed if $w = \eta$, though not in general a solution of the equation, provides at $z = b$ a value of w' which, at the point, satisfies the equation.

II.   When $w = \eta$ is a singular solution of the equation, then all the integrals of the equation, which become equal to $\eta(b)$ at $z = b$, where b is an ordinary point of η, have the form

$$w - \eta = P(z - b),$$

where P is a series of positive powers of $z - b$, the indices being either integers or commensurable fractions; the lowest index in P is always greater than unity.

III.   When $w = \eta$ is a particular solution of the equation, then all the integrals of the equation, which are equal to $\eta(b)$ at $z = b$, are given by

$$w = \eta.$$

(The analytical distinction between the cases II. and III. is afforded by the expansion

$$\frac{d(w - \eta)}{dz} = g_0 (w - \eta)^{\frac{\kappa}{a}} + \dots$$

The solution $w = \eta$ is *singular*, in relation to those integrals of the equation for which $\kappa < \alpha$; it is *particular*, in relation to those integrals of the equation for which $\kappa \geqslant \alpha$.)

IV.   It may happen that $w = \eta$ is a singular solution in relation to some integrals of the equation and is a particular solution in relation to other integrals; in that case, $w = \eta$ gives more than one distinct repeated root of

$$F(z, \eta, w') = 0,$$

though this result is not, of itself, sufficient to secure the double relation of the solution $w = \eta$.

262 IRRESOLUBLE FACTORS [109.

109. The discriminant may provide several roots. If it is an irresoluble function of w, these roots form one set. If it is a resoluble function of w, let it be expressed in the form $\Delta_1{}^{m_1}\Delta_2{}^{m_2}\Delta_3{}^{m_3}\ldots$, where Δ_1, Δ_2, Δ_3, ... are irresoluble; then the roots of $\Delta_1 = 0$ form one set, those of $\Delta_2 = 0$ another set, and so on. Then we have the proposition, due to Hamburger*, that, *if one root of a set is a solution, every root of that set is also a solution of the same kind, singular or particular;* so that, in fact, we can then regard $\Delta = 0$, if Δ is irresoluble, or $\Delta_1 = 0$, $\Delta_2 = 0$, ..., if Δ is resoluble, as solutions of the differential equation in the respective cases; and these solutions are singular or particular, according as any one root is singular or particular.

To prove this, let $w = \eta$ be a root of an irreducible equation of degree m in w, say $\Theta = 0$; and let it be a solution of the original differential equation, so that

$$F\left(z,\ \eta,\ \frac{d\eta}{dz}\right) = 0,$$

is satisfied. Now

$$\frac{d\eta}{dz} = -\frac{\dfrac{\partial\Theta}{\partial z}}{\dfrac{\partial\Theta}{\partial \eta}};$$

this quantity, by means of $\Theta = 0$, can be expressed in the form $\alpha_0 + \alpha_1\eta + \ldots + \alpha_{m-1}\eta^{m-1}$. Also, every power of $\dfrac{d\eta}{dz}$ can be expressed in this form: and every power of η, of index higher than $m - 1$, can be expressed in this form, so that $F\left(z,\ \eta,\ \dfrac{d\eta}{dz}\right) = 0$ can be expressed in the form

$$f_0 + f_1\eta + \ldots + f_{m-1}\eta^{m-1} = 0,$$

where the coefficients f do not involve η. Now $w = \eta$ is to satisfy the equation, either identically or as an algebraical consequence of $\Theta = 0$. The equation $\Theta = 0$ is irreducible, and therefore no equation of lower degree is satisfied in virtue of $\Theta = 0$; hence

---

* *Crelle*, t. cxii (1893), p. 220.

the above equation, as the final form of $F\left(z,\, \eta,\, \dfrac{d\eta}{dz}\right) = 0$, can be satisfied only if

$$f_0 = 0,\ f_1 = 0, \dots,\ f_{m-1} = 0.$$

In virtue of these relations, the substitution of any other root of $\Theta = 0$ for η will also lead to an equation which is satisfied: that is, every root of the irreducible equation $\Theta = 0$ is a solution if any one root is a solution.

The distinction between a singular solution and a particular solution is based upon the character of the expansion of $w' - \dfrac{d\eta}{dx} = v$, say, in powers of $w - \eta = u$, say. This expansion arises out of the equation

$$v^p\, h\,(z,\, \eta,\, u,\, v) + u^q\, k\,(z,\, \eta,\, u,\, v) = 0,$$

and is of the form

$$w' - \frac{d\eta}{dx} = v = g_0\,(z,\, \eta)\, u^{\frac{\kappa}{\alpha}} + g_1\,(z,\, \eta)\, u^{\frac{\kappa+1}{\alpha}} + \dots,$$

for the root η of $\Theta = 0$. But as $\Theta = 0$ is irreducible, the analysis for a root η_1, instead of a root η, in connection with the corresponding expansion, is precisely the same, step by step, with only the substitution of η_1 for η; so that, for any other root, we have

$$w' - \frac{d\eta_1}{dx} = g_0\,(z,\, \eta_1)\,(w - \eta_1)^{\frac{\kappa}{\alpha}} + g_1\,(z,\, \eta_1)\,(w - \eta_1)^{\frac{\kappa+1}{\alpha}} + \dots,$$

where the integers κ and α are the same as before.

Consequently any two roots of the irreducible equation $\Theta = 0$, are singular together, or are particular together, or are singular together in relation to one set of integrals and particular together in relation to some other set of integrals.

In these circumstances, the irreducible equation $\Theta = 0$ can be regarded as a solution which is either singular, or particular, or both singular and particular.

Note. In all these discussions as to singular solutions, and the relation of roots of the discriminant-equation to integrals of the differential equation, the functional character for completely unrestricted variation of the variables has been regarded as the most important element; and the introduction of the geometrical interpretation, wherein both the variables are restricted to real values, has been made only for the sake of incidental illustration.

On the subject of what may be called the geometry of singular solutions, there is a great amount of literature, the most important part of which dates from the publication, in 1873, of the well-known papers by Cayley and Darboux. It is not, however, my purpose to discuss the plane curves represented by differential equations of the first order (§ 60, Ex. 1): and therefore the discussion of the geometry of singular solutions, in regard to propositions regarding cusps, tac-loci and the like, connected with the discriminant of the differential equation and the discriminant of the equation which is the integral equivalent, is omitted. It is more properly associable with the discussion, in differential geometry, of the curves represented by differential equations of the first order.

Ex. 1. Consider the example

$$F = w'^3 - 4zww' + 8w^2 = 0.$$

The discriminant is

$$\Delta = 64 \left(w^4 - \tfrac{4}{27} z^3 w^3 \right) ;$$

and therefore the roots of $\Delta = 0$ which arise for consideration are

$$w = \eta = 0, \quad w = \eta = \tfrac{4}{27} z^3.$$

Evidently $w = 0$ satisfies the differential equation ; and $w = \tfrac{4}{27} z^3$ is easily seen to satisfy the equation.

First, as to the nature of the solution $w = \eta = 0$. It is necessary to obtain the expansion of $\dfrac{d}{dz}(w - \eta)$ in powers of $w - \eta$, that is, of w' in powers of w, from the equation $F = 0$; and the three values are found to be

$$w' = \frac{2}{z} w + \left(\frac{2}{z^4} + \frac{2}{z} \right) w^2 + \ldots,$$

$$w' = 2z^{\frac{1}{2}} w^{\frac{1}{2}} - \frac{w}{z} + \ldots,$$

$$w' = -2z^{\frac{1}{2}} w^{\frac{1}{2}} - \frac{w}{z} + \ldots.$$

As regards the first of these, the lowest index of powers of w in the expansion is unity ; for this value of w', we have $\kappa = a$; and the integral $w = \eta = 0$ is therefore a particular solution of the equation, in connection with this value of w'.

As regards the second and the third, the lowest index of powers of w is $\tfrac{1}{2}$; for these two values of w', we have $\kappa < a$; and the integral $w = \eta = 0$ is therefore a singular solution of the equation, in connection with these values of w'.

Secondly, as to the nature of the solution $w = \eta = \frac{4}{27} z^3$. It is necessary to obtain the expansion of $\frac{d}{dz}(w - \eta)$ in powers of $w - \eta$, that is, of $w' - \frac{4}{9} z^2$ in powers of $w - \frac{4}{27} z^3$, from the equation $F = 0$. Let

$$w - \tfrac{4}{27} z^3 = u, \quad w' - \tfrac{4}{9} z^2 = v ;$$

then

$$v^3 + \tfrac{4}{9}^2 v^2 z^2 - 4zuv + \tfrac{1}{2}\tfrac{6}{7} uz^3 + 8u^2 = 0.$$

There are two values of v which vanish with u; they are given by

$$v = \quad \tfrac{2}{3} i z^{\frac{1}{2}} u^{\frac{1}{2}} + \dots,$$

$$v = - \tfrac{2}{3} i z^{\frac{1}{2}} u^{\frac{1}{2}} + \dots.$$

The lowest index of powers of u in the expansion is $\frac{1}{2}$; for these two values of $w' - \frac{4}{9} z^2$, we have $\kappa < a$; and the integral $w = \eta = \frac{4}{27} z^3$ is therefore a singular solution of the equation in connection with these values of w'.

The general integral of the equation is

$$w = A (z - A)^2,$$

where A is determined by the equation

$$a = A (a - A)^2,$$

if $w = a$, $z = a$ are the assigned initial values.

The customary geometrical interpretation, obtained by regarding $z \, (= x)$ as the abscissa and $w \, (= y)$ as the ordinate of a point in a plane, illustrates the relation between the various integrals.

The equation

$$y = A (x - A)^2$$

represents a series of parabolas having their axes parallel to the axis of y, their vertices on the axis of x: for positive values of A, they are turned in the positive direction of the axis of y; for negative values of A, they are turned in the negative direction of the axis. They all touch the line $y = 0$; hence $y = 0$ is a singular solution, as being an envelope of integral curves. But one of the parabolas, viz. for $A = 0$, is the line $y = 0$; that is, $y = 0$ is a particular solution.

The curve

$$y = \tfrac{4}{27} x^3$$

is an envelope of the parabolas, touching the curve

$$y = A (x - A)^2$$

at the point $x = 3A$, $y = 4A^3$; thus $y = \frac{4}{27} x^3$ is a singular solution as being an envelope of integral curves, and the value of dy/dx for the common tangent at the point $x = 3A$, $y = 4A^3$, is $4A^2$. The three values at the point are the roots of

$$p^3 - 48A^4 p + 128A^6 = 0,$$

that is, $4A^2$ repeated and $-8A^2$; so that $y = \frac{4}{27} x^3$ is an envelope connected with two touching integral curves, but it is not thus connected with the third integral curve through the point.

Ex. 2. Discuss similarly the equations :—

(i) $z \left(\dfrac{dw}{dz} \right)^2 - 2w \dfrac{dw}{dz} + 4z = 0$;

(ii) $\left(\dfrac{dw}{dz} \right)^4 - 4w \left(z \dfrac{dw}{dz} - 2w \right)^2 = 0$;

(iii) $\left(\dfrac{dw}{dz} \right)^2 + 2z \dfrac{dw}{dz} - w = 0$;

(iv) $(1 - z^2) \left(\dfrac{dw}{dz} \right)^2 - 1 + w^2 = 0$;

(v) $z \left(\dfrac{dw}{dz} \right)^2 + \phi \left(2z \dfrac{dw}{dz} - w \right) = 0$,

where ϕ is any regular function ;

(vi) $\left(\dfrac{dw}{dz} \right)^3 - az \dfrac{dw}{dz} + z^3 = 0$.

PAINLEVÉ'S THEOREM.

110. In the case of the equation

$$\frac{dw}{dz} = f(w, z)$$

of the first order and the first degree, the function f being algebraical in w and uniform in z, it was proved (§ 90) that all the points of indeterminateness of the integral are fixed points determined by the equation itself. This theorem* applies also to the equation

$$F(z, w, w') = 0,$$

when F is algebraical in w', say of (finite) degree n in w', is algebraical† in w, and uniform in z.

---

* It was first enunciated by Painlevé, *Sur les lignes singulières des fonctions analytiques*, (Thèse, 1887), p. 41.

† The limitation, that the equation

$$F(z, w, w') = 0,$$

which is algebraical in w', should also be algebraical in w, if its points of indeterminateness are to be points fixed by the equation itself and not parametric points, is necessary, as may be seen from so simple an example as

$$w' = \frac{dw}{dz} = ae^{\frac{w}{a}},$$

the general integral of which is

$$w = a \log \frac{a}{a} - a \log (A - z),$$

where A is an arbitrary constant: the (parametric) point $z = A$ is a point of indeterminateness of the integral.

Much of the former proof, that applies to the equation of the first degree, is applicable to the present case. The equation $F = 0$ determines n values of w' as functions of w and z, say

$$w' = f_r(w, z), \qquad (r = 1, \ldots, n)$$

which are distinct from one another except for values of z and w satisfying

$$\Delta(z, w) = 0,$$

where Δ denotes the discriminant of $F = 0$ with regard to w'. Giving to z any arbitrary value z_0, the equation $\Delta = 0$ then provides a number, say N, of values of w, say $\eta_1, \ldots, \eta_N$; the preceding investigations shew that, in the immediate vicinity of $z = z_0$, $w = \eta_s$, some of the integrals of the equation can be expressed as regular functions of a fractional power of $z - z_0$, which acquire a value η_s at $z = z_0$, and the rest of the integrals can be expressed as regular functions of $z - z_0$, which acquire a value η_s at $z = z_0$. Hence the point $z = z_0$ can be a branch-point of integrals of the equation: but it is an algebraical critical point for them.

For particular values of z, such as $z = b$ (determined as a zero of a regular function of z occurring in an expansion such as that in § 105), there are values of w given by $\Delta(z, w) = 0$, say $w = \eta$, which are (or may be) points of indeterminateness of the integral; but these are fixed points, settled by the equation itself, and they are isolated points.

For values of z, such as $z = z_0$ or $z = b$, and for values of w which are distinct from those given by $\Delta(z, w) = 0$ for the value of z, the n values w' are distinct from one another. They remain distinct from one another so long as the variables do not acquire (or pass round) a simultaneous combination, which is effectively a branching-combination of values.

As regards the distinct branches, there may be combinations of values of w and z which can be accidental singularities of the first kind: these are parametric, but they are algebraical critical points; or there may be accidental singularities of the second kind: these can be points of indeterminateness, but they are fixed points; or there may be essential singularities of the function $f_r(w, z)$: but these again are fixed points. Accordingly, in the z-plane, we mark all the fixed points given by the equation as known points of indeterminateness: each of them is an isolated point; and therefore, beginning with any value of z, which is not one of these marked points, we can draw a curve in the plane so that it does not come within only an infinitesimal distance of any one of these points.

We then consider the aggregate of the n values of w', given by

$$w' = f_r(w, z), \qquad (r = 1, \ldots, n),$$

and, moving along the curve indicated, we proceed to form the continuations of the integrals that arise in connection with assigned initial values.

If we reach a point z on the curve which corresponds to a branching of the values, then, after z passes through that point, say to ζ, the aggregate of the n values is the same as if the curve had been drawn differently to ζ, though their distribution is different. As in the case of only one value of w', when the original equation is of the first degree, the continuations of the integrals can be carried beyond any point Z, unless a function $f_r(w, z)$ becomes either (i) infinite, (ii) indeterminate at Z, though uniform in the vicinity, (iii) indeterminate at Z, but multiform in the vicinity.

As regards infinite values, if the infinity be determinate, then the combination is an accidental singularity of the first kind for $w' = f_r(w, z)$; the corresponding integrals determined by that equation have a parametric critical point, but it is algebraical. If the infinity be not determinate, that is, if the value at the point be infinite but the point be a branch-point, the preceding investigations of this chapter shew that the corresponding integrals of the equations, all of which have the point for an infinity, have a parametric critical point there, but it is algebraical for every integral.

As regards a point Z which makes $f_r(w, z)$ indeterminate and leaves it uniform in the vicinity, the argument of § 90, which applies to the equation of the first degree, applies here practically word for word; and the inference is that the point Z is not one beyond which continuation of the value $w' = f_r(w, z)$ must cease: that is, the point is not a point of indeterminateness for the integral.

As regards a point Z which makes $f_r(w, z)$ indeterminate, but keeps it multiform in the vicinity, the multiformity arises solely from the original differential equation; the point is therefore a branch-point. (In such a case, algebraical transformations of the type

$$z - Z = \zeta^r, \qquad w - \eta = W^\kappa,$$

can be obtained, for integer values of r and κ, which make the expression of $f_r(w, z)$ uniform in the new vicinity of $\zeta = 0$ and $W = 0$: but it is unnecessary to make this change.) The integrals have been considered in the vicinity of every branch-point, the original equation being algebraical in w in every case; such branch-points as are points of indeterminateness lie off the curve; and for all others, the integrals are known to exist say in a vicinity round Z not large enough to extend to the nearest point of indeterminateness. As they exist in such a vicinity round Z, they can be continued beyond Z, that is, Z is not a point beyond which the continuation of the value $w' = f_r(w, z)$ must cease; the point, in fact, is not a point of indeterminateness.

Hence it follows that no point on any curve, drawn in the z-plane so as not to approach indefinitely near to the fixed points of indeterminateness, can be a point of indeterminateness. All such points, whether particular branch-points, or accidental singularities of the second kind, or essential singularities, of the functions $f_r(w, z)$, are fixed points determined by the equation itself; the only parametric singularities of the integrals are accidental singularities of the first kind, and algebraical branch-points which may give either finite or infinite values to the integrals.

Painlevé's theorem is thus established. Furthur, there is suggested the investigation of those equations of the first order and any degree such that all the exceptional points of the integrals of any kind are fixed points. It is known that the exceptional points, if parametric, can only be algebraical; and therefore it will be sufficient, for that purpose, to obtain the tests ensuring that an equation of the first order has no parametric exceptional points, which are algebraical in character.

CHAPTER IX.

DIFFERENTIAL EQUATIONS OF THE FIRST ORDER, THE INTEGRALS
OF WHICH HAVE NO PARAMETRIC BRANCH-POINTS*.

111. IT has been proved, in the preceding chapters, that the
exceptional points of the integral of an equation of the first order
$$F(z, w, w') = 0,$$
which is rational in w' and in w, and is uniform in z, belong to
one or other of three classes, viz.

(i)   poles, in the vicinity of which the integral is uniform;

(ii) branch-points, at which the branches of the integral may
      have finite or infinite values, and round which a number
      of branches circulate in one or more cycles: when the
      number of circulating branches is finite, the point is an
      algebraical critical point;

(iii) points of indeterminateness; these include essential
      singularities in the vicinity of which the integral is
      uniform: and other points, at which the values of the
      integral are unlimited in number and depend upon the
      method of approach of z to the point; in the vicinity
      of such points, there may also be branching of the
      integral, either definite or indefinite.

---

* The subject of this chapter originated with Fuchs's important Memoir,
"Ueber Differentialgleichungen deren Integrale feste Verzweigungspunkte besitzen," *Berl. Sitzungsber.* (1884), pp. 699—710. The results contained in this
memoir were amplified by Poincaré, *Acta Math.*, t. VII (1885), pp. 1—32. Another
method was devised by Picard for later developments; it is expounded, among
other places, in the second and third volumes of his *Cours d'Analyse*, having first
been given in the crowned "Mémoire sur les fonctions algébriques de deux
variables," *Liouville's Journal*, Sér. 4, t. V (1889), pp. 135—319.

In this connection, reference may also be made to Painlevé's crowned "Mémoire
sur les équations différentielles du premier ordre," *Annales de l'Ec. Norm.*,
Sér. 3me, t. VIII (1891), pp. 9—58, 103—140, 201—226, 267—284; to the memoir
by Wallenberg, cited in § 127, and to a brief memoir by the same writer, *Crelle*,
t. CXVI (1896), pp. 1—9.

It has further been proved (§ 110) that all the points of indeterminateness are fixed points, determined by the equation itself; but that the poles, and the algebraical critical points, of the integral are (or can be) parametric, and that, for the most general equation $F = 0$, some of these exceptional points are certainly parametric.

It is, however, conceivable that classes of equations exist for which all the exceptional points of the integral are fixed; it is further conceivable that more extensive classes of equations exist, for which some set of the exceptional points (though not all of them) belonging to the integral are fixed. The course of the general discussion shews that the occurrence of poles, in the vicinity of which the integral is uniform, is relatively rare and unimportant compared with the occurrence of algebraical critical points. Accordingly, we proceed to the investigation of the conditions, necessary and sufficient to secure that all the algebraical critical points of an equation

$$F(z, w, w') = 0$$

are fixed points, determined by the equation itself.

Equations of First Degree without Parametric Critical Points.

112.   We begin with the equation of the first degree; and we find that *the only equation of the first degree, in which all the algebraical critical (or branch) points are fixed, is Riccati's equation*

$$\frac{dw}{dz} = A_0 + A_1 w + A_2 w^2,$$

where A_0, A_1, A_2 are uniform functions of z.

The general equation of the first degree, which is rational in w, is

$$\frac{dw}{dz} = \frac{P(w, z)}{Q(w, z)},$$

where P and Q are integral polynomials in w, having no common factor.

First, we must have Q independent of w. (Then Q will be a function of z alone, and can be absorbed into the coefficients of powers of w in P; that is, we can take $Q = 1$.) If it were otherwise, let $w = a$, $z = a$, be a pair of values, which satisfy

$$Q(w, z) = 0,$$

and are such as to make $P(w, z)$ distinct from zero; we can take a arbitrarily, because P and Q have no common factor, and still find this condition satisfied. But taking $w = a$, $z = a$, as initial values for the equation, we know (§ 24) that $z = a$, where a is an arbitrary point, is an algebraical critical point for the equation. This possibility must be excluded; there must accordingly be no equation $Q = 0$, giving values of w and z; and therefore, as explained above, we take $Q = 1$.

Next, as P is a polynomial in w, let it be of degree m. The ground of the exclusion of variability for Q was on the score of a definitely infinite value for $\dfrac{dw}{dz}$, as connected with specific values of w and z. This possibility must still be excluded: and it can arise, in the new form of the equation, for infinite values of w, which therefore must be taken into account. For this purpose, let $w = \dfrac{1}{W}$, so that we have

$$\frac{dW}{dz} = -W^2 P\left(\frac{1}{W}, z\right) = -\frac{P_1(W, z)}{W^{m-2}}.$$

The critical value is now $W = 0$; the preceding case shews that

$$\frac{P_1(W, z)}{W^{m-2}}$$

must be an integral polynomial in W; and therefore

$$m - 2 \leqslant 0.$$

The function P is consequently of order not greater than 2 in w; and the original equation thus is of the form

$$\frac{dw}{dz} = A_0 + A_1 w + A_2 w^2,$$

where A_0, A_1, A_2 are functions of z, which are uniform because the expression for $\dfrac{dw}{dz}$ is uniform in z.

113. To discuss the integral of the equation, one convenient method is to make it depend upon a linear equation of the second order by the transformation

$$w = -\frac{1}{A_2}\frac{1}{u}\frac{du}{dz},$$

so that u is determined by

$$\frac{d^2u}{dz^2} - \left(\frac{1}{A_2}\frac{dA_2}{dz} + A_1\right)\frac{du}{dz} + A_0A_2u = 0.$$

The functions A_0, A_1, A_2 being uniform functions of z, let them be expressed in the form

$$A_0 = \frac{G_0}{G},\quad A_1 = \frac{G_1}{G},\quad A_2 = \frac{G_2}{G},$$

where the functions G, G_0, G_1, G_2 have no singularities for finite values of z; then the equation becomes

$$G^2G_2\frac{d^2u}{dz^2} - \left(GG_1G_2 + G^2\frac{dG_2}{dz} - G_2G\frac{dG}{dz}\right)\frac{du}{dz} + G_2^2G_0u = 0,$$

where now all the coefficients of derivatives of u are regular functions. The only points, which are possibly not ordinary points of the equation, are the roots of $G = 0$ and $G_2 = 0$, that is, they are fixed points of the original equation, being either zeros of A_2, or infinities of A_0 or A_1 or A_2. As regards the integral of the equation, compounded of a linear combination of two integrals, these (fixed) points may be algebraical branch-points, or branch-points of the same type as $(z - c)^\lambda$ where λ is incommensurable, or logarithmic singularities, or essential singularities (with or without branching), or more general points of indeterminateness.

Let this integral be

$$u = \alpha u_1 + \beta u_2,$$

where either u_1, or u_2, or both, must possess these various (fixed) exceptional points; then the integral w, which is

$$-\frac{1}{A_2}\frac{\alpha\dfrac{du_1}{dz} + \beta\dfrac{du_2}{dz}}{\alpha u_1 + \beta u_2},$$

contains the arbitrary parameter $\alpha \div \beta$. It is clear that, in general, any exceptional point of u_1 or u_2 of the classes indicated will be an exceptional point of w; all of these are fixed points. All other points in the plane, and therefore any parametric point, are ordinary for $\alpha u_1 + \beta u_2$ and therefore for $\alpha \dfrac{du_1}{dz} + \beta \dfrac{du_2}{dz}$; if a parametric point give a zero for $\alpha u_1 + \beta u_2$, it provides a pole for w; if the value of $\alpha u_1 + \beta u_2$ at the point is not zero, the parametric point is ordinary for w. Hence parametric points are either ordinary points or are poles of the integral w of Riccati's equation.

If the arbitrary parameter $\alpha \div \beta$ be determined by initial values $w = w_0$, $z = z_0$, where z_0 is an ordinary point of u_1 and u_2, then it is easy to see that

$$w = \frac{Sw_0 + T}{Uw_0 + V},$$

where S, U, V, T are functions of z and z_0, which have the fixed exceptional points of u_1 and u_2 as their (fixed) exceptional points.

114. In the discussion of the integral of the equation, it is also possible to introduce some simplification in the form by transformation.

Let

$$w = \frac{W}{A_2} - \theta,$$

choosing θ so that the term in W in the expression for $\dfrac{dW}{dz}$ may vanish; then

$$2\theta = \frac{A_1}{A_2} + \frac{1}{A_2{}^2} \frac{dA_2}{dz},$$

and the equation becomes

$$\frac{dW}{dz} = W^2 + J(z)$$

$$= W^2 + J,$$

say, where

$$4J = 4A_0 A_2 - A_1{}^2 + \frac{2}{A_2} \frac{d^2 A_2}{dz^2} - \frac{3}{A_2{}^2} \left(\frac{dA_2}{dz}\right)^2 + 2 \frac{dA_1}{dz} - \frac{2A_1}{A_2} \frac{dA_2}{dz}.$$

It is known that all the points of indeterminateness and all the critical points of the integral of the equation are fixed points; and therefore a parametric value of z can be only an ordinary point of the integral or a pole of the integral, and such a parametric value of z is an ordinary point of J.

When $z = \alpha$, where α is a parametric constant, is an ordinary point of the integral, then by Cauchy's existence-theorem an integral exists, which is a regular function of $z - \alpha$ and acquires an assigned value w_0 when $z = \alpha$; and the regular integral, thus determined, is unique.

When $z = \alpha$ is a pole of the integral, the initial condition attaching to w is that it shall acquire an infinite value when $z = \alpha$. Moreover, this point is neither a branch-point nor a point of indeterminateness: in the vicinity of $z = \alpha$, the integral is uniform. To determine the order of the infinity, let

$$W = \frac{\lambda}{(z - \alpha)^n} + \ldots,$$

the indices of the unexpressed powers of $z - \alpha$ being higher than $-n$, where n is a positive integer and λ is a (non-zero) constant. Substituting, so that the equation may be identically satisfied, we have

$$n + 1 = 2n, \quad -n\lambda = \lambda^2,$$

and therefore $n = 1$, $\lambda = -1$. Accordingly, we write

$$W = \frac{-1 + v}{z - \alpha},$$

where the new dependent variable v is a regular function of $z - \alpha$ vanishing when $z = \alpha$; it is determined by the equation

$$(z - \alpha) \frac{dv}{dz} = -v + v^2 + (z - \alpha)^2 J$$

$$= -v + v^2 + (z - \alpha)^2 \{J_0 + J_1(z - \alpha) + J_2(z - \alpha)^2 + \ldots\},$$

say. We know (§§ 64, 67) that only one integral of this equation exists which vanishes with $z - \alpha$, and that it is regular; it is easily found to be represented by the series

$$\tfrac{1}{3} J_0 (z - \alpha)^2 + \tfrac{1}{4} J_1 (z - \alpha)^3 + \tfrac{1}{5} (J_2 + \tfrac{1}{9} J_0^2) (z - \alpha)^4$$
$$+ \tfrac{1}{6} (J_3 + \tfrac{1}{6} J_0 J_1) (z - \alpha)^5 + \ldots.$$

To obtain a region of certain convergence of the series, we construct (§ 63) the dominant function V such that*

$$3V = -V + V^2 + (z - \alpha)^2 J;$$

we take the root of this quadratic which vanishes with $z = \alpha$, so that

$$V = 2 - \{4 - (z - \alpha)^2 J\}^{\frac{1}{2}}.$$

The region of convergence for V (and therefore also a certain region for v) is a circle of radius ρ having α for its centre, where ρ is the smaller of two quantities, one of them the distance of α from

---

* The factor 3 on the left-hand side is the value of θ in § 63; the quantities to be considered here are $2 - a$, $3 - a$, ..., where $a = -1$: the smallest of these is 3.

the nearest (fixed) singularity of J, the other of them a value of $|z - \alpha|$ determined by

$$|z - \alpha|^2|\, J\,| = 4.$$

We thus have the expression of the integral of the equation in the vicinity of an ordinary point of J, whether parametric or fixed; and its expression in the vicinity of a parametric point which is a singularity—a pole—of the integral. The expression of the integral in the vicinity of the other exceptional points, whether critical points or points of indeterminateness, (all of them being fixed), is obtained by the methods of Chapters IV and VI; and several special examples are there given.

115.   This form of equation, therefore, is the only one of the first order and the first degree that has its integral devoid of parametric exceptional points other than poles. It is, however, a question of an entirely different range of investigation to secure that the integral is devoid of critical fixed points and points of indeterminateness with branching, in other words, to secure that the integral is uniform.

Taking the equation in the form

$$\frac{dW}{dz} = W^2 + J,$$

where J is uniform, we see that, if the integral of this equation is uniform, so also is that of the original equation: and *vice versa*. Now if

$$W = -\frac{1}{U}\frac{dU}{dz},$$

then

$$\frac{d^2U}{dz^2} + UJ = 0,$$

so that, if U_1 and U_2 be two linearly independent solutions of this equation, then

$$W = \frac{\dfrac{dU_1}{dz} + c\,\dfrac{dU_2}{dz}}{U_1 + c\,U_2}.$$

If U_1 and U_2 are uniform functions, then also W is uniform; but not conversely, for U_1 and U_2 could have common irrational elements, or could have a common factor for which a fixed point is at once an essential singularity and a branch-point, and the point still could cease to be critical for W.

EQUATIONS OF ANY DEGREE, WITHOUT PARAMETRIC BRANCH-POINTS.

116. Proceeding next to find the conditions, which must be satisfied in order that all the critical points of the equation

$$F(z, w, w') = 0$$

may be fixed points, determined by the equation itself, we assume that $F = 0$ is irreducible in regard to w' and, being of degree m is expressible in the form

$$a_0 w'^m + a_1 w'^{m-1} + \ldots + a_{m-1} w' + a_m = 0,$$

where the coefficients a_0, a_1, ... , a_m are algebraical polynomials in w, are regular functions of z, and have no factor common to all.

From preceding investigations (Chap. VIII) it has appeared that an integral of the equation exists, which acquires an assigned value w_0 at an arbitrary point z_0 and is a regular function in the vicinity of z_0, provided every root w' of

$$F(z_0, w_0, w') = 0$$

is simple and finite; but if any root be infinite, or if any root be multiple, then z_0 is a branch-point for a number of integrals which have a common value w_0 at z_0.

Further, it was proved that, if $w = \eta$ is a root of the equation

$$\Delta(z, w) = 0,$$

where $\Delta(z, w)$ is the w'-discriminant of $F = 0$, and if $w' = \zeta$ be a multiple root of

$$F(z, \eta, \zeta) = 0,$$

so that it also satisfies

$$\frac{\partial}{\partial \zeta} F(z, \eta, \zeta) = 0;$$

and if

$$w = \eta + u, \quad w' = \zeta + v,$$

change the irreducible equation

$$F(z, w, w') = 0$$

into the (also irreducible) equation

$$F_1(z, u, v) = 0,$$

where F_1 is consequently not divisible by a power of v for all values of z and u; then any root v of this equation can be expressed in the form

$$v = g_0 u^{\frac{\kappa}{\alpha}} + g_1 u^{\frac{\kappa+1}{\alpha}} + \ldots,$$

where α is a positive integer, κ is an integer which may be positive, or zero, or negative, and where $g_0, g_1, \ldots$ are functions of z, the branch-points of which are fixed points; so that the parametric point z_0 is ordinary for each branch of these functions and therefore, within a domain round z_0 which does not extend so far as the nearest branch-point of any of them, the coefficients $g_0, g_1, \ldots$ can be expressed as regular functions of $z - z_0$. For such roots of $F = 0$ as are simple, α is unity; but some root (or roots) must be multiple, and then $\alpha \geqslant 2$. Taking

$$w - \eta = u = y^\alpha,$$

the equation for y is

$$\alpha y^{\alpha - 1} \frac{dy}{dz} = w' - \frac{d\eta}{dz}$$

$$= \zeta - \frac{d\eta}{dz} + g_0 y^\kappa + g_1 y^{\kappa+1} + \ldots.$$

117.   First, let the integer κ be negative; then

$$\frac{dz}{dy} = \epsilon_0 y^{\alpha - 1 - \kappa} + \epsilon_1 y^{\alpha - \kappa} + \epsilon_2 y^{\alpha - \kappa + 1} + \ldots,$$

where $\epsilon_0, = \dfrac{\alpha}{g_0}$, is a regular function of $z - z_0$, and all the other coefficients $\epsilon_1, \epsilon_2, \ldots$ are regular functions of $z - z_0$, in the region retained. The integer $\alpha \geqslant 1$, whether the root under consideration be simple or multiple, and κ is a negative integer, so that $\alpha - \kappa > 1$; hence (§ 24) y branches at z_0, being a regular function of $(z - z_0)^{\frac{1}{\alpha - \kappa}}$. Consequently $u, = y^\alpha$, and therefore also w, branches at z_0.

Now when κ is negative, $v = \infty$ when $u = 0$; and therefore *the equation*

$$F_1(z, u, v) = F(z, \eta + u, \zeta + v) = 0$$

must not have infinite roots for v when $u = 0$, if the original equation $F(z, w, w') = 0$ is to be devoid of parametric branch-points.

118. Secondly, let the integer κ be positive. There are two cases, according as $\zeta - \dfrac{d\eta}{dz}$ is not, or is, identically zero.

When $\zeta - \dfrac{d\eta}{dz}$ is not identically zero, its roots are fixed points; accordingly, the parametric point z_0 is not a root of $\zeta - \dfrac{d\eta}{dz}$, and we have

$$\frac{dz}{dy} = \gamma_0 y^{a-1} + \gamma_1 y^a + \cdots,$$

where γ_0, γ_1, ... are regular functions of $z - z_0$, of which γ_0 does not vanish when $z = z_0$. Then

$$z - z_0 = c_0 y^a + c_1 y^{a+1} + \cdots,$$

where $c_0 a$ is the value of γ_0 at $z = z_0$, that is, c_0 does not vanish. In the case of some roots of the original equation, $a \geqslant 2$; hence y is of the form

$$(z - z_0)^{\frac{1}{a}} P \{(z - z_0)^{\frac{1}{a}}\},$$

where $P(0)$ is not zero. Consequently

$$w - \eta = y^a$$
$$= (z - z_0) Q \{(z - z_0)^{\frac{1}{a}}\},$$

where Q is a regular function; and then $z = z_0$ is a branch-point for this part of the integral. If, then, the original equation $F = 0$ is to be devoid of parametric branch-points, $\zeta - \dfrac{d\eta}{dz}$ cannot differ from an identical zero; that is, η must be a solution of the equation.

Taking therefore $\zeta = \dfrac{d\eta}{dz}$, the form of the equation for y depends on the relation of $a - 1$ to κ.

(i) If $a - 1 > \kappa$, then

$$\frac{dz}{dy} = \frac{a}{g_0} y^{a-\kappa-1} + \cdots,$$

so that

$$z - z_0 = b_0 y^{a-\kappa} + b_1 y^{a-\kappa+1} + \cdots.$$

Since $\alpha - 1 > \kappa$ in this case, that is, $\alpha - \kappa > 1$, we have y a regular function of $(z - z_0)^{\frac{1}{\alpha - \kappa}}$, and therefore

$$w - \eta = y^\alpha$$
$$= (z - z_0)^{\frac{\alpha}{\alpha - \kappa}} Q\left\{(z - z_0)^{\frac{1}{\alpha - \kappa}}\right\},$$

so that $z = z_0$ is a branch-point for these integrals.

(ii) If $\alpha - 1 = \kappa$, then

$$\frac{dz}{dy} = \frac{\alpha}{g_0} + h_1 y + \ldots,$$

so that

$$z - z_0 = b_0 y + b_1 y^2 + \ldots.$$

Here y, and therefore also $w (= \eta + y^\alpha)$, has $z = z_0$ for an ordinary point.

(iii) If $\alpha - 1 < \kappa$, then

$$\frac{dy}{dz} = \frac{g_0}{\alpha} y^{\kappa - (\alpha - 1)} + \frac{g_1}{\alpha} y^{\kappa - \alpha + 2} + \ldots.$$

Now $z = z_0$ is a parametric point; all the points of indeterminateness of the original equation $F = 0$ are fixed points, determined by the equation itself; and therefore $z = z_0$ is not a point of indeterminateness either of $F = 0$ or of the deduced equation between y and z. Since $\kappa - (\alpha - 1)$ is a positive integer, it follows that

$$y = 0$$

is the only solution which vanishes when $z = z_0$; that is, $w = \eta$ is the solution of the original equation. (It is a particular solution.)

Hence, in order that z_0 may not be a branch-point, we must have $\alpha - 1 \leqslant \kappa$, and η must be a solution of the equation: consequently, *when w' is regarded as an algebraical function of w, given by $F(z, w, w') = 0$, and when, for an arbitrary value z_0 of z, $w = \eta$ is a branch-point of order $\alpha - 1$, the α branches having ζ for their common value, then $\zeta = \dfrac{d\eta}{dz}$, in order that the integral of the equation*

$$F(z, w, w') = 0$$

may not have z_0 as a branch-point, so that η is a solution of the equation; and $F\left(z, w, \dfrac{d\eta}{dz}\right) = 0$ must have $w = \eta$ as a root of multiplicity equal to, or greater than, $\alpha - 1$.

119. Thirdly, let $\kappa = 0$. Again there are two cases, according as $\zeta - \dfrac{d\eta}{dz}$ is not, or is, identically zero.

When $\zeta - \dfrac{d\eta}{dz}$ is not identically zero, the same course of analysis as was adopted (§ 118) when κ is a positive integer, shews that t_0 is a branch-point for those integrals associated with values of $\alpha \geqslant 2$; and there are such integrals. Since z_0 is not to be a branch-point of the integral of the equation, it follows that $\zeta - \dfrac{d\eta}{dz}$ must be identically zero, and therefore that η must be a solution of the original equation.

Now some of the roots of $\Delta(z, \eta) = 0$ are simple roots of $F\left(z, w, \dfrac{d\eta}{dz}\right) = 0$, regarded as an equation in w; but one at least of them is multiple, and $\alpha \geqslant 2$ for that root. Since $\kappa = 0$, the analysis for the case when κ is a positive integer applies in the instance $\alpha - 1 > \kappa$; and $z = z_0$ is a branch-point for those integrals. When $\kappa = 0$, the root

$$v = g_0 u^{\frac{\kappa}{\alpha}} + g_1 u^{\frac{\kappa+1}{\alpha}} + \ldots,$$

is not zero when $u = 0$; and this case must be excluded. Hence *the equation*

$$F_1(z, u, v) = F\left(z, \eta + u, \dfrac{d\eta}{dz} + v\right) = 0$$

must not have any multiple root v which is finite when $u = 0$, if the original equation is to be devoid of parametric branch-points.

120. When the equation is arranged in powers of w', it is

$$F(z, w, w') = A_0 w'^m + A_1 w'^{m-1} + \ldots + A_{m-1} w' + A_m,$$

where the coefficients A are integral polynomials in w.

If A_0 involves w, let $w = \xi$ be a value of w which makes A_0 vanish. As $F = 0$ is to have no parametric branch-points, the equation $w = \xi$, if it gives a solution of the original equation $F = 0$, does not violate the hypothesis. The alternative is that $w = \xi$ should not give a solution of $F = 0$: and then we take

$$w = \xi + u,$$

making this substitution in the coefficients A alone in the first place. Now

$$A_0(z, \ w) = A_0(z, \ \xi + u) = A_0{}',$$

say, where $A_0{}'$ is an integral polynomial in u, which vanishes with u because $A_0(z, \ \xi) = 0$. Others of the coefficients A, after this substitution, also may vanish with u; but not all of them can do so, because $w - \xi$ would then be a common factor of all—the possibility of existence of which has been excluded. Suppose that

$$A_i{}' = A_i(z, \ \xi + u) = u^{r_i} P_i(u),$$

where $P_i(u)$ is not zero for $u = 0$; also $r_0 > 0$, and some one or more of the integers $r_1, \ldots, r_n$ must vanish.

Then at least one root w' of the equation

$$A_0{}' w'^m + A_1{}' w'^{m-1} + \ldots + A'_{m-1} w' + A'_m = 0$$

is infinite when $u = 0$. To obtain the order of the infinity, construct a Puiseux diagram, marking in a plane referred to two perpendicular axes $O\xi$, $O\eta$ the points

$$(m, r_0); \ (m - 1, r_1); \ \ldots (1, r_{m-1}); \ (0, r_m).$$

A line drawn through the point $m - i, r_i$, making an angle $\tan^{-1} \mu$ with the positive direction of $O\xi$, is

$$y - r_i = \mu \left\{ x - (m - i) \right\},$$

so that the intercept on the axis $O\eta$ is $r_i - \mu(m - i)$; this is the index of the lowest term in the expansion of $A_i{}' w'^{m-i}$ in ascending powers of u for a value of w' which, when expanded also in ascending powers, begins with a term in $u^{-\mu}$. The point $(0, r_m)$ is on the axis $O\eta$; one of the points $(0, r_m); \ (1, r_{m-1}); \ \ldots; \ (m - 1, r_1)$ is on the axis $O\xi$, say $(m - j, r_j)$, so that $r_j = 0$; the point (m, r_0) is off the axis $O\xi$. Through $(0, r_m)$ take a line coinciding with $O\eta$; make it turn in the counterclockwise sense, till it meets some point or points in the tableau: then make it turn about the last of such points, until it meets others; and so on. At some stage it passes through a point or points on $O\xi$; on continuing to revolve, when it passes through the last of these on $O\xi$, (which has its abscissa $< m$), it will meet at least one point in the tableau, giving to its direction a positive inclination $\tan^{-1} \mu$ less than $\frac{1}{2}\pi$. Hence there will be certainly one portion of the line, and there may be more than one portion of the line, giving positive values.

to μ, where a term in $u^{-\mu}$ is the first term in the expansion of w' in ascending powers of u.

Further, if a value μ be determined by a portion of the line joining $(m - j, r_j)$ to $(m - i, r_i)$, where $m - j < m - i$ and $r_j = 0$, we have

$$r_j - (m - j)\,\mu = r_i - (m - i)\,\mu,$$

then

$$\mu = \frac{r_i}{j - i},$$

so that μ is a positive commensurable quantity.

Let a root w', which is infinite when u is zero, be obtained in the form

$$w' = g_0 u^{-\kappa} + g_1 u^{-\kappa+1} + g_2 u^{-\kappa+2} + \dots,$$

where κ is a positive commensurable quantity; the zeros of g_0 may be branch-points of g_0, but they are fixed points; a parametric point z_0 is ordinary for each branch of g_0, and likewise for the other coefficients g_1, g_2, ..., so that we may consider them expressed as regular series of powers in $z - z_0$, of which at least the coefficient g_0 does not vanish when $z = z_0$.

Now ξ is not a solution, so that $\dfrac{d\xi}{dz}$ is not infinite: thus

$$\frac{du}{dz} = w' - \frac{d\xi}{dz}$$

$$= g_0 u^{-\kappa} + \dots + \dots,$$

the term $-\dfrac{d\xi}{dz}$ being inserted in the series certainly later than the first term. As in previous cases, this gives

$$u = (z - z_0)^{\frac{1}{\kappa+1}}\, Q\{(z - z_0)^{\frac{1}{\kappa+1}}\},$$

where $Q(0)$ is not zero. Since κ is a positive commensurable quantity, it follows that $z = z_0$ is a branch-point for u and therefore for the integral w. The equation

$$F(z, w, w') = 0$$

is to be devoid of parametric branch-points; the hypothesis on which the preceding result has been deduced must be excluded; and therefore the coefficient A_0, which, at the utmost, is a polynomial in w, must not determine any root $w = \xi$; consequently, A_0 *must be independent of* w. It at once follows

(i) that, for a parametric value of z, no root w' can be infinite for finite values of w; and (ii) that A_0 can be taken as equal to 1: for it now is a function of z alone, and we can divide $F(z, w, w')$ throughout by A_0 without affecting the character of the coefficients.

Next, to take account of possibly infinite values of w, write

$$w = \frac{1}{W};$$

then, assuming $A_0 = 1$, the equation in W is

$$\left(\frac{dW}{dz}\right)^m - W^2 A_1\left(\frac{1}{W}, z\right)\left(\frac{dW}{dz}\right)^{m-1}$$

$$+ W^4 A_2\left(\frac{1}{W}, z\right)\left(\frac{dW}{dz}\right)^{m-2} - \ldots = 0.$$

By what has just been proved, it is necessary, if $F(z, w, w') = 0$ is to have no parametric branch-points, that $\dfrac{dW}{dz}$ should not, for a parametric value of z_0, have an infinite value for zero values of z. Hence all the coefficients $W^{2i} A_i\left(\dfrac{1}{W}, z\right)$ must be integral polynomials; in other words, the coefficient $A_i(w, z)$ is an algebraical polynomial in w of degree not greater than $2i$.

121. Combining these various results, we have the following theorem[*], due to Fuchs:—

The conditions, necessary and sufficient to secure that the differential equation

$$F(z, w, w') = 0,$$

of degree m, shall have no parametric branch-points, are:—

(A) *The equation must have the form*

$$w'^m + \psi_1 w'^{m-1} + \psi_2 w'^{m-2} + \ldots + \psi_m = 0,$$

where ψ_i (for $i = 1, \ldots, m$) is an algebraical polynomial in w of degree not higher than $2i$, the coefficients of the various powers of w being uniform functions of z;

(B) *If Δ denote the discriminant of F, and if $w = \eta$ be a root of $\Delta(z, w) = 0$ which leaves all the roots of*

$$F(z, \eta, w') = 0$$

* *Berl. Sitzungsber.*, (1884), p. 707.

distinct from one another, no condition need be imposed; but if it be a root of the discriminant-equation which makes one or more roots of

$$F(z, \eta, w') = 0$$

multiple, then η must be a solution of the original differential equation so that we must have $w' = \dfrac{d\eta}{dz}$ for those multiple roots.

(C) *If the root $w' = \dfrac{d\eta}{dz} = \zeta$ of the equation*

$$F(z, \eta, w') = 0$$

is of multiplicity α, then $w = \eta$ must be a root of

$$F(z, w, \zeta) = 0$$

of multiplicity equal to, or greater than, $\alpha - 1$.

In the course of the establishment of the theorem, it has been proved

(i) that w' cannot become infinite for finite values of w when z has a parametric value, though it may be so when z has a particular value (or any one of a number of particular values) determined by the equation itself:

(ii) that, if $wW = 1$, then $\dfrac{dW}{dz}$ cannot become infinite when $W = 0$ if z has a parametric value, though it may be so when z has a particular value (or any one of a number of particular values) determined by the equation itself:

(iii) that the values of w and w', determined by

$$F = 0, \quad \frac{\partial F}{\partial w'} = 0,$$

constitute a solution (singular, or particular, or both singular and particular) of the equation.

Ex. The conditions, that the critical points of the equation

$$\tfrac{1}{4}\lambda^2 \left(\frac{dw}{dz}\right)^2 = (\mu^2 + \lambda w)\left(2\mu \frac{dw}{dz} + w^2\right)$$

should be fixed, are satisfied : so that the only parametric points, which are exceptional for the integral, are poles.

Let a be a pole of the integral of order n, so that w can be expressed in the form

$$w = \frac{\theta}{(z-a)^n} P(z-a),$$

where P is a regular function in the vicinity of $z=a$ and $P(0)=1$. When substitution takes place, a is a pole of order $2n+2$ on the left-hand side; for λw^3 it is of order $3n$; for $2\lambda\mu w \dfrac{dw}{dz}$, it is of order $2n+1$; and for other terms, it is of order less than $2n+2$. Hence

$$2n+2=3n\,;$$

and equating coefficients of the terms of highest order in $(z-a)^{-1}$, we have

$$\tfrac{1}{4}\lambda^2\theta^2 n^2=\lambda\theta^3,$$

that is, we have

$$w=\frac{\lambda}{(z-a)^2}P(z-a).$$

As a matter of fact,

$$P(z-a)=1+2\mu(z-a)\,;$$

so that the integral of the equation is a rational function of z.

122.   The equation

$$F(z,\,w,\,w')=0$$

is rational and integral of degree m in w'; and it has been proved to be rational and integral of degree not greater than $2m$ in w, in order that the desired property may be possessed: the coefficients of the powers of w' being of limited degrees in w.   The variable z is of a parametric character when the equation is regarded as an algebraical relation between w and w'.

It thus is natural to associate with the equation a Riemann surface having m sheets.   The branch-points of the surface are the places

$$w=\eta,\quad w'=\frac{d\eta}{dz},$$

given by $F=0$, $\dfrac{\partial F}{\partial w'}=0$; and the surface is such that if, at any place $w=\eta$, $w'=\dfrac{d\eta}{dz}$, there are α sheets which branch there, then we associate with that branch-place a root w of

$$F\left(z,\,w,\,\frac{d\eta}{dz}\right)=0,$$

which is of multiplicity at least $\alpha-1$.   Let $2p+1$ denote the connectivity* of the Riemann surface, so that p denotes the class (or genus) of the surface and also the class (or genus) of the equation.

* *Th. Fns.*, § 178.

Thus far, the only limitation upon the explicit occurrence of z in the equation

$$F(z, w, w') = 0$$

is that the coefficients of the various algebraical combinations of w and w' are uniform functions of z; and the branch-points of the equation are fixed points, being roots of such functions as g_0 in the expansions in § 120.

If these uniform functions are transcendental functions of z, then the number of roots of g_0 may be unlimited; and so the integral of the equation could have an unlimited number of branch-points, all of them fixed.

If all th uniform functions are rational functions of z, so that, in effect, $F(z, w, w') = 0$ is algebraical in w', in w, and in z, then the number of branch-points is limited; and all of them are determined by the equation itself, being fixed points.

123.   Proceeding now to the various cases that can arise, we begin with the simplest, viz. when the genus of the equation is zero, so that $p = 0$. Then it is known * that both the quantities which are regarded as variables in the equation can be expressed rationally as functions of a new variable t, say

$$w = \frac{\phi_1(t)}{\phi_0(t)}, \quad w' = \frac{\phi_2(t)}{\phi_0(t)},$$

where ϕ_0, ϕ_1, ϕ_2 are algebraical integral polynomials of degree not higher than $2m$ in t; in the present case, their coefficients are functions of z. Hence

$$\frac{dt}{dz} = \frac{\phi_0\phi_2 - \phi_0\dfrac{\partial\phi_1}{\partial z} + \phi_1\dfrac{\partial\phi_0}{\partial z}}{\phi_0\dfrac{\partial\phi_1}{\partial t} - \phi_1\dfrac{\partial\phi_0}{\partial t}},$$

where the right-hand side is a rational function of t. The branch-points of the original equation are to be fixed points; they manifestly can arise only through branch-points of t, for w is uniform in z and is rational in t; therefore the branch-points

---

* Salmon, *Higher Plane Curves*, § 44.

of t are fixed points and consequently (§ 112) the equation for t must have the form

$$\frac{dt}{dz} = A_0 + A_1 t + A_2 t^2,$$

where A_0, A_1, A_2 are uniform functions of z.

There thus arises no substantially new class of equations of genus zero; they all are rational transformations of Riccati's equation. When given, they can be transformed to Riccati's equation.

124. Next, consider equations of genus unity, so that $p = 1$. It is known* that both the quantities, which are regarded as variables in the equation, can be expressed rationally as uniform doubly-periodic functions of a single argument; or, restricting the expression of the variables to algebraical functions, they can be expressed in the form

$$w = \frac{\phi_1 + \psi_1 R^{\frac{1}{2}}}{\phi_0 + \psi_0 R^{\frac{1}{2}}}, \quad w' = \frac{\phi_2 + \psi_2 R^{\frac{1}{2}}}{\phi_0 + \psi_0 R^{\frac{1}{2}}},$$

where ϕ_0, ϕ_1, ϕ_2 are rational integral functions of t of degree $k \leqslant \frac{1}{2}m$, ψ_0, ψ_1, ψ_2 are rational integral functions of t of degree $h \leqslant \frac{1}{2}m$, such that $h + k = m - 2$, and R is a quartic polynomial in t: the coefficients in every case being functions of z.

When the two equations are combined, we obtain, after substituting for w' in the derivative of w, an equation for $\frac{dt}{dz}$ which, when rationalised, is of the second order. As the quantity w is not to have parametric branch-points, it is clear that t may not have parametric branch-points; hence, by the condition (A) of § 121, the equation for $\frac{dt}{dz}$ must be of the form

$$\left(\frac{dt}{dz}\right)^2 - 2\psi_1 \frac{dt}{dz} + \psi_2 = 0,$$

where ψ_1 and ψ_2 are polynomials of degrees not higher than two and four respectively, the coefficients being functions of z.

---

* Clebsch, *Crelle*, t. LXIV (1865), pp. 210—270; Cayley, *Coll. Math. Papers*, vol. VIII, pp. 181—187, where such equations are called bicursal.

Solving for $\dfrac{dt}{dz}$, we have

$$\frac{dt}{dz} = A_0 + A_1 t + A_2 t^2 + \lambda R^{\frac{1}{2}},$$

where A_0, A_1, A_2, λ are functions of z: and $\lambda^2 R$, $= \psi_1{}^2 - \psi_2$, is a quartic polynomial in t.

Further, the condition (B) of § 121 must be satisfied. The discriminant-equation is $R = 0$; any root of this equation must satisfy the differential equation. Now the value of $\dfrac{dt}{dz}$ is given by

$$\frac{\partial R}{\partial t}\frac{dt}{dz} + \frac{\partial R}{\partial z} = 0$$

for the particular root of R: and also, for this root, by

$$\frac{dt}{dz} = A_0 + A_1 t + A_2 t^2.$$

Hence we have

$$\frac{\partial R}{\partial z} + (A_0 + A_1 t + A_2 t^2)\frac{\partial R}{\partial t} = 0,$$

satisfied by any root of $R = 0$, so that the left-hand side vanishes for each root of $R = 0$, that is, when $R = 0$: and it is of degree in t greater than R. It must therefore contain R as a factor, and consequently

$$\frac{\partial R}{\partial z} + (A_0 + A_1 t + A_2 t^2)\frac{\partial R}{\partial t} = (B_0 + B_1 t) R,$$

where B_0 and B_1 may be functions of z: the quantity R must satisfy this equation. Also R is of degree 4 in t; and since a factor λ^2 is associated with R in the expression for $\psi_1{}^2 - \psi_2$, we may assume that, in R, the coefficient of t^4 is unity, say

$$R = (t - \alpha)(t - \beta)(t - \gamma)(t - \delta).$$

In the expressions for w and w' in terms of t, the general character and the particular conditions are conserved when t is subjected to a homographic substitution

$$t = \frac{au + b}{cu + d},$$

where a, b, c, d are functions of z such that

$$ad - bc = 1.$$

It therefore is to be expected that the corresponding equations, determining u and the associated quartic expression, will be of the same general form as those for t. To verify this, we have

$$\frac{du}{dz} = (cu + d)^2 \frac{dt}{dz}$$

$$= A_0(cu + d)^2 + A_1(cu + d)(au + b) + A_2(au + b)^2 + \lambda_1 R'^{\frac{1}{2}},$$

where $R' = (u - \alpha')(u - \beta')(u - \gamma')(u - \delta')$,

$$\alpha = \frac{a\alpha' + b}{c\alpha' + d}, \quad \beta = \frac{a\beta' + b}{c\beta' + d}, \cdots,$$

and $\lambda_1 = \lambda \{(a - c\alpha)(a - c\beta)(a - c\gamma)(a - c\delta)\}^{\frac{1}{2}}$;

that is, $\dfrac{du}{dz} = A_0' + A_1'u + A_2'u^2 + \lambda_1 R'^{\frac{1}{2}}.$

Similarly, we find

$$\frac{\partial R'}{\partial z} + (A_0' + A_1'u + A_2'u^2)\frac{\partial R'}{\partial t} = (B_0' + B_1'u).R',$$

so that the equations are covariantive for homographic transformation.

We can select a normal form by taking an appropriate homographic substitution. As the quantities a, b, c, d are at our disposal, subject to the condition $ad - bc = 1$, choose them so that $\alpha' = 0$, $\beta' = 1$, $\gamma' = -1$; then

$$R' = u(u^2 - 1)(u - \delta'),$$

where δ', if variable, is the only function of z in R'. It follows that the system can always be transformed so that the quartic polynomial in the radical, which occurs in the expression for $\dfrac{dt}{dz}$, can be made to occur in the form

$$R = R(t) = t(t^2 - 1)(t - \delta),$$

where δ may be a function of z. Substituting in

$$\frac{\partial R}{\partial z} + (A_0 + A_1 t + A_2 t^2)\frac{\partial R}{\partial t} = (B_0 + B_1 t)R,$$

which must be satisfied identically, we have

$$t(1 - t^2)\frac{d\delta}{dz} + (A_0 + A_1 t + A_2 t^2)\frac{\partial R}{\partial t} = (B_0 + B_1 t)R.$$

Now make $t = -1, 0, 1$ in succession, each of which causes R and the coefficient of $\dfrac{d\delta}{dz}$ to vanish : hence for these three values

$$(A_0 + A_1 t + A_2 t^2)\frac{\partial R}{\partial t}$$

must vanish.   Moreover, $\dfrac{\partial R}{\partial t}$ cannot vanish for $t = -1, 0, 1$; for in the respective cases, the factor $t+1$, t, or $t-1$ would be a repeated factor of R, and the genus of the original equation would then be $p = 0$, not $p = 1$.

Consequently, $A_0 + A_1 t + A_2 t^2$ vanishes for the three values of t, and therefore $A_0 = 0$, $A_1 = 0$, $A_2 = 0$; thus

$$t(1 - t^2)\frac{d\delta}{dz} = (B_0 + B_1 t)\, R,$$

which is to be satisfied identically.   It gives

$$\frac{d\delta}{dz} = (B_0 + B_1 t)(t - \delta),$$

which can be satisfied identically, only if

$$\frac{d\delta}{dz} = 0, \quad B_0 = 0, \quad B_1 = 0,$$

that is, δ is a constant and does not depend on z.

We now have

$$\frac{dt}{dz} = \lambda R^{\frac{1}{2}},$$

that is,

$$R^{-\frac{1}{2}} dt = \lambda dz = d\mu,$$

say, where μ is a function of z.   Hence, after quadratures, we can express t as a uniform doubly-periodic function of $\mu + A$, where μ is a function of z; the critical points of t are those of μ, where $\dfrac{d\mu}{dz} = \lambda$, that is, they are fixed points.

Hence, when the genus of the equation $F = 0$ is unity and the other conditions of § 121 are satisfied, its integral can be obtained by algebraical transformations and a quadrature.

125.   When the genus p of the equation

$$F(z, w, w') = 0$$

is greater than unity, and the integral of the equation is required to have all its critical points fixed, then the integral can be obtained by algebraical processes, and it occurs in the form

$$w = A(z),$$

where A is algebraical in the functions of z that are coefficients of w and w' in F, and A also involves initially assigned values of w and w'. This theorem was first enunciated* by Poincaré, having been obtained by considerations associated mainly with the theory of automorphic functions: but the general result, as distinguished from details and from the actual derivation of the integral, can be established more simply by a proof due to Picard†, based upon considerations associated with the theory of birational transformation of Riemann surfaces.

Let x denote an arbitrary initial value of z, and let y denote a corresponding arbitrary initial value of w; then the values of w' (denoted, say, by y'), which can be taken as initial values, satisfy the equation

$$F(x, y, y') = 0$$

of precisely the same form as the given equation. Manifestly when this is regarded as an algebraical relation between x, y, y', any two of the three quantities can be considered as independent; so that, regarding the equation as one between y and y', involving a parameter x, we have the genus of the equation equal to p. Moreover, the values of y' are different from one another unless y is such as to make the discriminant of F vanish, that is, unless only a limited number of values of y be excluded.

Now let the independent variable pass from x to z by any path not passing through any of the fixed critical points: there are supposed to be no parametric critical points, so that the path is not otherwise restricted. Taking x, y, y' as a set of initial values and proceeding along the path, we obtain at the end definite final values z, w, w'; also, because there are no parametric critical points, the path can, without changing these final values, be deformed in the plane, provided only it is not made to pass over any of the fixed critical points. Thus the final values w and w' depend upon the extremities of the path, upon the initial values

* *Acta Math.* t. vii (1885), pp. 1—32: this memoir should be consulted also with reference to § 124.

† *Cours d'Analyse*, t. iii. pp. 62, 81—87.

y and y', and only to a limited definite extent upon the path; in other words, w and w' can be expressed in terms of y and y', as well as of z and x, in the form

$$\begin{aligned} w &= g\,(z,\,x,\,y,\,y') \\ w' &= G\,(z,\,x,\,y,\,y') \end{aligned}\Bigg\},$$

where the functions g and G are uniform functions of y and y'. (For the purposes of the verification of the equation, we should have $G = \partial g/\partial z$; but it is not necessary to introduce this relation at the present stage of the argument.)

The quantities y and y' are parametric and not special numerical values. Hence the path can be reversed, the argument applied with z, w, w' as initial values, and equally general deformations of the path are possible; hence y and y' can be expressed in terms of w and w', as well as of z and x, in the form

$$\begin{aligned} y &= h\,(z,\,x,\,w,\,w') \\ y' &= H\,(z,\,x,\,w,\,w') \end{aligned}\Bigg\},$$

where the functions h and H are uniform functions of w and w'. Combining these results, it follows that there exists a bi-uniform transformation between the set of variables w, w' and the set of variables y, y'. Further, the equation $F(z,\,w,\,w') = 0$ is rational in w and w'; and the equation $F(x,\,y,\,y') = 0$ is rational in y and y'. Hence the bi-uniform transformations are rational in the variables that are transformed: in other words, there exists a birational transformation which transforms the equation $F = 0$ into itself.

Such a birational transformation is, in general, not unique. It is known that, when $p = 0$, birational transformations exist involving three arbitrary parameters: and, when $p = 1$, birational transformations exist involving one arbitrary parameter. But when $p > 1$ (the case which at present is under consideration), the number of birational transformations is known to be limited.

Some of the limited number of birational transformations, which exist when $p > 1$, can be deduced from the properties of integrals of the first kind belonging to the associated Riemann surface. Let such an integral, appertaining to the equation

$$F\,(z,\,w,\,w') = 0,$$

be denoted by

$$\int I_m\,(z,\,w,\,w')\,dw;$$

there are p such integrals in all, and they can be represented by giving to m the values 1, 2, ..., p. When the variables are transformed by the birational transformations that are known to exist, the subject of integration becomes a function of y and y' ; and as the integral is everywhere finite on the surface, it is still an integral of the first kind, and therefore* is expressible as a linear combination of the p normal integrals of the first kind appertaining to the equation

$$F(x, y, y') = 0.$$

Hence

$$\int I_m(z, w, w')\,dw \equiv \Sigma A_{m,\kappa} \int I_\kappa(x, y, y')\,dy,$$

the moduli of the congruences being the periods of the integrals, and the coefficients A, independent of w, w', y, y', being possibly functions of z, x and constants; and therefore

$$I_m(z, w, w')\,dw = \Sigma A_{m\kappa} I_\kappa(x, y, y')\,dy,$$

a differential relation which subsists in virtue of the birational transformation.

The same argument applies to each of the p normal integrals of the first kind; and therefore taking any other of them, say $\int I_n(z, w, w')\,dw$, we deduce

$$I_n(z, w, w')\,dw = \Sigma A_{n\kappa} I_\kappa(x, y, y')\,dy,$$

another differential relation subsisting in virtue of the (same) birational transformation. As these two relations are consistent with one another, it follows that the birational transformation is implied in the deduced equation

$$\frac{I_m(z, w, w')}{I_n(z, w, w')} = \frac{\Sigma A_{m\kappa} I_\kappa(x, y, y')}{\Sigma A_{n\kappa} I_\kappa(x, y, y')}.$$

To determine it more explicitly, we should in the first instance take an equation of this kind with (parametric) coefficients A, the quantities I being normal integrals of the first kind; and then the coefficients A must be obtained in connection with the fundamental equations

$$F(z, w, w') = 0, \quad F(x, y, y') = 0.$$

Even without this more explicit determination, the general character of the new relation is known: it is an algebraical relation

* *Th. Fns.*, § 234.

between w, w' and the initial parametric quantities y, y' : that is, the integral of the equation

$$F(z, w, w') = 0,$$

when it is of genus greater than unity and it has all its critical points fixed, is algebraical.

It is manifest that, in order to secure the existence of the birational transformation deduced through integrals of the first kind, limitations upon the forms of the functions I_m must be satisfied; these in turn will impose limitations upon the form of the equation $F = 0$, which must be satisfied in order that the assumed hypothesis as to the critical points may be justified. Moreover, in constructing the equation of transformation, the original equation $F = 0$ has been regarded merely as an algebraical equation between w and w' : that the results obtained may constitute a solution of the differential equation, the further relation

$$w' = \frac{dw}{dz}$$

between the quantities given by the birational transformation must be satisfied; when this is the case, the equations represent an algebraical integral.

126.   Summarising the results, we have the following addition to Fuchs's theorem, made by Poincaré:

When Fuchs's conditions that the critical points of the differential equation

$$F(z, w, w') = 0$$

should be fixed points and not parametric are satisfied, then, when the genus of the equation (regarded as algebraical in w and w') is zero, it can be transformed to Riccati's equation; when the genus is unity, it is integrable by transformation and a quadrature; when the genus is greater than unity, the integral of the equation is algebraical.

The appropriate reduction, when $p = 0$, is given by the customary unicursal expressions for w and w'. The reduction to the quadratable form, when $p = 1$, is derivable from the corresponding expressions for w and w'. When $p > 1$ and the conditions of § 121 are satisfied, several of these are useful in the actual construction of the integrals of the first order; the further analysis for the derivation of the algebraical integral is then within an algebraic range. Moreover, by the birational transformation in the last case, the surface associated with the equation

$$F(z, w, w') = 0$$

is transformed into itself. But such a surface has $3p - 3$ moduli (if $p > 1$), which are invariable through such a transformation; hence as the moduli, which might involve z alone for the first form and x alone for the second form, are unaltered, they must be pure constants—a result due to Poincaré.

BINOMIAL EQUATIONS WITH FIXED CRITICAL POINTS.

127.   As a special class, consider binomial equations* of the form

$$w'^m = R(w, z),$$

where R is an algebraical polynomial of degree not greater than $2m$, in accordance with condition (A) of § 121.

Let $w = \eta$ be a root of $R(w, z) = 0$, which is the discriminant-equation; then it makes the m roots w' equal to one another and zero in value. If $w = \eta$ is not a branch-point for the values of w', no condition attaches to η (B, § 121): in order that this may be the case, the index of $w - \eta$ in R must be an integer multiple of m, that is, it must be either $2m$ or m. If $w = \eta$ is a branch-point for sets of values of w', then (B, § 121) η is a solution of the equation; hence $\dfrac{d\eta}{dz} = 0$, and therefore η is a constant, say a. As regards the index of $w - a$ in R, let it be n; and suppose that s is the greatest common measure of m and n. Then the point $w = a$ gives s cycles, each of $\dfrac{m}{s}$ members w' branching there; and in the expression of these members, each cycle is given by an equation of the form

$$w'^{\frac{m}{s}} = (w - a)^{\frac{n}{s}} P(w - a),$$

where P is a regular function such that $P(0)$ is not zero.

---

* Wallenberg, *Schlöm. Zeitschr.*, t. xxxv (1890), pp. 193 et seq.

Hence (C, § 121)

$$\frac{n}{s} \geqslant \frac{m}{s} - 1,$$

that is,

$$n \geqslant m - s.$$

Now whether factors of R be of the form $w - a$ (where a is constant) or of the form $w - \eta$ (where η is variable), we have

$$2m \geqslant \Sigma n$$
$$\geqslant \Sigma (m - s).$$

Moreover, for the Riemann surface connected with the equation involving w' and w as variables, the ramification Ω is given[*] by

$$\Omega = \Sigma s \left(\frac{m}{s} - 1 \right)$$
$$= \Sigma (m - s),$$

so that

$$\Omega \leqslant 2m.$$

But if p be the class (or genus) of the surface,

$$\Omega = 2m + 2(p - 1);$$

hence $p = 0$ or $p = 1$.

When $p = 0$, the equation becomes a Riccati equation: when $p = 1$, the integral can be obtained by quadratures: in each case, it may be, after algebraical transformations. Thus the integration of every binomial differential equation, all the critical points of which are fixed points, can be made to depend upon the integration of a Riccati equation or can be effected by quadratures.

128. We proceed to obtain all the irreducible binomial equations which have all their critical points fixed.

Denote by $w - \eta$ a factor of R such that η may be a function of z. Then $R(w, z)$ may contain either

    (i)   a single power $(w - \eta)^{2m}$: or

    (ii)   two single powers $(w - \eta_1)^m (w - \eta_2)^m$: or

    (iii) a single power $(w - \eta)^m$: or

    (iv) no factor of the form $w - \eta$;

and in addition, it can contain a factor λ, where λ is a function of z.

[*] *Th. Fns.*, § 178.

(i)   The first two of these cases can be dealt with at once. In the first, we have

$$w'^m = \lambda\,(w - \eta)^{2m},$$

so that, if $\lambda = \mu^m$, we have

$$w' = \mu\,(w - \eta)^2,$$

a special form of Riccati's equation.

(ii)   Similarly, the second is

$$w'^m = \lambda\,(w - \eta_1)^m\,(w - \eta_2)^m,$$

which is reducible to

$$w' = \mu\,(w - \eta_1)\,(w - \eta_2),$$

another form of Riccati's equation.

(iii)   As regards the third form, if $R\,(w,\,z)$, which can be (but need not be) of degree $2m$ in w, contains no factors other than λ and $(w - \eta)^m$, the equation is

$$w'^m = \lambda\,(w - \eta)^m;$$

this is reducible to

$$w' = \mu\,(w - \eta),$$

and can be integrated by quadratures. Accordingly, for the further discussion of the case when $R\,(w,\,z)$ contains a factor $(w - \eta)^m$, it will be assumed that it contains* factors of the form $w - a$.

When $R\,(w,\,z)$ contains a single factor $(w - \eta)^m$, let the equation be

$$w'^m = \lambda\,(w - \eta)^m\,P\,(w),$$

where $P\,(w)$ is a product of factors of the form $w - a$, and in w is of degree not higher than m. Let such a factor be $(w - a_1)^{n_1}$; and let s_1 be the greatest common measure of m and n_1, so that (C, § 121)

$$n_1 \geqslant m - s_1.$$

Hence

$$m \geqslant \Sigma n_1$$

$$\geqslant \Sigma\,(m - s_1);$$

and therefore

$$\Sigma \left(1 - \frac{s_1}{m}\right) \leqslant 1.$$

---

* It hardly needs to be remarked that the quantity η may be a constant in any particular instance; the most general case is that in which η is a function of the independent variable z.

Now as s_1 is a factor of m, $\dfrac{s_1}{m} \leqslant \tfrac{1}{2}$, and therefore $1 - \dfrac{s_1}{m} \geqslant \tfrac{1}{2}$. Consequently, in the preceding summation, there can be only two terms or only one term, corresponding to two different factors $w - a$ or to one such factor.

When there are two terms in the summation, then

$$1 - \frac{s_1}{m} = \tfrac{1}{2}$$

for each of them, that is, $\dfrac{s_1}{m} = \tfrac{1}{2}$, and so $n_1 = \tfrac{1}{2}m$, since s_1 is the greatest common measure of n_1 and m. The equation is

$$w'^m = \lambda \, (w - \eta)^m \, (w - a_1)^{\frac{1}{2}m} \, (w - a_2)^{\frac{1}{2}m},$$

which is reducible to

$$w'^2 = \mu \, (w - \eta)^2 \, (w - a_1) \, (w - a_2).$$

When there is only one term in the summation, then (C, § 121)

$$n_1 \geqslant m - s_1,$$

where s_1 is the greatest common measure of m and n_1. The specified property must be possessed also for infinite values of w; so that writing $wW = 1$, we have

$$(-1)^m \, W'^m = \lambda \, (1 - W\eta)^m \, (1 - a_1 W)^{n_1} \, W^{m - n_1}.$$

Hence as s_1 is still the greatest common measure of m and $m - n_1$, we have (C, § 121)

$$m - n_1 \geqslant m - s_1.$$

From the former case,

$$m - n_1 \leqslant s_1,$$

so that

$$s_1 \geqslant m - s_1$$

and therefore

$$s_1 \geqslant \tfrac{1}{2}m.$$

But s_1 is a factor of m. Were it equal to m, the equation would be

$$w'^m = \lambda \, (w - \eta)^m \, (w - a_1)^m,$$

a special form of an equation already considered (ii). Hence

$$s_1 = \tfrac{1}{2}m,$$

and the equation is

$$w'^m = \lambda \, (w - \eta)^m \, (w - a_1)^{\frac{1}{2}m},$$

the reduced form of which is

$$w'^2 = \mu \, (w - \eta)^2 \, (w - a_1).$$

(iv) Now, suppose that $R(w, z)$ contains no factor of the type $w - \eta$, so that all its factors which involve w are of the type $w - a$; let the equation be

$$w'^m = \lambda \, (w - a_1)^{n_1} (w - a_2)^{n_2} \ldots (w - a_k)^{n_k},$$

where

$$n_1 + n_2 + \ldots + n_k \leqslant 2m,$$

on account of the degree of R in w. If s_i denote the greatest common measure of m and n_i, then (C § 121)

$$n_i \geqslant m - s_i,$$

which secures the property for the values $w = a_i$. To secure the property for infinite values of w, we similarly must have

$$\theta \geqslant m - s,$$

where $\theta = 2m - (n_1 + n_2 + \ldots + n_k)$, and s is the greatest common measure of θ and m.

From the former conditions, we have

$$\Sigma \, (m - s_i) \leqslant \Sigma n_i \leqslant 2m,$$

and therefore

$$\Sigma \left(1 - \frac{s_i}{m} \right) \leqslant 2.$$

There are various cases, according to the number of distinct factors in $R(w)$.

The integer s_i is m, or $\tfrac{1}{2}m$, or $\tfrac{1}{3}m$, ..., according to the value of n_i, of which it is a factor. Unless one (or more than one) of the values of s_i be m, then

$$\frac{s_i}{m} \leqslant \tfrac{1}{2},$$

so that

$$1 - \frac{s_i}{m} \geqslant \tfrac{1}{2} :$$

consequently as

$$\Sigma \left(1 - \frac{s_i}{m} \right) \leqslant 2,$$

there cannot be more than four terms in the summation.

If one of the quantities s_i be m, so that $n_i = m$, then for the remaining factors

$$\Sigma \left(1 - \frac{s_i}{m}\right) \leqslant 1:$$

that is, there cannot then be more than two other factors.

If two quantities s_i be m, there are no other factors.

The last two cases are special instances of the more general forms already dealt with; the first, therefore, supplies a new result.

129. First, suppose that there is only one distinct factor, say $(w - a)^n$, so that

$$w'^m = \lambda (w - a)^n.$$

If m and n have any common factor s, the equation is reducible to one of degree $\dfrac{m}{s}$ and of the same form: hence we can take $s = 1$. Then

$$n \geqslant m - 1 = m - 1 + q,$$

say. To take account of infinite values, write $wW = 1$ and then

$$(-W')^m = \lambda (1 - aW)^n W^{2m-n},$$

and in order to secure the required property, we have

$$2m - n \geqslant m - 1,$$

that is,

$$m + 1 - q \geqslant m - 1,$$

and therefore

$$q \leqslant 2.$$

When $q = 0$, we have the equation

$$w'^m = \lambda (w - a)^{m-1}.$$

When $q = 1$, the equation reduces to the form $w' = \mu (w - a)$, which already has been considered.

When $q = 2$, we have the equation

$$w'^m = \lambda (w - a)^{m+1}.$$

Thus there are two new forms when $R(w)$ contains a single factor.

Secondly, suppose that there are two distinct factors, so that

$$w'^m = \lambda (w - a_1)^{n_1} (w - a_2)^{n_2},$$

where

$$2m - n_1 - n_2 = \theta \geqslant 0;$$

and assume that m, n_1, n_2 have no common factor: the equation would otherwise be reducible to another, for which this condition is satisfied. Let s_1 be the greatest common factor of m and n_1, s_2 that of m and n_2, so that s_1 and s_2 are prime to one another: and let s be that of m and θ, when $\theta > 0$. Then (C, § 121)

$$n_1 \geqslant m - s_1,$$
$$n_2 \geqslant m - s_2,$$
$$\theta \geqslant m - s,$$

so that

$$2m \geqslant 3m - s - s_1 - s_2,$$

and therefore

$$m \leqslant s + s_1 + s_2.$$

Now s, when it occurs, s_1, and s_2, are factors of m. Let $\dfrac{s}{m} = \dfrac{1}{\sigma}$, $\dfrac{s_1}{m} = \dfrac{1}{\sigma_1}$, $\dfrac{s_2}{m} = \dfrac{1}{\sigma_2}$: so that

$$1 \leqslant \frac{1}{\sigma} + \frac{1}{\sigma_1} + \frac{1}{\sigma_2}.$$

Let $\theta = 0$; then

$$n_1 + n_2 = 2m,$$

so that either both integers are equal to m, in which case the equation is reducible to

$$w' = \mu (w - a)(w - b),$$

a special instance of a form already retained: or, one of the integers n_1 being greater than m, say $m + \gamma$, the other is $m - \gamma$. Since s_2 divides n_2 and m, it divides γ and therefore also n_1; the equation can be reduced to

$$w'^{\frac{m}{s_2}} = \lambda^{\frac{1}{s_2}} (w - a_1)^{\frac{n_1}{s_2}} (w - a_2)^{\frac{n_2}{s_2}}.$$

Now $\dfrac{n_2}{s_2}$ and $\dfrac{m}{s_2}$ have no common factor, so that (C, § 121)

$$\frac{n_2}{s_2} \geqslant \frac{m}{s_2} - 1;$$

and $n_2 = m - \gamma$, so that

$$\frac{n_2}{s_2} < \frac{m}{s_2}.$$

Hence

$$\frac{n_2}{s_2} = \frac{m}{s_2} - 1;$$

and therefore

$$\frac{n_1}{s_2} = \frac{m}{s_2} + 1.$$

Thus the equation reduces to

$$w'^t = \mu (w - a_1)^{t+1} (w - a_2)^{t-1},$$

where t is an integer.

Let $\theta > 0$; then

$$1 \leqslant \frac{1}{\sigma} + \frac{1}{\sigma_1} + \frac{1}{\sigma_2}.$$

We manifestly can leave on one side the case when n_1 or n_2 is m; for we then have a special instance of the equation containing a factor $(w - \eta)^m$: that is, we do not take $\sigma_1 = 1$ or $\sigma_2 = 1$.

The inequality is satisfied if $\sigma = 1$, that is, $s = m$, so that $\theta = 2m$ or $\theta = m$. If $\theta = 2m$, the equation is

$$w'^m = \lambda,$$

which is reducible to

$$w' = \mu.$$

If $\theta = m$, the equation is

$$w'^m = \lambda (w - a_1)^{n_1} (w - a_2)^{n_2},$$

where now $n_1 + n_2 = m$. Also

$$n_1 \geqslant m - s_1, \quad n_2 \geqslant m - s_2.$$

But $s_1 \leqslant \frac{1}{2}m$, $s_2 \leqslant \frac{1}{2}m$, so that

$$n_1 \geqslant \frac{1}{2}m, \quad n_2 \geqslant \frac{1}{2}m;$$

and therefore, because $n_1 + n_2 = m$, we have

$$n_1 = \frac{1}{2}m, \quad n_2 = \frac{1}{2}m.$$

The equation is

$$w'^m = \lambda (w - a_1)^{\frac{1}{2}m} (w - a_2)^{\frac{1}{2}m},$$

which is reducible to

$$w'^2 = \mu (w - a_1)(w - a_2).$$

Hence we now may take $\sigma > 1$.

Each of the numbers n_1 and n_2 is less than m. If not, let n_1 (which is not equal to m) be greater than m, say $m + \gamma$; then

$$n_2 < m,$$

so that

$$\frac{n_2}{s_2} < \frac{m}{s_2},$$

and

$$\frac{n_2}{s_2} \geqslant \frac{m}{s_2} - 1;$$

therefore

$$n_2 = m - s_2.$$

Also $\qquad\qquad\qquad\qquad \theta < m,$

so that $\qquad\qquad\qquad\qquad \dfrac{\theta}{s} < \dfrac{m}{s},$

and $\qquad\qquad\qquad\qquad \dfrac{\theta}{s} \geqslant \dfrac{m}{s} - 1 \, ;$

therefore $\qquad\qquad\qquad \theta = m - s.$

Now $s \leqslant \tfrac{1}{2}m$; so that
$$\theta \geqslant \tfrac{1}{2}m.$$

Also $\qquad\qquad\qquad\qquad \theta = 2m - n_1 - n_2$

$$= s_2 - \gamma$$

$$< s_2.$$

And s_2 is a factor of m, so that $s_2 \leqslant \tfrac{1}{2}m$: hence
$$\theta < \tfrac{1}{2}m,$$

contradicting the former result. Hence both the indices n_1, n_2 are less than m: and θ is less than m; so that now, as above,

$$n_1 = m - s_1, \quad n_2 = m - s_2, \quad \theta = m - s,$$

and we have
$$m = s_1 + s_2 + s,$$

and therefore
$$\frac{1}{\sigma} + \frac{1}{\sigma_1} + \frac{1}{\sigma_2} = 1.$$

The possible solutions are

$$\sigma, \, \sigma_1, \, \sigma_2 = 2, \, 3, \, 6 \, ;$$

$$2, \, 4, \, 4 \, ;$$

$$3, \, 3, \, 3 \, ;$$

giving
$$\sigma_1, \, \sigma_2 = 2, \, 6 \, ; \; 2, \, 3 \, ; \; 3, \, 6 \, ; \; 2, \, 4 \, ; \; 4, \, 4 \, ; \; 3, \, 3.$$

Now
$$n_1 = m\left(1 - \frac{1}{\sigma_1}\right), \quad n_2 = m\left(1 - \frac{1}{\sigma_2}\right) \, ;$$

hence the reduced forms of the equation

$$w'^m = \lambda \, (w - a_1)^{n_1} \, (w - a_2)^{n_2}$$

in the respective cases are

$$w'^6 = \mu\,(w - a_1)^3\,(w - a_2)^5$$
$$w'^6 = \mu\,(w - a_1)^3\,(w - a_2)^4$$
$$w'^6 = \mu\,(w - a_1)^4\,(w - a_2)^5$$
$$w'^4 = \mu\,(w - a_1)^2\,(w - a_2)^3$$
$$w'^4 = \mu\,(w - a_1)^3\,(w - a_2)^3$$
$$w'^3 = \mu\,(w - a_1)^2\,(w - a_2)^2$$

Thirdly, suppose that there are three distinct factors, so that the equation is

$$w'^m = \lambda\,(w - a_1)^{n_1}\,(w - a_2)^{n_2}\,(w - a_3)^{n_3},$$

where

$$n_1 + n_2 + n_3 \leqslant 2m.$$

We may leave on one side the case when one of the integers n is equal to m, because it has already been discussed in the more general form where the quantity a is a function of z.

Moreover, no one of the indices can be greater than m: for if one of them, say n_1, is greater than m, then

$$n_2 + n_3 < m.$$

Now, as in other instances,

$$\frac{n_2}{m} = 1 - \frac{s_2}{m} \geqslant \tfrac{1}{2},$$

and similarly

$$\frac{n_3}{m} \geqslant \tfrac{1}{2}:$$

which render $n_2 + n_3 < m$ an impossibility. Accordingly, n_1, n_2, n_3 are, each of them, less than m; and we have

$$n_i = m - s_i, \qquad\qquad (i = 1, 2, 3),$$

where s_i is the greatest common measure of m and n_i.

Let

$$2m - n_1 - n_2 - n_3 = \theta,$$

and first, suppose that θ is not zero. Then θ may not be $2m$; nor can $\theta = m$, for then

$$m = \Sigma\,(m - s_i)$$
$$= m\Sigma\left(1 - \frac{s_i}{m}\right)$$
$$\geqslant \tfrac{3}{2}m.$$

Hence, if θ is not zero, we have

$$\theta = m - s,$$

where s is the greatest common measure of m and θ. In this case

$$2m = m - s + \Sigma(m - s_i),$$

so that

$$2 = 1 - \frac{s}{m} + \Sigma\left(1 - \frac{s_i}{m}\right).$$

Each of the quantities $1 - \dfrac{s}{m}$, $1 - \dfrac{s_i}{m}$, is equal to, or greater than, $\frac{1}{2}$; as there are four of them, each of them is $\frac{1}{2}$, so that

$$n_i = \tfrac{1}{2}m.$$

The equation then is

$$w'^m = \lambda (w - a_1)^{\frac{1}{2}m} (w - a_2)^{\frac{1}{2}m} (w - a_3)^{\frac{1}{2}m},$$

which reduces to

$$w'^2 = \lambda (w - a_1)(w - a_2)(w - a_3).$$

When θ is zero, so that

$$n_1 + n_2 + n_3 = 2m,$$

then

$$\Sigma\left(1 - \frac{s_i}{m}\right) = 2,$$

and therefore

$$\Sigma \frac{s_i}{m} = 1 ;$$

that is, with the same notation as in the case of two factors, we have

$$\frac{1}{\sigma_1} + \frac{1}{\sigma_2} + \frac{1}{\sigma_3} = 1.$$

The possible solutions are

$$\sigma_1, \sigma_2, \sigma_3 = 2, 3, 6 ; \ 2, 4, 4 ; \ 3, 3, 3 ;$$

and

$$n_i = m\left(1 - \frac{1}{\sigma_i}\right),$$

for $i = 1, 2, 3$. The reduced forms of the equation

$$w'^m = \lambda (w - a_1)^{n_1} (w - a_2)^{n_2} (w - a_3)^{n_3}$$

in the respective cases are

$$\left.\begin{aligned}
w'^6 &= \mu (w - a_1)^3 (w - a_2)^4 (w - a_3)^5 \\
w'^4 &= \mu (w - a_1)^2 (w - a_2)^3 (w - a_3)^3 \\
w'^3 &= \mu (w - a_1)^2 (w - a_2)^2 (w - a_3)^2
\end{aligned}\right\}.$$

Fourthly, let there be four distinct factors in $R(w)$, which is the greatest number of distinct factors that $R(w)$ can have. The equation is

$$w'^m = \lambda (w - a_1)^{n_1} (w - a_2)^{n_2} (w - a_3)^{n_3} (w - a_4)^{n_4}.$$

The preceding analysis, for the instance $n_1 + n_2 + n_3 < 2m$, can be applied here: and it is easy to prove that $n_i = \frac{1}{2}m$, $(i = 1, 2, 3, 4)$, so that the reduced equation is

$$w'^2 = \mu (w - a_1)(w - a_2)(w - a_3)(w - a_4).$$

The equations obtained constitute the aggregate of the binomial equations, all the critical points of which are fixed.

130.   As is remarked by Wallenberg, these equations are not independent of one another.   Thus, taking

$$w - a_4 = \frac{1}{u},$$

the equation

$$w'^2 = \mu (w - a_1)(w - a_2)(w - a_3)(w - a_4)$$

becomes

$$u'^2 = \mu_1 (u - \alpha_1)(u - \alpha_2)(u - \alpha_3),$$

on changing μ_1 and the constants α.   Similarly, for the others. Accordingly, omitting all such forms as can be derived from others that are retained, we have the system

$$w' = \mu (w - \eta_1)(w - \eta_2),$$
$$w'^2 = \mu (w - \eta)^2 (w - a_1)(w - a_2),$$
$$w'^t = \mu (w - a_1)^{t+1} (w - a_2)^{t-1},$$
$$w'^6 = \mu (w - a_1)^3 (w - a_2)^4 (w - a_3)^5,$$
$$w'^4 = \mu (w - a_1)^2 (w - a_2)^3 (w - a_3)^3,$$
$$w'^3 = \mu (w - a_1)^2 (w - a_2)^2 (w - a_3)^2,$$
$$w'^2 = \mu (w - a_1)(w - a_2)(w - a_3)(w - a_4),$$

which are independent of one another; all the others can be deduced by transformation of these forms or by transformation of special cases of these forms.

The first of these is a Riccati equation.

For the second, let

$$\frac{w - a_1}{w - a_2} = \frac{y^2}{\mu (a_2 - \eta)^2},$$

where y is the new dependent variable; the equation for y is

$$2\frac{dy}{dz} = \mu(a_1 - \eta)(a_2 - \eta) + y\left(\frac{\mu'}{\mu} - \frac{2\dot{\eta}'}{a_2 - \eta}\right) - y^2,$$

a Riccati form.

For the third, let

$$\frac{w - a_1}{w - a_2} = y;$$

the equation for y is

$$\left(\frac{dy}{dz}\right)^t = (a_1 - a_2)^t\, y^{t+1},$$

the integral of which can be obtained by quadrature.

For each of the others, the genus (or class) p is unity: the integral depends upon algebraical transformation and quadratures, and as will be seen later (Chap. x), is in each case expressible by means of uniform doubly-periodic functions.

Ex. Shew that, if z does not explicitly occur in the equation

$$w'^m + a_1 w'^{m-1} + \ldots + a_{m-1} w' + a_m = 0,$$

the only equations which have all their critical points fixed are those of genus 0 or 1.

Note. If z does not occur explicitly in the original differential equation, the latter has the form

$$f(w, w') = 0.$$

Manifestly, when z occurs in the integral, it must occur in the form $z - a$, where a is an arbitrary constant; and therefore every critical point of the integral is parametric. If then the integral is to have all its critical points fixed, it follows that there are **no** critical points: that is, the integral is a uniform function.

The discussion of the necessary conditions and the determination of the various forms of equations, the integrals of which are thus limited, will be resumed in the next chapter.

Ex. 1. As an example of an equation for which $p > 1$, consider*

$$F = (4zw' - w)^4 - z^3 w'^4 + b = 0.$$

Forming the discriminant of F and taking any of its roots—this can be effected by treating $F = 0$, $\partial F/\partial w' = 0$ as simultaneous equations—, we find

$$w = \eta = \{(256z)^{\frac{1}{3}} - 1\}^{\frac{3}{4}},$$

giving twelve values in all.   Each of these is a solution of the equation.

---

* Wallenberg, *l.c.*, p. 352.

The branching, in the immediate vicinity of each of the twelve simultaneous values of η and $\dfrac{d\eta}{dz}$, is simple; the ramification is $12(2-1)=12$, and the genus of the equation is given by $2p+1=12-8+3$, so that $p=3$.

All the Fuchsian conditions, that the critical points should be fixed, are satisfied. Since the genus is greater than unity, the integral of the equation is algebraical (§ 125); and it can be derived through a transformation, that is birational between the variables w, w' and y, y' in the equations

$$(4zw' - w)^4 - z^3 w'^4 + b = 0, \quad (4xy' - y)^4 - x^3 y'^4 + b = 0.$$

Such a birational transformation is clearly given by

$$4zw' - w = 4xy' - y, \quad z^{\frac{3}{4}} w' = x^{\frac{3}{4}} y',$$

the irrational character so far as concerns the (parametric) quantities x, z not affecting the general character of the transformation. Hence

$$w = 4zw' - (4xy' - y)$$
$$= 4z^{\frac{1}{4}} x^{\frac{3}{4}} y' - (4xy' - y)$$
$$= 4z^{\frac{1}{4}} x^{\frac{3}{4}} y' - (x^3 y'^4 - b)^{\frac{1}{4}}.$$

Let $x^3 y'^4 - b$, which is connected with the initial arbitrary values of the variables, be denoted by a^4; then

$$w = 4z^{\frac{1}{4}} (a^4 + b)^{\frac{1}{4}} - a,$$

which is the integral. The critical points of the integral are $z = 0$, $z = \infty$, both fixed; and the integral manifestly is algebraical.

Ex. 2.　Shew that each of the equations

　(i)　$w'^3 - 4zww' + 8w^2 = 0$:

　(ii)　$w'^3 - \dfrac{4}{z} ww'^2 + \dfrac{4}{z^2} w^2 w' - \dfrac{1}{z} = 0$:

　(iii)　$z^3 w'^3 - (c^4 + 4z^2 w) w'^2 + (4zw^2 - 2c^2 z) w' - z^2 = 0$:

has all its critical points fixed; and obtain the integral in each case.

(Wallenberg.)

Note. The memoir by Wallenberg, which has been quoted and from which the preceding examples are taken, contains a number of interesting discussions relating to differential equations of the class under consideration in this chapter. In the same connection, Briot and Bouquet's *Théorie des fonctions elliptiques*, (2nd ed.), Book v, Chap. iv, may be consulted with advantage.

CHAPTER X.

EQUATIONS OF THE FIRST ORDER WITH UNIFORM INTEGRALS, AND WITH ALGEBRAICAL INTEGRALS*.

EQUATIONS WITH UNIFORM INTEGRALS†.

131. THE question of determining whether the general solution of the equation

$$F(z, w, w') = 0$$

is an algebraic equation or whether the integral is a uniform function, is one of much greater complexity than the preceding investigation; but if the differential equation be free from explicit occurrence of the independent variable, so that it has the form

$$F(w, w') = 0,$$

then similar analysis leads to the conditions necessary and sufficient to secure that the integral function w is a uniform function of z. The remark in § 130, Note, indicates that the last question must be largely included in the earlier investigation; but in spite of some repetition, a full discussion will be given here, so as not to leave an important lacuna in the theory. It

---

* For reasons that will appear in the course of the chapter, the theory of equations of the first order with algebraical integrals is discussed only very slightly. The methods belong to a range of ideas outside those which it is my chief aim to expound in this place: moreover, they appear to me not yet to have received that complete discussion or that sufficiently final form which compels their full admission into a text-book.

† Various references are given, in the course of §§ 131—137, to some of the more important memoirs used in giving an account of the method adopted. Other authorities, that may be consulted on the subject of differential equations of the form $F(W, w) = 0$, are Jordan, *Cours d'Analyse*, t. III, pp. 122—136; Picard, *Cours d'Analyse*, t. III, ch. IV; Phragmén, *Stockh. Öfv.*, t. XLVIII (1891), pp. 623—668.

also is desirable to expound the somewhat different and earlier method, due to Briot and Bouquet.

Let the equation be of the mth degree in $\dfrac{dw}{dz}$, supposed irreducible; when arranged in powers of the derivative, it takes the form

$$\left(\frac{dw}{dz}\right)^m + \left(\frac{dw}{dz}\right)^{m-1} f_1(w) + \left(\frac{dw}{dz}\right)^{m-2} f_2(w) + \ldots = 0.$$

Because w is a uniform function of z, it has, quà function of z, no branch-points; and $\dfrac{dw}{dz}$ has, quà function of z, no branch-points. Hence infinities of w are infinities of $\dfrac{dw}{dz}$, and *vice versa*; and therefore $\dfrac{dw}{dz}$ cannot become infinite for a finite value of w. It follows that the coefficients $f_1(w)$, $f_2(w)$, ... of the various powers of the derivative are integral functions of w; they are already known, by the character of the equation, to be rational.

Moreover, all the general properties possessed by w are possessed by its reciprocal $u = \dfrac{1}{w}$. When u is made the dependent variable, we have

$$\left(\frac{du}{dz}\right)^m - \left(\frac{du}{dz}\right)^{m-1} u^2 f_1\left(\frac{1}{u}\right) + \left(\frac{du}{dz}\right)^{m-2} u^4 f_2\left(\frac{1}{u}\right) - \ldots = 0$$

as the equation determining u. Now $\dfrac{du}{dz}$ cannot become infinite except for infinite values of u, since u is a uniform function of z; hence the coefficients of powers of $\dfrac{du}{dz}$ must be rational integral functions of u. This condition can be satisfied only if $f_s(w)$ be of degree in w not higher than $2s$.

Hence, denoting $\dfrac{dw}{dz}$ by W and $\dfrac{du}{dz}$ by U, we have the theorem :—

I. *The differential equation*

$$F(W, w) = W^m + W^{m-1} f_1(w) + W^{m-2} f_2(w) + \ldots = 0$$

cannot determine w as a uniform function of z, unless the co-efficients

$$f_1(w), \quad f_2(w), \quad f_3(w), \ldots$$

are rational integral functions of w of degrees not higher than 2, 4, 6, ... *respectively: and when this condition is satisfied, it is satisfied also for the equation*

$$U^m - U^{m-1} u^2 f_1\left(\frac{1}{u}\right) + U^{m-2} u^4 f_2\left(\frac{1}{u}\right) - \ldots = 0,$$

which determines u, the reciprocal of w.

132. The equation, in the first instance, determines W as a function of w; and values of w may be ordinary points or may be branch-points for W, quà function of w. In the vicinity of such points, it is necessary to secure that w, as depending upon z, shall be uniform.

First, consider finite values for w: let $w = \gamma$. For points in the immediate vicinity of that value, the values of W are not infinite: they may be

   (i)   distinct from one another, and no one of them zero at the point; or

   (ii)   distinct from one another, and at least one of them zero at the point; or

   (iii)   not distinct from one another, so that $w = \gamma$ is then a branch-point of the function.

   (i)   Let any value Γ, a constant different from zero, be the value of W for $w = \gamma$. Then in the vicinity we have

$$\frac{dw}{dz} = \Gamma\{1 + \lambda(w - \gamma) + \mu(w - \gamma)^2 + \ldots\},$$

and therefore

$$\Gamma dz = \frac{dw}{1 + \lambda(w - \gamma) + \mu(w - \gamma)^2 + \ldots}$$

$$= \{1 + 2\lambda'(w - \gamma) + 3\mu'(w - \gamma)^2 + \ldots\} dw,$$

where λ', μ', ... are constants. Hence, if z_0 be the value of z when $w = \gamma$, we have

$$\Gamma(z - z_0) = w - \gamma + \lambda'(w - \gamma)^2 + \mu'(w - \gamma)^3 + \ldots,$$

and the inversion of this equation gives

$$w - \gamma = \Gamma(z - z_0) + P(z - z_0),$$

that is, w is then a uniform function of z in the vicinity of z_0. No new condition, attaching to the original equation, arises.

(ii) Since the values are distinct from one another, and at least one of them is zero for $w = \gamma$, we must have

$$\frac{dw}{dz} = a\,(w - \gamma)^n \{1 + b\,(w - \gamma) + c\,(w - \gamma)^2 + \ldots\}$$

for at least one of the values of W; and n is a positive integer, as γ is not a branch-point.

First, if n be unity, we have

$$\frac{dw}{w - \gamma} \{1 + b'\,(w - \gamma) + c'\,(w - \gamma)^2 + \ldots\} = a\,dz,$$

so that

$$\log(w - \gamma) + P\,(w - \gamma) = az,$$

the constant of integration being absorbed in $P\,(w - \gamma)$. This gives

$$(w - \gamma)\,e^{P(w-\gamma)} = e^{az},$$

and therefore, inverting the functional relation,

$$w - \gamma = e^{az}\,Q\,(e^{az}),$$

that is, w is a uniform function in the vicinity of its own value γ, but it can acquire this value only for logarithmically infinite values of z. No new condition, attaching to the original equation, arises.

Secondly, if n be 2, so that

$$\frac{dw}{dz} = a\,(w - \gamma)^2 \{1 + b\,(w - \gamma) + c\,(w - \gamma)^2 + \ldots\},$$

then, proceeding as before, we have

$$\frac{1}{w - \gamma}\,b \log(w - \gamma) + R\,(w - \gamma) = az.$$

If b be different from zero, then, as on pp. 315, 316, it can be proved that w is not uniform in the vicinity of $z = \infty$. Hence b must be zero, so that

$$w - \gamma = -\frac{1}{az}\,S\left(\frac{1}{az}\right),$$

giving w as a uniform function of z in the vicinity of its own value γ; and w can acquire the value γ only for algebraically infinite values of z. The new condition, attaching to the original equation, will be included in a subsequent case (III., § 133).

When $n > 2$, similar analysis shews that $z = \infty$ is a branch-point of w; that is, w is not then a uniform function of z. Thus the only values of n are 1, 2.

(iii) If $w = \gamma$ be a branch-point, then two cases arise according as W is not, or is, zero : it cannot be infinite, because γ is not infinite.

If W be not zero, we have the value of W in the form

$$W = a \{1 + b (w - \gamma)^{\frac{1}{p}} + c (w - \gamma)^{\frac{2}{p}} + ...\},$$

where p is a positive integer. The integral of this equation is of the form

$$(w - \gamma) \{1 + b' (w - \gamma)^{\frac{1}{p}} + c' (w - \gamma)^{\frac{2}{p}} + ...\} = a (z - \alpha),$$

and this makes w uniform in the vicinity of $z = \alpha$, only if powers of $w - \gamma$ with non-integer indices be absent from the last equation and therefore also from the former. When the fractional powers are absent from the former, the implication is that $w = \gamma$ is really not a branch-point for W, quà function of w, but only that more than one of its values are equal to a; then w is a uniform function of z, and therefore W is a uniform function of w, and *vice versa*.

If however W be zero at the branch-point, then its value in the vicinity takes the form

$$W = a (w - \gamma)^{\frac{q}{p}} + b (w - \gamma)^{\frac{q+1}{p}} + c (w - \gamma)^{\frac{q+2}{p}} + ... ;$$

and, as W cannot be infinite for a finite value of w, the fraction q/p is positive. It may be less than 1, equal to 1, or greater than 1. Hence :—

II. *If any finite value γ of w be a branch-point of W regarded as a function of w, then, in order that w may be uniform, all the values of W affected by the point must be zero for $w = \gamma$.*

133. If $q/p < 1$, the integration of the equation leads to a relation of the form

$$z - \alpha = a' (w - \gamma)^{\frac{p-q}{p}} + b' (w - \gamma)^{\frac{p-q+1}{p}} + ...$$

in which all the indices are positive. The inversion of this relation makes w uniform in the vicinity of $z = \alpha$, only if $p - q$ be unity, that is, if *the zero of W as a function of w be of degree $1 - \frac{1}{p}$, when the degree is less than unity ;* and the value of z is finite.

If $q/p = 1$, then we have

$$W = a(w - \gamma) + b(w - \gamma)^{1+\frac{1}{p}} + c(w - \gamma)^{1+\frac{2}{p}} + \ldots,$$

and therefore

$$a\,dz = \frac{dw}{w - \gamma}\{1 + a'(w - \gamma)^{\frac{1}{p}} + b'(w - \gamma)^{\frac{2}{p}} + \ldots\},$$

so that

$$az = \log(w - \gamma) + a''(w - \gamma)^{\frac{1}{p}} + b''(w - \gamma)^{\frac{2}{p}} + \ldots.$$

Let $w - \gamma = v^p$, $Z = e^{\frac{az}{p}}$; then this equation becomes

$$p\log Z = p\log v + a''v + b''v^2 + \ldots,$$

that is, $$Z = ve^{\lambda v + \mu v^2 + \cdots} = vP(v);$$

whence, by inversion, we have a relation of the form

$$v = ZQ(Z),$$

so that $$w - \gamma = e^{az}Q(e^{\frac{az}{p}}),$$

shewing that w is uniform for values in the vicinity of $w = \gamma$: it is simply-periodic in that vicinity, the period being $\dfrac{2p\pi i}{a}$, and it can acquire the value γ only for (logarithmically) infinite values of z.

If $q/p > 1$, let $q = p + n$, where n and p are prime to one another; then we have

$$W = a(w - \gamma)^{1+\frac{n}{p}} + b(w - \gamma)^{1+\frac{n+1}{p}} + \ldots,$$

so that

$$a\,dz = \{(w - \gamma)^{-1-\frac{n}{p}} + b'(w - \gamma)^{-1-\frac{n-1}{p}} + c'(w - \gamma)^{-1-\frac{n-2}{p}} + \ldots\}\,dw,$$

or $$z = \alpha(w - \gamma)^{-\frac{n}{p}} + \beta(w - \gamma)^{-\frac{n-1}{p}} + \ldots$$

$$+ \delta(w - \gamma)^{-\frac{1}{p}} + \epsilon\log(w - \gamma) + P\{(w - \gamma)^{\frac{1}{p}}\}.$$

Hence w can acquire its value γ only for (algebraically) infinite values of z.

As a first condition for uniformity, the coefficient ϵ must vanish, that is, in the expansion of $\dfrac{dz}{dw}$ in powers of $(w - \gamma)^{\frac{1}{p}}$, there must be no term involving $(w - \gamma)^{-1}$. For let

$$z = Z^{-n}, \quad w - \gamma = v^p,$$

so that
$$v^n = Z^n \{\alpha + \beta v + \dots + \delta v^{n-1} + \epsilon v^n \log v + v^n P(v)\}.$$
Then, if
$$v = uZ,$$
we have
$$u^n = Q(uZ) + \epsilon u^n Z^n (\log u + \log Z),$$
where Q is a series of integral powers of uZ converging for sufficiently small values of $|uZ|$.

Since z is infinitely large for sufficiently small values of $|w - \gamma|$, we have Z infinitesimally small. When $Z = 0$, the value of $Z^n \log Z$ is zero; but for values of Z that are not zero, the quantity has an infinite number of different values of the form

$$Z^n (\operatorname{Log} Z + 2m\pi i),$$

and there will then be an infinite number of distinct equations determining u, one corresponding to each of the values of m. Hence u (and therefore v, and therefore also $w - \gamma$), in that case, has an infinite number of distinct branches in the vicinity of $Z = 0$; then w is not uniform in the vicinity of $Z = 0$. As a first condition for uniformity, we must therefore have $\epsilon = 0$.

We take $\epsilon = 0$: then the equation between z and v, where $w - \gamma = v^p$, is

$$z = v^{-n} \{\alpha + \beta v + \gamma v^2 + \dots\},$$

the inversion of which can give v (and therefore can give $w - \gamma$) as a uniform function of z, only if $n = 1$. When $n = 1$, we have $w - \gamma$ uniform; and w can obtain its value γ only for algebraically infinite values of z.

Combining these results, we have the theorem :

III.   *If for a finite value γ of w, which is a branch-point of W, the equation in W has a zero for p branches, then, in order that w may be uniform, the degree of that zero is of one of the forms*

$$1 - \frac{1}{p}, \ 1, \ and \ 1 + \frac{1}{p}; \ and \ if \ it \ be \ of \ the \ form^* \ 1 + \frac{1}{p}, \ the \ term \ in$$

$(w - \gamma)^{-1}$ *must be absent from the expression of* $\dfrac{dz}{dw}$ *in powers of* $w - \gamma$.

134.   Only finite values of w have been considered. For the consideration of infinite values of w, we pass to the equation in u:

---

* The case $p = 1$ occurs in (ii), § 132 : it now is included in III.

and only zero values of u need be taken into account. If w be uniform, u also is uniform and *vice versa*; hence:—

IV. *In order that the function w may be uniform when its value tends to become infinitely large, the conditions in* II. *and* III. *must apply to the equation in u for the value $u = 0$.*

The branch-points of W, regarded as a function of w, as well as points where the roots though equal are distinct as in II., are (in addition possibly to $u = 0$) the common roots of the equations

$$f(W, w) = 0, \quad \frac{\partial f(W, w)}{\partial W} = 0.$$

If, then, the conditions in II. *and* III. *be satisfied for all these points, and if the conditions in* IV. *be satisfied for $u = 0$, that is, for infinite values of w, then the integral of the equation*

$$\left(\frac{dw}{dz}\right)^m + f_1(w)\left(\frac{dw}{dz}\right)^{m-1} + \ldots + f_{m-1}(w)\frac{dw}{dz} + f_m(w) = 0$$

is a uniform function of z.

135.   The classes of uniform functions of z can be obtained as follows.

The function, inverse to w, is given by the equation $\dfrac{dz}{dw} = W^{-1}$, and therefore

$$z = \int \frac{dw}{W}.$$

Let the Riemann's surface for the algebraical equation

$$f(W, w) = 0,$$

regarded as an equation between a dependent variable W and an independent variable w capable of assuming all values, be constructed; and let its connectivity be $2P + 1$. Then $\int W^{-1} dw$ is the integral of a uniform function of position on the surface; and if w_0 be a value at any point, then all other values at that point differ from w_0 by integer multiples of

   (i)   the moduli of the integral at the $2P$ cross-cuts,

   (ii) the moduli of the integral at such other cross-cuts as may be necessary on account of the expression of the subject of integration as a function of w.

Hence the argument of w, a uniform function of z, is of the form $z + \Sigma m\Omega$, where the coefficients m are integers and the quantities Ω are constant.

It is known* that uniform functions of z with more than two linearly independent periods cannot exist; hence there may be two moduli, or only one modulus, or none. In the last case, as there are m values of W for one of w, there are m values of z for one of w; and no value of w provides an essential singularity for z. Thus z is an algebraical function of w: and conversely w is an algebraical function of z which, being uniform, is rational. It therefore follows that *the uniform function of z is either*

(i)   *a doubly-periodic function of z; or*

(ii) *a simply-periodic function of z; or*

(iii) *a rational function of z.*

Further†, *the class of the Riemann's surface for the equation* $f(W, w) = 0$ *is either unity or zero;* for in what precedes, the value of P is not greater than unity, when the limitations as to the possible number of periods are assigned.

It is now easy to assign the criteria ·determining the class of functions to which w belongs, when it is known to be a uniform function of z satisfying the differential equation.

(i) Let w be a uniform doubly-periodic function. Take any parallelogram of periods in the finite part of the plane: all values of z within the parallelogram are finite, and all possible values of w are acquired within the parallelogram.

Let γ be a finite value of w for a point $z = c$; then, since the function is uniform, we have

$$w - \gamma = (z - c)^m P (z - c),$$

where m is an integer and $P(z - c)$ does not vanish for $z = c$: and, by inversion, we also have

$$z - c = (w - \gamma)^{\frac{1}{m}} Q \{(w - \gamma)^{\frac{1}{m}}\},$$

where Q is finite but does not vanish for $w = \gamma$.

Now

$$\frac{dw}{dz} = (z - c)^{m-1} \{mP(z - c) + (z - c) P'(z - c)\}$$

$$= (w - \gamma)^{1 - \frac{1}{m}} Q_1 \{(w - \gamma)^{\frac{1}{m}}\},$$

where Q_1 does not vanish for $w = \gamma$.

* *Th. Fns.*, § 108.

† This result is due to Hermite, and is stated by him in a letter to Cayley, *Proc. Lond. Math. Soc.*, t. IV (1873), pp. 343—345. The limitation of the class to zero or unity is not, in itself, sufficient to ensure that w is a uniform function of z.

If $m = 1$, then γ is an ordinary point for $\dfrac{dw}{dz}$.

If $m > 1$, then γ is a zero branch-point for W, of index-degree equal to

$$1 - \frac{1}{m}.$$

If, in the vicinity of $z = b$, w be infinitely large of order q, then $z = b$ is a zero of u of order q, so that we have

$$u = (z - b)^q\, P_1 (z - b)\,;$$

as in the first of these cases, it follows that

$$\frac{du}{dz} = u^{1 - \frac{1}{q}} P_2 (u^{\frac{1}{q}}),$$

where P_2 does not vanish for $u = 0$.

Hence it follows that if, for finite or for infinite values of w, all the branch-points for W be zeros and each of them have its degree less than unity, the index of the degree being of the form $1 - \dfrac{1}{p}$, where p is the number of branches of W affected, then the uniform function w is doubly-periodic.

(ii)   Let w be a uniform simply-periodic function, of period ω; then it is known* that w can be expressed in the form

$$w = f(e^{\frac{2\pi z i}{\omega}}) = f(Z).$$

Take any strip† in the z-plane as for a simply-periodic function, bounded by lines whose inclination to the axis of real quantity is $\frac{1}{2}\pi + \arg.\ \omega$: in this strip the function acquires all its values. The variable Z is finite in the strip except at the infinite limits; at one infinite limit we have $z = ki\omega$, where k is positive and infinitely great, and then $Z = e^{-2\pi k} = 0$, and at the other we can take $z = - ki\omega$ and then $Z = e^{2\pi k} = \infty$; so that $Z = 0$ and $Z = \infty$ at the infinite limits.

Let γ be a finite value of w for a finite point $z = c$ and let $C = e^{\frac{2\pi c i}{\omega}}$: then we have

$$w - \gamma = f(Z) - f(C) = (Z - C)^q\, g\,(Z - C),$$

---

* *Th. Fns.*, § 113.   † *Th. Fns.*, § 111.

where $g(Z - C)$ does not vanish for $Z = C$ and q is a positive integer.

When $q = 1$, we have

$$Z - C = (w - \gamma)\, G\,(w - \gamma),$$

where G does not vanish for $w = \gamma$; and then

$$\frac{dw}{dz} = \frac{2\pi i}{\omega}\, Z\, \{g\,(Z - C) + (Z - C)\, g'\,(Z - C)\}$$

$$= H\,(w - \gamma),$$

where H does not vanish for $w = \gamma$; the point $w = \gamma$ is an ordinary point for $\dfrac{dw}{dz}$.

When $q > 1$, we have

$$Z - C = (w - \gamma)^{\frac{1}{q}}\, G\, \{(w - \gamma)^{\frac{1}{q}}\},$$

where G does not vanish for $w = \gamma$; and then

$$\frac{dw}{dz} = \frac{2\pi i}{\omega}\, Z\, (Z - C)^{q-1}\, \{qg\,(Z - C) + (Z - C)\, g'\,(Z - C)\}$$

$$= (w - \gamma)^{1 - \frac{1}{q}}\, h\, \{(w - \gamma)^{\frac{1}{q}}\},$$

where h does not vanish for $w = \gamma$. Such a point is a branch-zero for q branches of W, and its index-degree is $1 - \dfrac{1}{q}$.

If the value of w be infinite for the finite point $z = c$, then we have

$$u = (Z - C)^q\, g\,(Z - C).$$

If $q = 1$, the point is an ordinary point for $\dfrac{du}{dz}$; if $q > 1$, it is a branch-zero for q branches of $\dfrac{du}{dz}$, and its index-degree is $1 - \dfrac{1}{q}$.

When $z = \infty$, then $Z = 0$ or $Z = \infty$. The value of the function w for infinite values of z is either finite or infinite.

Let w be a finite quantity γ, for infinitely large values of z. When Z is very small, we have

$$w - \gamma = Z^q f(Z),$$

where q is a positive integer and f does not vanish for $Z = 0$; and then

$$Z = (w - \gamma)^{\frac{1}{q}}\, g\, \{(w - \gamma)^{\frac{1}{q}}\},$$

where g does not vanish for $w = \gamma$. Then

$$\frac{dw}{dz} = \frac{2\pi i}{\omega} Z Z^{q-1} \{qf(Z) + Zf'(Z)\}$$

$$= Z^q h(Z),$$

where h does not vanish when $Z = 0$; and therefore

$$\frac{dw}{dz} = (w - \gamma) P_1 \{(w - \gamma)^{\frac{1}{q}}\},$$

or the point $w = \gamma$ is a branch-zero of q branches of $\dfrac{dw}{dz}$, and its index-degree is unity.

When Z is very large, we have

$$w - \gamma = Z^{-q} f_1\left(\frac{1}{Z}\right),$$

where q is a positive integer, and f_1 is finite and not zero for $Z = \infty$. As before, it is easy to see that

$$\frac{dw}{dz} = (w - \gamma) P_2 \{(w - \gamma)^{\frac{1}{q}}\},$$

or the point $w = \gamma$ is a branch-zero of q branches of $\dfrac{dw}{dz}$, and its index-degree is unity.

If, however, the value of w be infinite for infinitely large values of z, then we have

$$u = Z^q f_1(Z)$$

when Z is very small, and

$$u = Z^{-q} f_2\left(\frac{1}{Z}\right)$$

when Z is very large. As before, the point $u = 0$ is then, in each case, a branch-zero of q branches of $\dfrac{du}{dz}$, and its index-degree is unity.

Hence if all the branch-points of W are zeros: if, moreover, one of them has its degree equal to unity, and if all the other branch-zeros are of index-degree less than unity, the index of the degree being of the form $1 - \dfrac{1}{p}$, where p is the number of branches of W affected : then the uniform function w determined by the equation $f(W, w) = 0$ is simply-periodic.

(iii)   Let w be a rational function of z; then it can be expressed in the form

$$w = \frac{f_1(z)}{f_2(z)},$$

where f_1 and f_2 are rational integral functions of z.

Finite values of w can arise from values of z in the vicinity of a zero of $f_1(z)$, say $z = c$, or an infinity of $f_2(z)$.   For the former, we have, if γ denote the value of z,

$$w - \gamma = (z - c)^m F(z - c),$$

where F does not vanish for $z = c$:   and then, inverting the functional relation,

$$z - c = (w - \gamma)^{\frac{1}{m}} P(w - \gamma),$$

where m is a positive integer which may be 1 or greater than 1.

Now

$$\frac{dw}{dz} = (z - c)^{m-1} \{mF(z - c) + (z - c) F'(z - c)\},$$

so that, if $m = 1$, we have

$$\frac{dw}{dz} = Q(w - \gamma),$$

where Q does not vanish when $w = \gamma$; and, if $m > 1$, we have

$$\frac{dw}{dz} = (w - \gamma)^{1 - \frac{1}{m}} Q_1 \{(w - \gamma)^{\frac{1}{m}}\},$$

where Q_1 does not vanish when $w = \gamma$.   Hence $w = \gamma$ is either an ordinary point for W: or it is a branch-point at which m branches vanish, the index-degree of the zero being $1 - \dfrac{1}{m}$.

For an infinity of $f_2(z)$, we must have $z = \infty$; and therefore, for infinitely large values of z, we have

$$w - \gamma = z^{-\lambda} F\left(\frac{1}{z}\right),$$

where F does not vanish when $z = \infty$.   Proceeding as before, we have

$$\frac{dw}{dz} = (w - \gamma)^{1 + \frac{1}{\lambda}} F_1 \{(w - \gamma)^{\frac{1}{\lambda}}\},$$

where F_1 does not vanish when $w = \gamma$.   If $\lambda = 1$, $w = \gamma$ is an ordinary point, a case which has been considered; if $\lambda > 1$, $w = \gamma$

is a branch-point for W at which λ branches vanish, and the index-degree of the zero is $1 + \dfrac{1}{\lambda}$.

Infinite values of w can arise from values of z that are infinitely large—in connection with $f_1(z)$—or from values of z that are zeros of the denominator. For the former, we have

$$u = z^{-\lambda} F\left(\frac{1}{z}\right),$$

where λ is a positive integer and F does not vanish for $z = \infty$; and then proceeding as before, we have

$$\frac{du}{dz} = u^{1+\frac{1}{\lambda}} F(u^{\frac{1}{\lambda}}),$$

so that, if $\lambda = 1$, $u = 0$ is an ordinary point, a case of which account has already been taken; and if $\lambda > 1$, $u = 0$ (that is, $w = \infty$) is a branch-point for U at which λ branches vanish, and the index-degree of the zero is $1 + \dfrac{1}{\lambda}$.

Moreover, as w is a rational function, we do not have both $w = \gamma$ and $u = 0$ for infinite values of z.

It thus appears that, when w is a rational function, there is only one value of w which, being a branch-point for W, gives m branches vanishing, and has the index-degree of the zero equal to $1 + \dfrac{1}{m}$; all other branch-points of W give zeros that are of index-degree less than unity, each being of the form $1 - \dfrac{1}{n}$, where n is the number of branches that vanish at the point.

136. The following is a summary of the results that have been obtained :—

I. In order that an irreducible differential equation of the first order may have a uniform function for its integral, it must be of the form

$$F(W, w) = \left(\frac{dw}{dz}\right)^m + \left(\frac{dw}{dz}\right)^{m-1} f_1(w) + \ldots + f_m(w) = 0,$$

where $f_1(w)$, $f_2(w)$, ..., $f_m(w)$ are rational integral functions of w of degrees not higher than 2, 4, 6, ..., $2m$ respectively:

and this condition as to degree is then satisfied for the equation

$$G(U, u) = F\left(-\frac{1}{u^2}U, \frac{1}{u}\right)$$

$$= \left(\frac{du}{dz}\right)^m - \left(\frac{du}{dz}\right)^{m-1} u^2 f_1\left(\frac{1}{u}\right) + \dots \pm u^{2m} f_m\left(\frac{1}{u}\right) = 0.$$

II. If any finite value of w be a branch-point of W, when regarded as a function of w determined by the equation

$$F(W, w) = 0,$$

then all the affected values of W must be zero for that value of w; and likewise for the value $u = 0$ in connection with the equation

$$G(U, u) = 0.$$

III. If for a value of w, which is a branch-point of W when regarded as a function of w, there be a multiple root of $F(W, w) = 0$ which is zero for n branches, the index-degree for each of those branches is of one of the forms $1 - \frac{1}{n}$, 1, $1 + \frac{1}{n}$; and likewise for the value $u = 0$ in connection with the equation

$$G(U, u) = 0.$$

IV. The class of the equation $F(W, w) = 0$, and therefore the class of the Riemann's surface associated with the equation, is either zero or unity.

V. If all the multiple zero-roots of W, for finite values or for an infinite value of w, be of index-degree less than unity, each of them being of the form $1 - \frac{1}{n}$, then w is a uniform doubly-periodic function of z.

VI. If, for some value of w, there be a single set of m multiple zero-roots of index-degree equal to unity, and if, for finite values or for an infinite value of w, all the other sets of multiple zero-roots have their respective index-degrees less than unity and of the form $1 - \frac{1}{n}$, then w is a uniform singly-periodic function of z.

VII. If, for some value of w, there be a single set of m multiple zero-roots the index-degree of which is equal to $1 + \frac{1}{m}$, and if, for other values of w, all the other sets of multiple zero-roots have their respective index-degrees less than unity and of the form $1 - \dfrac{1}{n}$, then w is a rational function of z.

In all other cases the equation, supposed irreducible, cannot have a uniform function of z for its integral. If the equation have a uniform function of z for its integral, and the preceding conditions in V., VI. or VII., be not satisfied, the equation is reducible*, that is, it can be replaced by rational equations of lower degree to which the criteria apply.

Note. The preceding method may be considered as essentially due to Briot and Bouquet.

There is another method of proceeding, which leads to the same result. It is based upon Hermite's theorem (§ 135), proved independently ; and its development will be found in memoirs by Fuchs† and Raffy‡. A reference to the memoirs which have been quoted shews that the equation $F(W, w) = 0$, when it is satisfied by a uniform function of z, can be associated with the theory of unicursal curves and of bicursal curves.

137. The preceding general results will now be applied to the particular equation

$$\left(\frac{dw}{dz}\right)^s = f(w),$$

where f is a rational, integral, algebraical function of degree not greater than $2s$.

Let

$$f(w) = \lambda^s (w-a)^l (w-b)^m \ldots,$$

where $\lambda, a, b, \ldots$ are constants and $l, m, \ldots$ are integers, and

$$l + m + \ldots \leqslant 2s.$$

The equation in $u \left(= \dfrac{1}{w} \right)$ and $\dfrac{du}{dz}$ is

$$(-1)^s \left(\frac{du}{dz}\right)^s = \lambda^s u^{2s-l-m-\cdots} (1 - au)^l (1 - bu)^m \ldots;$$

---

* This investigation is based upon two memoirs by Briot and Bouquet, *Journ. de l'Éc. Polytechnique*, t. xxi, Cah. xxxvi (1856), pp. 134—198, 199—254; and upon their *Traité des fonctions elliptiques*, pp. 341—350, 376—392. A memoir by Cayley, *Proc. Lond. Math. Soc.*, vol. xviii (1887), pp. 314—324, may also be consulted.

† *Comptes Rendus*, t. xciii (1881), pp. 1063—1065; *Sitzungsber. d. Akad. d. Wiss. zu Berlin*, 1884, pp. 709, 710.

‡ *Annales de l'Éc. Norm.*, 2me Sér., t. xii (1883), pp. 105—190; *ib.*, 3me Sér., t. ii.(1885), pp. 99—112.

thus the values of $\dfrac{dw}{dz}$ and $\dfrac{du}{dz}$ are respectively

$$\frac{dw}{dz} = \lambda \, (w-a)^{\frac{l}{s}} (w-b)^{\frac{m}{s}} \dots,$$

$$-\frac{du}{dz} = \lambda u^{2 - \frac{l}{s} - \frac{m}{s} - \cdots} (1 - au)^{\frac{l}{s}} (1 - bu)^{\frac{m}{s}} \dots.$$

Because the integral of the equation must be uniform, each of the indices $2 - \dfrac{l}{s} - \dfrac{m}{s} - \dots, \dfrac{l}{s}, \dfrac{m}{s}, \dots$ must be of one of the forms $1 - \dfrac{1}{p}$, 1, or $1 + \dfrac{1}{p}$; and p may be 1, but the point is then not a branch-point. Then the smallest value of p is 2, and the least index is therefore $\frac{1}{2}$; hence, as

$$\frac{l}{s} + \frac{m}{s} + \dots \leqslant 2,$$

there cannot be more than *four* distinct (that is, non-repeated) factors in $f(w)$. Hence

(i)   if one of the indices $\dfrac{l}{s}$, $\dfrac{m}{s}$, ..., be greater than 1, each of the other indices must be less than 1, unless it be 2 when all the others are zero ;

(ii)   if one of the indices $\dfrac{l}{s}$, $\dfrac{m}{s}$, ..., be equal to 1, then either each of the other indices must be less than 1, or one other is equal to 1, and then there is no remaining index ;

(iii)   if each of the indices $\dfrac{l}{s}$, $\dfrac{m}{s}$, ..., be less than 1, then $2 - \dfrac{l}{s} - \dfrac{m}{s} - \dots$ may be less than 1, or equal to 1, or greater than 1.

These cases, associated with the possible numbers of factors, will be taken in order.

I.   Let there be a single factor ; the equation is

$$\left(\frac{dw}{dz}\right)^{s} = \lambda^{s} \, (w-a)^{l},$$

and therefore

$$\left(-\frac{du}{dz}\right)^{s} = \lambda^{s} u^{2s-l} (1 - au)^{l}.$$

Now $\dfrac{l}{s}$, not being 2, is either $1 - \dfrac{1}{s}$, 1, $1 + \dfrac{1}{s}$; and these forms cover also the necessary forms of $2 - \dfrac{l}{s}$.

If $l = s - 1$, then one index (for $w = a$) is equal to $1 - \dfrac{1}{s}$, and the other (for $u = 0$) is equal to $1 + \dfrac{1}{s}$: the function w is rational in z, and z is infinite only when $w = \infty$: hence the integral w is a *rational integral* function of z.

If $l=s+1$, the reasoning is similar; and the integral is a *rational meromorphic* function of z.

If $l=s$, the indices are each equal to unity : the integral is a *simply-periodic* function of z. The equation is reducible.

If $l=2s$, the equation is reducible ; the integral is *rational*.

The equations in the respective cases are

$$\left(\frac{dw}{dz}\right)^s = \lambda^s (w-a)^{s-1} \quad\text{.........................} \quad \text{(A.)},$$

$$\left(\frac{dw}{dz}\right)^s = \lambda^s (w-a)^{s+1} \quad\text{..........................} \quad \text{(A.)},$$

$$\left(\frac{dw}{dz}\right) = \lambda \ (w-a) \quad\text{...............................} \quad \text{(S. P.)},$$

$$\frac{dw}{dz} = \lambda \ (w-a)^2 \quad\text{................................} \quad \text{(A.)},$$

where (A.) implies that the uniform integral is a rational function of z, and (S. P.) implies that it is a simply-periodic function ; the letters (D. P.) will be used to imply that the uniform integral is a doubly-periodic function.

II. Let there be two distinct factors ; then the equation is

$$\left(\frac{dw}{dz}\right)^s = \lambda^s (w-a)^l (w-b)^m.$$

First, let one of the indices in the expression for $\dfrac{dw}{dz}$ be greater than 1, say $\dfrac{l}{s}$. It is not necessarily in its lowest terms ; when reduced to its lowest terms, let

$$\frac{l}{s} = 1 + \frac{1}{\rho}.$$

Then $\dfrac{m}{s}$ must be less than 1 ; when reduced to its lowest terms, let

$$\frac{m}{s} = 1 - \frac{1}{\sigma},$$

which is the necessary form. And $2 - \dfrac{l}{s} - \dfrac{m}{s} - \ldots$ must be less than 1, and it must be expressible in the form $1 - \dfrac{1}{\tau}$: hence

$$2 - \left(1 + \frac{1}{\rho}\right) - \left(1 - \frac{1}{\sigma}\right) = 1 - \frac{1}{\tau},$$

and therefore

$$1 + \frac{1}{\rho} = \frac{1}{\sigma} + \frac{1}{\tau},$$

where ρ and σ are each greater than unity. If $\tau > 1$, the right-hand side is manifestly less than the left ; and therefore we must have $\tau = 1$, $\rho = \sigma$; and the common value of ρ and σ is s. The integral is then a rational function of z.

Secondly, let one of the indices in the expression for $\dfrac{dw}{dz}$ be equal to 1, say

$l=s$. Then $\dfrac{m}{s}$ is either 1 or of the form $1-\dfrac{1}{\sigma}$.

If $\dfrac{m}{s}=1$, the exponent of u in the expression for $\dfrac{du}{dz}$ is zero : the equation is

$$\left(\frac{dw}{dz}\right)^s = \lambda^s (w-a)^s (w-b)^s,$$

which is reducible ; it has a simply-periodic function for its integral.

If $\dfrac{m}{s}=1-\dfrac{1}{\sigma}$, the exponent of u in the expression for $\dfrac{du}{dz}$ is $\dfrac{1}{\sigma}$. This must be of the form $1-\dfrac{1}{\rho}$, so that

$$\frac{1}{\sigma}+\frac{1}{\rho}=1 ;$$

hence, as σ and ρ are each greater than 1, each must be 2. The equation is

$$\left(\frac{dw}{dz}\right)^s = \lambda^s (w-a)^s (w-b)^{\frac{1}{2}s},$$

which is reducible ; and the integral is a simply-periodic function.

Thirdly, let each of the indices in the expression for $\dfrac{dw}{dz}$ be less than 1 ; as they are not necessarily in their lowest terms, let $\dfrac{l}{s}=1-\dfrac{1}{\rho}$, $\dfrac{m}{s}=1-\dfrac{1}{\sigma}$. Then the index of u in the expression for $\dfrac{du}{dz}$ is $\dfrac{1}{\rho}+\dfrac{1}{\sigma}$; because ρ and σ are each greater than 1, this index cannot be greater than 1.

If $\dfrac{1}{\rho}+\dfrac{1}{\sigma}=1$, the only possible values are $\rho=2$, $\sigma=2$; the equation is

$$\left(\frac{dw}{dz}\right)^s = \lambda^s (w-a)^{\frac{1}{2}s} (w-b)^{\frac{1}{2}s},$$

which is reducible ; the integral is a simply-periodic function of z.

If $\dfrac{1}{\rho}+\dfrac{1}{\sigma}$ be less than 1, then, as it is the index of u in the expression for $\dfrac{du}{dz}$, it must be of the form $1-\dfrac{1}{\tau}$, where τ is greater than 1 : thus

$$\frac{1}{\rho}+\frac{1}{\sigma}+\frac{1}{\tau}=1,$$

and then all the indices in the expressions for $\dfrac{dw}{dz}$ and $\dfrac{du}{dz}$ are less than 1. Hence for such equations as exist, the integrals will be doubly-periodic functions.

In this equation the interchange of ρ and σ gives no essentially new arrangement. We must have $\tau > 1$: the solutions for values of τ greater than 1 are:—

(a) $\tau = 2$; then $\dfrac{1}{\rho} + \dfrac{1}{\sigma} = \dfrac{1}{2}$, so that $\rho = 3$, $\sigma = 6$; $\rho = 4$, $\sigma = 4$.

(b) $\tau = 3$; then $\dfrac{1}{\rho} + \dfrac{1}{\sigma} = \dfrac{2}{3}$, so that $\rho = 2$, $\sigma = 6$; $\rho = 3$, $\sigma = 3$.

(c) $\tau = 4$; then $\dfrac{1}{\rho} + \dfrac{1}{\sigma} = \dfrac{3}{4}$, so that $\rho = 2$, $\sigma = 4$.

(d) $\tau = 5$ gives no solution.

(e) $\tau = 6$; then $\dfrac{1}{\rho} + \dfrac{1}{\sigma} = \dfrac{5}{6}$, so that $\rho = 2$, $\sigma = 3$.

And no higher value of τ gives solutions.

Hence the whole system of equations, satisfied by a uniform function of z and having two distinct factors in $f(w)$, is:—

$$\left(\frac{dw}{dz}\right)^s = \lambda^s (w-a)^{s-1} (w-b)^{s+1} \ldots\ldots\ldots\text{(A.)},$$

$$\left(\frac{dw}{dz}\right) = \lambda \ (w-a) \quad (w-b) \quad \ldots\ldots\ldots\text{(S. P.)},$$

$$\left(\frac{dw}{dz}\right)^2 = \lambda^2 (w-a)^2 \quad (w-b) \quad \ldots\ldots\ldots\text{(S. P.)},$$

$$\left(\frac{dw}{dz}\right)^2 = \lambda^2 (w-a) \quad (w-b) \quad \ldots\ldots\ldots\text{(S. P.)},$$

$$\left(\frac{dw}{dz}\right)^6 = \lambda^6 (w-a)^4 \quad (w-b)^5 \quad \ldots\ldots\ldots\text{(D. P.)}, \ (1),$$

$$\left(\frac{dw}{dz}\right)^4 = \lambda^4 (w-a)^3 \quad (w-b)^3 \quad \ldots\ldots\ldots\text{(D. P.)}, \ (2),$$

$$\left(\frac{dw}{dz}\right)^6 = \lambda^6 (w-a)^3 \quad (w-b)^5 \quad \ldots\ldots\ldots\text{(D. P.)}, \ (3),$$

$$\left(\frac{dw}{dz}\right)^3 = \lambda^3 (w-a)^2 \quad (w-b)^2 \quad \ldots\ldots\ldots\text{(D. P.)}, \ (4),$$

$$\left(\frac{dw}{dz}\right)^4 = \lambda^4 (w-a)^2 \quad (w-b)^3 \quad \ldots\ldots\ldots\text{(D. P.)}, \ (5),$$

$$\left(\frac{dw}{dz}\right)^6 = \lambda^6 (w-a)^3 \quad (w-b)^4 \quad \ldots\ldots\ldots\text{(D. P.)}, \ (6).$$

III. Let there be three distinct factors: then the equation is

$$\left(\frac{dw}{dz}\right)^s = \lambda^s (w-a)^l (w-b)^m (w-c)^n,$$

and therefore

$$\left(-\frac{du}{dz}\right)^s = \lambda^s u^{2s-l-m-n} (1-au)^l (1-bu)^m (1-cu)^n.$$

If one of the indices in the expression for $\dfrac{dw}{dz}$ be greater than 1, say $\dfrac{l}{s} = 1 + \dfrac{1}{\rho}$,

then $\dfrac{m}{s}$, $\dfrac{n}{s}$ must be of the form $1 - \dfrac{1}{\sigma}$, $1 - \dfrac{1}{\tau}$, where σ and τ are each greater than 1.

The index of u in the expression for $\dfrac{du}{dz}$ is then $\dfrac{1}{\sigma} + \dfrac{1}{\tau} - \dfrac{1}{\rho} - 1$, a quantity which is necessarily negative, for ρ is finite; and the index should either be zero or be of a form $1 - \dfrac{1}{\mu}$. Hence no one of the indices $\dfrac{l}{s}$, $\dfrac{m}{s}$, $\dfrac{n}{s}$ can be greater than 1.

Secondly, let one of the indices in the expression for $\dfrac{dw}{dz}$ be equal to 1, say $l = s$. Then since $m + n \leqslant s$, only one of the indices is unity; and therefore $\dfrac{m}{s}$, $\dfrac{n}{s}$ are of the form $1 - \dfrac{1}{\rho}$, $1 - \dfrac{1}{\sigma}$, where ρ and σ are each greater than 1. The index of u in the expression for $\dfrac{du}{dz}$ is then $\dfrac{1}{\rho} + \dfrac{1}{\sigma} - 1$, and it cannot be negative; hence the only possible values are $\rho = 2 = \sigma$, and they make the index zero. There is thus one index equal to 1, and the others are less than 1: the integral of the equation is a simply-periodic function of z.

Thirdly, let all the indices in the expression for $\dfrac{dw}{dz}$ be less than 1: then they are of the forms $1 - \dfrac{1}{\rho}$, $1 - \dfrac{1}{\sigma}$, $1 - \dfrac{1}{\tau}$, where ρ, σ, τ are greater than 1; and the index of u in the expression for $\dfrac{du}{dz}$ is $\dfrac{1}{\rho} + \dfrac{1}{\sigma} + \dfrac{1}{\tau} - 1$. Because the smallest value of ρ, σ, τ is 2, this last index is not greater than $\frac{1}{2}$; hence it must be $1 - \dfrac{1}{\mu}$, where, because this quantity is the index of u, μ is equal to 1 or to 2. In either case, all the indices are less than 1; and therefore the integrals of the corresponding equations are doubly-periodic functions of z.

If
$$\dfrac{1}{\rho} + \dfrac{1}{\sigma} + \dfrac{1}{\tau} - 1 = 1 - \dfrac{1}{2},$$

so that $\dfrac{1}{\rho} + \dfrac{1}{\sigma} + \dfrac{1}{\tau} = \dfrac{3}{2}$, the only possible solution is

$$\rho, \ \sigma, \ \tau = 2, \ 2, \ 2.$$

If $\dfrac{1}{\rho} + \dfrac{1}{\sigma} + \dfrac{1}{\tau} = 1$, the only possible solutions are

$$\rho, \ \sigma, \ \tau = 2, \ 3, \ 6;$$

$$2, \ 4, \ 4;$$

$$3, \ 3, \ 3.$$

Hence the whole system of equations, satisfied by a uniform function of z and having three distinct factors in $f(w)$, is :—

$$\left(\frac{dw}{dz}\right)^2 = \lambda^2 (w-a)^2 (w-b) (w-c) \quad \ldots \ldots \text{(S. P.)},$$

$$\left(\frac{dw}{dz}\right)^2 = \lambda^2 (w-a) (w-b) (w-c) \quad \ldots \ldots \text{(D. P.)}, \; (7),$$

$$\left(\frac{dw}{dz}\right)^6 = \lambda^6 (w-a)^3 (w-b)^4 (w-c)^5 \quad \ldots \ldots \text{(D. P.)}, \; (8),$$

$$\left(\frac{dw}{dz}\right)^4 = \lambda^4 (w-a)^2 (w-b)^3 (w-c)^3 \quad \ldots \ldots \text{(D. P.)}, \; (9),$$

$$\left(\frac{dw}{dz}\right)^3 = \lambda^3 (w-a)^2 (w-b)^2 (w-c)^2 \quad \ldots \ldots \text{(D. P.)}, \; (10).$$

IV.   Let there be four distinct factors ; then the equation is

$$\left(\frac{dw}{dz}\right)^s = \lambda^s (w-a)^l (w-b)^m (w-c)^n (w-d)^p.$$

Since $\dfrac{l}{s}, \dfrac{m}{s}, \dfrac{n}{s}, \dfrac{p}{s}$ are each of a form $1 - \dfrac{1}{\rho}$, and their sum is not greater than 2, it is easy to see that the only possible solution is given by $\dfrac{l}{s} = \dfrac{m}{s} = \dfrac{n}{s} = \dfrac{p}{s} = \dfrac{1}{2}$; each index is less than 1, and the integral is a doubly-periodic function.

Hence the single equation, satisfied by a uniform function of z and having four distinct factors in $f(w)$, is

$$\left(\frac{dw}{dz}\right)^2 = \lambda^2 (w-a) (w-b) (w-c) (w-d) \ldots \ldots \text{(D. P.)}, \; (11).$$

Those of the complete system of equations, which have their integrals either rational functions of z or simply-periodic functions of z, are easily integrated.   The remainder, which have uniform doubly-periodic functions of z for their integrals, are most easily integrated by first determining the irreducible infinities of the functions and their orders : and then, by using the properties of doubly-periodic functions, the integral can be constructed.

The irreducible infinities can be determined as follows.   In the equation for $\dfrac{du}{dz}$, let the index of u be $1 - \dfrac{1}{\rho}$; and let $s = \sigma\rho$.   Then the equation which determines u is

$$\left(\frac{du}{dz}\right)^s = \lambda^s u^{\sigma(\rho-1)} (1 - au)^l \ldots,$$

so that, for very small values of u, we have

$$\left\{u^{-1+\frac{1}{\rho}} + \ldots\right\} du = a\lambda dz,$$

where a is a primitive sth root of unity.   Hence

$$a\lambda (z-c) = \rho u^{\frac{1}{\rho}} + \ldots,$$

and therefore

$$\frac{1}{w} = u = a^\rho \lambda^\rho (z-c)^\rho + \ldots.$$

It thus appears that the accidental singularity of w at $z=c$ is of order ρ ; and, as there are σ distinct values of a^ρ, there are σ distinct accidental singularities to be associated with the respective values.

Applying these to the equations which, having doubly-periodic functions for the integrals, are numbered (1) to (11), we have the following results, where σ is the number of distinct irreducible accidental singularities and ρ is the order of each of these singularities :

number of equation	(1)	(2)	(3)	(4)	(5)	(6)	(7)	(8)	(9)	(10)	(11)
number of singularities $=\sigma$	3	2	2	1	1	1	1	6	4	3	2
order of singularity $=\rho$	2	2	3	3	4	6	2	1	1	1	1

All the binomial equations, which have uniform functions of z for their integrals, have been obtained. The general results, summarised in § 136, can be applied to other equations ; the application to trinomial equations will be found in the treatise by Briot and Bouquet (cited p. 325, note).

Note. The binomial equations can be treated otherwise, by forming the equation

$$z - a = \int (w-a)^{-\frac{l}{s}} (w-b)^{-\frac{m}{s}} \ldots \ldots dw \; ;$$

but, as indicated at the beginning of § 137, the method in the text is adopted in order to illustrate the general results of § 136. (See also Note, § 136.)

Ex. 1. Prove that the integral of the equation

$$\left(\frac{dw}{dz}\right)^3 - \left(\frac{dw}{dz}\right)^2 + 4w^3 - 27w^6 = 0$$

is a rational function of z ; that the integral of

$$\left(\frac{dw}{dz}\right)^3 - \left(\frac{dw}{dz}\right)^2 - 4w^2 - 27w^4 = 0$$

is a simply-periodic function of z ; and that the integral of

$$\left(\frac{dw}{dz}\right)^3 + 3\left(\frac{dw}{dz}\right)^2 + w^6 - 4 = 0$$

is a doubly-periodic function of z.

Find the infinities of each of the functions : and integrate the equations.

(Briot and Bouquet.)

Ex. 2. Shew that, if an irreducible trinomial equation of the form

$$\left(\frac{dw}{dz}\right)^m + \left(\frac{dw}{dz}\right)^{m-1} f_1(w) + f_m(w) = 0$$

have a uniform integral, then m may not be greater than 5 ; and that, if m be 4 or 5, the uniform integral is a doubly-periodic function.

Apply this result to the discussion of the equation

$$\left(\frac{dw}{dz}\right)^5 + \left(\frac{dw}{dz}\right)^4 (w^2-1) - \frac{4^4}{5^5} w^2 (w^2-1)^4 = 0.$$

(Briot and Bouquet.)

Ex. 3. Shew that the integral of the equation

$$\left(\frac{dw}{dz}\right)^6 = \lambda\,(w-a)^2\,(w-b)^5\,(w-c)^5$$

is a two-valued doubly-periodic function of z.   (Schwarz.)

Ex. 4. Shew that, if a function w be determined by a differential equation

$$F\left(\frac{dw}{dz},\ w\right) = 0,$$

where F is a rational integral algebraic function of w and $\frac{dw}{dz}$, of degree m in $\frac{dw}{dz}$, and does not contain z explicitly, then to each value of w there correspond m series of values of z, the terms in each series differing from one another by multiples of periods. Prove further that, if the integral w have only a limited number of values for each value of z, then it is determined by[an algebraical relation between w and u, where u may be z, or $e^{\frac{2\pi z i}{\omega}}$, or $\wp\,(z)$.

(Briot and Bouquet.)

These results should be compared with the results relative to functions which possess an algebraical addition-theorem*.

General Considerations on Equations compatible with a given Equation†.

138. After having discussed equations, all the critical points of which are fixed, we proceed to consider for more general cases the formal relations between the equation, say

$$F\,(w',\,w,\,z) = 0,$$

---

* *Th. Fns.*, Chap. xiii.

† In addition to the references given in § 140 dealing with the general theory of equations having algebraical integrals, it is proper to refer to the investigations of Darboux, *Bull. des Sciences*, 2me Sér., t. ii (1878), pp. 60—96, 123—144, 151—200, and Poincaré, *Palermo Rend.*, t. v (1891), pp. 161—191; also to a series of papers by Autonne, *Journ. de l'Éc. Polytech.*, t. xlii (1891), pp. 35—122, t. xliii (1892), pp. 47—180, t. xliv (1893), pp. 79—183, t. xlv (1894), pp. 1—53, 2me Sér., t. ii (1897), pp. 51—169, t. iii (1897), pp. 1—74, and a memoir, *Annales de Lyon*, t. iii (1892), 1er Fascicule. All of these relate to equations of the first order and the first degree.

There are also some papers dealing with equations

$$y' = \frac{P\,(y,\,x)}{Q\,(y,\,x)},$$

having integrals of the form

$$X\,(y-x_1)^{m_1}\,(y-x_2)^{m_2}\,\ldots\,(y-x_n)^{m_n} = \text{constant},$$

where X, x_1, ..., x_n are variable; see Korkine, *Math. Ann.*, t. xlviii (1897), pp. 317—364, and *Comptes Rendus*, t. cxxiii (1896), pp. 38—40; also Painlevé, *Ann. de Toulouse*, t. x (1896), G, *Comptes Rendus*, t. cxxii (1896), pp. 1319—1322, *ib.*, t. cxxiii (1896), pp. 88—91.

and any integral; the equation $F = 0$ being rational in w' and in w and, for many purposes, limited to be rational in z also.

In order to solve such an equation, or indeed an equation of any order, some other equation must be obtained which is compatible with it but is not a mere algebraical equivalent of it. When this new equation does not involve w', so that it is of the form

$$g(w, z) = 0,$$

it is called an integral of the equation; and if it involve an arbitrary element, connected with either a parameter or with initial values of the variables, it will be a general integral of the equation. When the new equation involves not merely z and w but also derivatives of w up to any order, it is of the form

$$h(z, w, w', w'', \ldots) = 0.$$

Now it is possible to deduce from $F = 0$ all the derivatives of order higher than the first by the process of successive differentiation; and each such derivative is uniquely obtained. When the values of these derivatives have been substituted in $h = 0$, the latter takes the form

$$r(w', w, z) = 0,$$

which may or may not involve an arbitrary constant. This last form is accordingly the most general form of equation compatible with but algebraically independent of $F = 0$; it manifestly includes the case of a compatible equation explicitly independent of w'.

The analytical expression of compatibility is easily constructed as follows. The values of all the successive derivatives of w, deduced from the one equation in series and from the other equation in series, must be the same either identically or in virtue of $F = 0$, $r = 0$; and this will be the case, without added conditions, solely in virtue of $F = 0$, $r = 0$ if they exist together. The two equations $F = 0$, $r = 0$ can be regarded as determining w' and w in terms of z; when these determined values are substituted, each of the two (compatible) equations becomes an identity. Accordingly, we have

$$\frac{\partial F}{\partial w'} w'' + \frac{\partial F}{\partial w} w' + \frac{\partial F}{\partial z} = 0,$$

$$\frac{\partial r}{\partial w'} w'' + \frac{\partial r}{\partial w} w' + \frac{\partial r}{\partial z} = 0,$$

on differentiating; and **therefore,** eliminating the quantity w'' which must have a value common to these two relations, we find

$$- \frac{\partial r}{\partial w'} \left(\frac{\partial F}{\partial w} w' + \frac{\partial F}{\partial z} \right) + \frac{\partial r}{\partial w} w' \frac{\partial F}{\partial w'} + \frac{\partial r}{\partial z} \frac{\partial F}{\partial w'} = 0:$$

an equation that is satisfied simultaneously with $F = 0$, $r = 0$.

If our object be the determination of the most general relation that is compatible with $F = 0$, the preceding equation is manifestly a partial differential equation of the first, order characteristic of the form of such compatible relation. Moreover, if we assume that $F = 0$ is irreducible—an assumption that is justifiable—then that characteristic equation is not satisfied solely on account of $F = 0$; that is to say, the partial differential equation determines the form of the compatible relation. Further, as the singular solutions (if any) can be discussed separately and have already been dealt with in principle, we shall assume that $\frac{\partial F}{\partial w'}$ does not vanish concurrently with F.

139. The subsidiary system of the partial differential equation is

$$\frac{dw'}{w' \frac{\partial F}{\partial w} + \frac{\partial F}{\partial z}} = \frac{- dw}{w' \frac{\partial F}{\partial w'}} = \frac{- dz}{\frac{\partial F}{\partial w'}}.$$

It is easy to see that $F = 0$ is satisfied in virtue of these equations. Any other integral of this system will give a compatible equation; denoting it by $r(w', w, z)$, then

$$r(w', w, z) = a,$$

where a is an arbitrary constant, is an equation compatible with $F = 0$. The elimination of w' between $F = 0$ and $r = a$ leads to a general integral of the original differential equation.

Let another equation compatible with $F = 0$ be given by

$$s(w', w, z) = c,$$

where c is an arbitrary constant. Then since the relation

$$\frac{\partial u}{\partial w'} \left(w' \frac{\partial F}{\partial w} + \frac{\partial F}{\partial z} \right) - \frac{\partial u}{\partial w} w' \frac{\partial F}{\partial w'} - \frac{\partial u}{\partial z} \frac{\partial F}{\partial w'} = 0$$

is satisfied when $u = F$, when $u = r$, and when $u = s$, we have

$$J \left(\frac{F, r, s}{w', w, z} \right) = 0,$$

by determinantal elimination from the three equations, it being assumed that $\dfrac{\partial F}{\partial w'}$ is not zero. Now $J = 0$ manifestly is not satisfied in virtue of $r = a$ or of $s = c$, for the arbitrary constants a and c occur only on the right-hand sides of these compatible equations, and do not occur in $J = 0$. Further, the whole investigation is formally the same whether the original irreducible equation be taken in the form $F = 0$ or in a form $F = b$, where b is any arbitrary constant; hence $J = 0$ is not satisfied in virtue of $F = 0$ for, owing to the identity of analytical form, $J = 0$ would then be satisfied in virtue of $F = b$, an impossibility excluded for reasons similar to those adduced in connection with $r = a$, $s = c$ Hence $J = 0$ is satisfied identically; and therefore between the three quantities F, r, s, regarded as involving three arguments w', w, z, there exists a functional relation

$$\Theta\,(F,\ r,\ s) = 0,$$

the coefficients of which are free from w', w, z and involve only the permanent constants in F, r, s. For our present purposes, $F = 0$ is a permanent equation, $r = a$ is an equation compatible with $F = 0$, $s = c$ is another such equation : hence

$$\Theta\,(0,\ a,\ c) = 0,$$

say

$$\phi\,(a,\ c) = 0.$$

It therefore appears that if $r = a$, $s = c$ be two equations compatible with $F = 0$, some functional relation

$$\phi\,(a,\ c) = 0$$

exists between the parameters a and c, the form of ϕ depending upon the form of r and s.

Conversely, if two compatible equations $r = a$, $s = c$ are known, such that the functional relation

$$\phi\,(a,\ c) = 0$$

is satisfied, then any arbitrary functional combination, say

$$u = \psi\,(r,\ s),$$

where ψ is an arbitrary function at choice, leads to another compatible equation

$$u = k.$$

Since r and s are, each of them, solutions of the homogeneous linear partial differential equation, which is characteristic of the

compatible equations, it is at once evident that the foregoing
function also leads to a solution. Also

$$k = \psi(a, c).$$

It thus appears that, when one compatible equation has been
obtained in the form $r = a$, an unlimited number of other compatible equations can be deduced from the forms

$$\left. \begin{aligned} \Theta(F, r, s) &= 0 \\ u &= \psi(r, s) \end{aligned} \right\};$$

and the question naturally arises as to the simplest forms that can
be chosen as forms of reference.

140. Without entering upon the discussion of the general
question thus suggested, it will be enough to limit the discussion
to one class of equations, viz. those which have algebraical integrals.
And here it is necessary to take account of the fact, established by
Abel's theorem, that a relation among transcendental integrals
may be equivalent to an algebraic relation between the arguments
of the integrals.   Thus the equation

$$\left(\frac{dw}{dz}\right)^2 = \frac{1 - w^2}{1 - z^2}$$

has an integral

$$\sin^{-1} w - \sin^{-1} z = \cos^{-1} a,$$

which is transcendental in form; but it also has an integral

$$w^2 - 2awz + z^2 = 1 - a^2,$$

which is an analytical equivalent of the transcendental relation.
Also, it is possible that an equation should appear to involve a
parameter in a transcendental form when a transformation can be
made so that a (new) parameter occurs only algebraically: thus
the equation

$$w = z \tan \alpha + c \sec \alpha$$

is effectively included in the equation

$$(w - \gamma z)^2 = c^2 (1 + \gamma^2),$$

α being the parameter in the first and γ in the second.

As the original differential equation $F = 0$ is algebraical in w
and w', it will be assumed that, whenever a compatible equation
(or the general integral) contains transcendental forms and can be
replaced by an equivalent equation containing algebraical forms,

this change is effected; but of course, such a change from transcendental to algebraical forms is possible only under limiting conditions as to the mode of occurrence and the character of the transcendental functions. Without entering upon the discussion of these conditions—a discussion which involves the significance of Abel's theorem in relation to the comparison of transcendental functions—we assume, as already stated, that the equation adopted as compatible with $F = 0$ is made algebraical whenever this is possible; and we might proceed to the consideration of equations which have algebraical integrals. Conditions necessary and sufficient to secure that the integral of an equation of the first order shall be a uniform function of the independent variable have already been obtained: and accordingly such equations will be regarded as completely discussed. The method appropriate for that discussion requires a development as yet not effected, before it can be made suitable for this more general question; and the range of ideas employed for the development of the corresponding theory lies outside the scheme of this section of the present work. Accordingly, we shall deal only with some elements of that theory: and for the present, shall merely refer to the investigations of Picard* and Painlevé†, based so largely upon the theory of rational transformation of curves and of surfaces‡.

Equations with Algebraical Integrals.

141.   Suppose then that an equation

$$r(w', w, z) = a$$

is compatible with the original equation

$$F(w', w, z) = 0,$$

taken to be algebraical in w' and w: and let the function $r(w', w, z)$ be designated an integral of $F = 0$. The elimination of w' between $F = 0$ and $r = a$ leads to an equation

$$G(w, z, a) = 0,$$

---

* *Cours d'Analyse*, t. ii, ch. xii; t. iii, ch. iv.

† See the memoir cited at the beginning of Chapter ix: and the *Stockholm Lectures*, frequently quoted in preceding chapters.

‡ It should be added that the first attempt at effecting relations between differential equations, of the first order and any degree, and the Riemann theory of algebraic functions was made by Clebsch, *Math. Ann.*, t. vi (1873), pp. 211—213.

which, as it involves an arbitrary parameter a, is the general solution of $F=0$. Owing to this relation between $G=0$ and $F=0$, the elimination of a between

$$G=0, \quad G_1 = w'\frac{\partial G}{\partial w} + \frac{\partial G}{\partial z} = 0$$

involves the relation $F=0$; and the relation $r=a$ is satisfied in virtue of these same equations, $F=0$, $G=0$.

Now assume that the function G is integral and rational so far as concerns w, and that F is integral and rational so far as concerns w' and w; then G_1 also is integral and rational so far as concerns w' and w. By the ordinary theory of elimination, we have

$$G = A'F + B'r,$$

where A', B' are functions of w', w, z, rational in w' and w; and therefore

$$r = AF + BG,$$

where A, B are functions similar to A', B'. Now in virtue of $F=0$, $G=0$, every rational function of w' and w, and therefore both A and B, can be made polynomials in w' and w, the coefficients being functions of z that may have a meromorphic form. To effect the change to this form in any given instance, we first use $F=0$ to make A and B polynomials in w' of degree not so high as F: and then use $G=0$ to make the coefficients of the various powers of w' polynomials in w of degree not so high as G. When this transformation is made, then r is the sum of two terms, each of which is the product of two polynomials in w and w'; hence r is itself a polynomial in w and w'. It therefore follows that, *if the general integral of the equation $F=0$ is rational in the dependent variable w, $F=0$ itself being rational in w' and w, then every integral, such as $r(w', w, z)$, compatible with $F=0$, can be expressed in a form that is rational in w' and w.*

Further, *if $F=0$ and its general solution are rational also in z, then $r(w', w, z)$ also is rational in z.* For interchanging the independent and the dependent variables, the equation

$$F(w', w, z) = 0$$

becomes

$$F\left(\frac{1}{z'}, w, z\right) = 0,$$

which is rational in z and z'; by what has just been proved, every integral compatible with this equation is rational in z and z', that is,

$$r\left(\frac{1}{z'}, w, z\right)$$

is rational in z.

Again, the equation

$$r(w', w, z) = a,$$

which is compatible with $F = 0$, is rational in w and w' when F has that character. Moreover by using $F = 0$, the function

$$r(w', w, z)$$

can be made an integral function of w'; and by repeated substitution from $F = 0$, the degree of this integral function in w' can be made less than the degree of $F = 0$: that is, *if the differential equation*

$$F(w', w, z) = 0$$

be of degree m and have its general solution rational, every integral compatible with $F = 0$ can be expressed as a polynomial in w' of degree $\leqslant m - 1$, the coefficients of the various powers of w' being rational functions of w.

It is an immediate corollary that the arbitrary constant enters linearly into the general solution of

$$\frac{dw}{dz} = R(w, z),$$

where R is rational in w and the equation has a rational integral.

Lastly, since r is rational in w', as also is F, and since $G(w, z, a)$ is the eliminant of $r = a$, $F = 0$, it follows that G, the general solution, is rational in the parameter a.

142. Next, consider solutions of the equation defined by arbitrarily assigned initial values. Let w_0 be assigned as a value of w when $z = z_0$; let w_0' denote one of the m associated values of the derivative, given by say

$$F_0 = F(w_0', w_0, z_0) = 0 ;$$

and assume that these are not connected with the singular integral (if any) of the equation. Then when the variable describes a path from z_0 to z, not passing through any of the singularities of the equation, the function $r(w_0', w_0, z_0)$ becomes $r(w', w, z)$ at the end of the path; or since $r = a$, we have

$$r(w', w, z) = a = r(w_0', w_0, z_0).$$

The general solution of the equation is known to have the form

$$G(w, z, a) = 0,$$

so that it can be expressed in the form

$$G\{w, z, r(w_0', w_0, z_0)\} = 0,$$

G being rational in w and a, and r being rational in w_0' : that is, G is rational in w_0'. Now w_0' is any one of the m values satisfying $F_0 = 0$; hence eliminating w_0' between $G = 0$, $F_0 = 0$, we have an equation

$$\mathfrak{G}(w, z, w_0, z_0) = 0,$$

which is rational in w and in w_0. If $F = 0$ and its general solution be rational in z, then also $\mathfrak{G}$ is rational in z and z_0.

Note. One remark may be made in passing. It might be thought that, if an integral function acquires only a limited number of values when the variable z describes any paths (whether round fixed critical points or round parametric critical points), then the algebraical equation of which it is a root is necessarily rational in z, when the original differential equation is rational in z. That this is not the case can be seen from a simple example. The equation

$$\tfrac{1}{4} w'^2 = (w + 1)(w' - w)$$

is rational in w' and w: its primitive is

$$w = 2ae^z + a^2 e^{2z},$$

which is rational in w but not in z; and if the parameter a be determined by the condition that $w = w_0$ when $z = z_0$, then $\mathfrak{G} = 0$ is the equivalent of

$$\frac{(w + 1)^{\frac{1}{2}} - 1}{(w_0 + 1)^{\frac{1}{2}} - 1} = e^{z - z_0},$$

when the latter is freed from radicals. Manifestly $\mathfrak{G}$ is rational in w and w_0, and is transcendental in z and z_0.

143. It has been seen that an unlimited number of integral functions are compatible with the original differential equation: and it is desirable to determine whether the original differential equation possesses a general integral in the form of an algebraical equation. The following considerations indicate a possible method. Let $r(w', w, z)$ and $s(w', w, z)$ be two integral functions of the compatible character, say

$$r(w', w, z) = X, \quad s(w', w, z) = Y;$$

and denote by

$$\phi(X, Y) = 0$$

the relation between X and Y, satisfied in virtue of $F = 0$. Then we can regard

$$r = X, \quad s = Y$$

as a rational transformation which, when applied to the equation

$$F(w', w, z) = 0,$$

leads to an algebraical equation

$$\phi(X, Y) = 0.$$

According to the form of the transformation, there will be a relation between the genus, say p, of F and the genus, say p', of ϕ: in particular, consider as possible the case when the transformation is birational, so that p and p' are equal.

Let $I_1, \ldots, I_p$ denote the p normal elementary integrals of the first kind associated with F; and let

$$\int \frac{Q(X, Y)}{\frac{\partial \phi}{\partial Y}} dX$$

be any integral of the first kind associated with ϕ. When the variables are transformed, the integral still remains everywhere finite on the Riemann surface associated with F, that is, it is an integral of the first kind; and therefore it is expressible in the form

$$\lambda_1 I_1 + \ldots + \lambda_p I_p,$$

where $\lambda_1, \ldots, \lambda_p$ are parametric in this relation. If

$$I_k = \int \frac{Q_k(w, w')}{\frac{\partial F}{\partial w'}} dw, \qquad (k = 1, \ldots, p),$$

we have

$$\int \frac{Q(X, Y)}{\frac{\partial \phi}{\partial Y}} dX \equiv \Sigma \lambda_k I_k,$$

the moduli of the congruence being the periods : and therefore, as the quantities λ_k are parametric,

$$\frac{Q\,(X,\ Y)}{\dfrac{\partial \phi}{\partial Y}}\,dX = \Sigma\,\frac{\lambda_k Q_k\,(w,\ w')}{\dfrac{\partial F}{\partial w'}}\,dw.$$

If we conceive the equation $F = 0$ resolved so as to express w' (irrationally) in terms of w and z, and the resulting expression substituted in $r = X$, the new equation is a resolved form of the general primitive. It can be regarded as the integral of

$$dw - w'dz = 0\ ;$$

that is, if Θ be an integrating factor, we have

$$\Theta\,(dw - w'dz)$$

as a perfect differential. But we know (§ 139) that any combination of r and s, that is, X and Y, is an integral compatible with $F = 0$; and so

$$\frac{Q\,(X,\ Y)}{\dfrac{\partial \phi}{\partial Y}}$$

is an (irrational) function of X and so is a compatible integral. Hence as $dX = 0$ concurrently with $dw - w'dz = 0$, and therefore also

$$\frac{Q\,(X,\ Y)}{\dfrac{\partial \phi}{\partial Y}}\,dX = 0$$

concurrently with $dw - w'dz = 0$, it follows that

$$\frac{Q\,(X,\ Y)}{\dfrac{\partial \phi}{\partial Y}}\,\frac{dX}{dw}$$

is an integrating factor of the equation. Denote by M the quantity

$$\sum_{k=1}^{p}\,\frac{\lambda_k Q_k\,(w,\ w')}{\dfrac{\partial F}{\partial w'}},$$

so that M is an integrating factor of

$$dw - w'dz = 0,$$

when w' is the appropriate function of w and z. Hence

$$M\,(dw - w'dz) = \text{perfect differential},$$

so that

$$-\frac{\partial M}{\partial z} = \frac{\partial}{\partial w}(Mw')$$

$$= w'\frac{\partial M}{\partial w} + M\frac{\partial w'}{\partial w},$$

that is,

$$\frac{\partial F}{\partial w'}\left(\frac{\partial M}{\partial z} + w'\frac{\partial M}{\partial w}\right) = M\frac{\partial F}{\partial w}.$$

Now, on the hypothesis that the integral is rational, the form of M is known, save as to the (parametric) coefficients λ: that is, now taking account of the variation of z as well as of w and w' (which have been considered rather in relation to the general properties of $F = 0$ and the associated Riemann surface), when

$$\sum_{k=1}^{p} \frac{\lambda_k Q_k(w, w')}{\frac{\partial F}{\partial w'}}$$

is substituted for M in the differential equation which it satisfies, the latter must become an identity. The quantities Q_k are known in connection with the equation $F = 0$ of genus p; and therefore the quantities λ_k, functions of z, must satisfy a number of relations in order that the equation for M may be identically satisfied. These relations will impose conditions on the form of F which, in fact, are the conditions that $F = 0$ should have a rational integral.

144.   It is manifest that this is rather a descriptive indication than a convenient method: any attempt at actual expression of the conditions would be extremely laborious, even if it could be completed. No other method, however, seems generally available.

But if it does not promise to be an effective method for the expression of conditions or the actual construction of the integral, there is a manifest possibility of developing general properties of the equations. With the reference to Painlevé's investigations that already (§ 140) has been made, we shall cease any further discussion of this portion of the subject, as being outside the range proposed for this volume.